T0337835

Wireless Multi-Antenna Channels

Wiley Series on Wireless Communications and Mobile Computing

Series Editors: Dr Xuemin (Sherman) Shen, *University of Waterloo, Canada*
Dr Yi Pan, *Georgia State University, USA*

The "Wiley Series on Wireless Communications and Mobile Computing" is a series of comprehensive, practical and timely books on wireless communication and network systems. The series focuses on topics ranging from wireless communication and coding theory to wireless applications and pervasive computing. The books provide engineers and other technical professionals, researchers, educators, and advanced students in these fields with invaluable insight into the latest developments and cutting-edge research.

Other titles in the series:

Misic and Misic: *Wireless Personal Area Networks: Performance, Interconnection, and Security with IEEE 802.15.4*, January 2008, 978-0-470-51847-2

Takagi and Walke: *Spectrum Requirement Planning in Wireless Communications: Model and Methodology for IMT-Advanced*, April 2008, 978-0-470-98647-9

Pérez-Fontán and Espiñeira: *Modeling the Wireless Propagation Channel: A Simulation Approach with MATLAB®*, August 2008, 978-0-470-72785-0

Ippolito: *Satellite Communications Systems Engineering: Atmospheric Effects, Satellite Link Design and System Performance*, August 2008, 978-0-470-72527-6

Lin and Sou: *Charging for Mobile All-IP Telecommunications*, September 2008, 978-0-470-77565-3

Myung and Goodman: *Single Carrier FDMA: A New Air Interface for Long Term Evolution*, October 2008, 978-0-470-72449-1

Wang, Kondi, Luthra and Ci: *4G Wireless Video Communications*, April 2009, 978-0-470-77307-9

Cai, Shen and Mark: *Multimedia Services in Wireless Internet: Modeling and Analysis*, June 2009, 978-0-470-77065-8

Stojmenovic: *Wireless Sensor and Actuator Networks: Algorithms and Protocols for Scalable Coordination and Data Communication*, February 2010, 978-0-470-17082-3

Liu and Weiss, *Wideband Beamforming: Concepts and Techniques*, March 2010, 978-0-470-71392-1

Riccharia and Westbrook, *Satellite Systems for Personal Applications: Concepts and Technology*, July 2010, 978-0-470-71428-7

Zeng, Lou and Li, *Multihop Wireless Networks: Opportunistic Routing*, July 2011, 978-0-470-66617-3

Qian, Muller and Chen: *Security in Wireless Networks and Systems*, February 2013, 978-0-470-512128

Wireless Multi-Antenna Channels

Modeling and Simulation

Serguei L. Primak

The University of Western Ontario, Canada

Valeri Kontorovich

CINVESTAV, Mexico

A John Wiley & Sons, Ltd., Publication

Library of Congress Cataloging-in-Publication Data

Primak, L. Serguei.
  Wireless multi-antenna channels : modeling and simulation / Serguei L. Primak, Valeri Kontorovich.
    p. cm. – (Wireless communications and mobile computing)
  Includes bibliographical references and index.
  ISBN 978-0-470-69720-7 (hardback)
  1. Roaming (Telecommunication)–Mathematical models.   2. MIMO systems–Mathematical models.
3. Antenna radiation patterns–Mathematical models.   4. Antenna arrays–Mathematical models.
5. Adaptive antennas–Mathematical models.   I. Kontorovich, V. IA. (Valeri Kontorovich)   II. Title.
  TK5103.4874.P75 2011
  621.3845′6–dc23

                                                                                        2011025940

A catalogue record for this book is available from the British Library.

ISBN: 978-0-470-69720-7 (H/B)
ISBN: 978-1-119-95471-2 (ePDF)
ISBN: 978-1-119-95472-9 (oBook)
ISBN: 978-1-119-96086-7 (ePub)
ISBN: 978-1-119-96087-4 (eMobi)

Typeset in 10/12pt Times by Laserwords Private Limited, Chennai, India.

Contents

About the Series Editors

Xuemin (Sherman) Shen (M'97-SM'02) received the B.Sc degree in electrical engineering from Dalian Maritime University, China in 1982, and the M.Sc. and Ph.D. degrees (both in electrical engineering) from Rutgers University, New Jersey, USA, in 1987 and 1990 respectively. He is a Professor and University Research Chair, and the Associate Chair for Graduate Studies, Department of Electrical and Computer Engineering, University of Waterloo, Canada. His research focuses on mobility and resource management in interconnected wireless/wired networks, UWB wireless communications systems, wireless security, and ad hoc and sensor networks. He is a co-author of three books, and has published more than 300 papers and book chapters in wireless communications and networks, control and filtering. Dr. Shen serves as a Founding Area Editor for *IEEE Transactions on Wireless Communications*; Editor-in-Chief for *Peer-to-Peer Networking and Application*; Associate Editor for *IEEE Transactions on Vehicular Technology; KICS/IEEE Journal of Communications and Networks, Computer Networks; ACM/Wireless Networks; and Wireless Communications and Mobile Computing (Wiley)*, etc. He has also served as Guest Editor for *IEEE JSAC, IEEE Wireless Communications*, and *IEEE Communications Magazine*. Dr. Shen received the Excellent Graduate Supervision Award in 2006, and the Outstanding Performance Award in 2004 from the University of Waterloo, the Premier's Research Excellence Award (PREA) in 2003 from the Province of Ontario, Canada, and the Distinguished Performance Award in 2002 from the Faculty of Engineering, University of Waterloo. Dr. Shen is a registered Professional Engineer of Ontario, Canada.

Dr. Yi Pan is the Chair and a Professor in the Department of Computer Science at Georgia State University, USA. Dr. Pan received his B.Eng. and M.Eng. degrees in computer engineering from Tsinghua University, China, in 1982 and 1984, respectively, and his Ph.D. degree in computer science from the University of Pittsburgh, USA, in 1991. Dr. Pan's research interests include parallel and distributed computing, optical networks, wireless networks, and bioinformatics. Dr. Pan has published more than 100 journal papers with over 30 papers published in various IEEE journals. In addition, he has published over 130 papers in refereed conferences (including IPDPS, ICPP, ICDCS, INFOCOM, and GLOBECOM). He has also co-edited over 30 books. Dr. Pan has served as an editor-in-chief or an editorial board member for 15 journals including five IEEE *Transactions* and has organized many international conferences and workshops. Dr. Pan has delivered over 10 keynote speeches at many international conferences. Dr. Pan is an IEEE Distinguished Speaker (2000–2002), a Yamacraw Distinguished Speaker (2002), and a Shell Oil Colloquium Speaker (2002). He is listed in Men of Achievement, Who's Who in America, Who's Who in American Education, Who's Who in Computational Science and Engineering, and Who's Who of Asian Americans.

1

Introduction

1.1 General remarks

The explosion in demand for wireless services experienced over the past 20 years has put significant pressure on system designers to increase the capacity of the systems being deployed. While the spectral resource is very scarce and practically exhausted, the biggest possibilities are predicted to be in the areas of spectral reuse by unlicensed users or in exploiting the spatial dimension of the wireless channels. The former approach is now under intense development and is known as the cognitive radio approach (Haykin 2005). The latter approach is as old as communication systems themselves and is known mostly through the receive diversity techniques, well studied in both Western (Middleton 1960; Simon and Alouini 2000) and former USSR literature (Fink 1970; Klovski 1982a). These techniques are mainly used to improve the signal to noise ratio in the receiver in fading environments. In order to exploit the additional (spatial) dimension of the wireless channel, a number of technologies were suggested in early 1990s, including smart antennas. The development of this antenna technology mainly focused on the development of the estimation of the angle of arrival, optimal beamforming, and space-time signal processing. However, these techniques offered only a limited increase in the channel capacity.

In recent years, however, development of the multiple-input multiple-output (MIMO) system has emerged as the most potent technique for increasing the capacity of wireless channels. This technique exploits sampling in the spatial dimension on both sides of the communication links, combined in such a way that they either create virtual, multiple, parallel spatial data pipes to increase capacity linearly with the number of pipes and/or to add diversity to improve the quality of the links. A large number of research articles and monographs have treated different aspects of this subject (Correia, Ed. 2007; Paulraj et al. 2003; Tse and Viswanath 2005;

Wireless Multi-Antenna Channels: Modeling and Simulation, First Edition.
Serguei L. Primak and Valeri Kontorovich.
© 2012 John Wiley & Sons, Ltd. Published 2012 by John Wiley & Sons, Ltd.

Verdu 1998), including channel modeling, modulation, diversity-multiplexing trade-off, and so on. Since initial papers by Teletar (Teletar 1999), MIMO technologies have been included in many existing standards of 4G communications. Today MIMO technology appears to be the natural candidate for most large-scale commercial wireless products.

Most of the researchers are focusing on investigation of MIMO systems under the important assumption that fading is Rayleigh, channel state information is perfectly known, and the scattering is reach. Such results provide good limiting estimates for capacity, performance, and delay. However, it often provides over optimistic results. The main contribution of this book lies in addressing the following issues:

- Suggestion of the generalized Gaussian model of MIMO wireless channels.

- Investigation of performance of different coding schemes in generalized Gaussian channels.

- Suggestion of an efficient simulator of MIMO wireless channels based on a geometric-based modeling paradigm.

- In-depth studying of the effect of channel estimation on performance in MIMO systems.

- Introduction of the multitaper approach to channel estimation.

- Investigation of the second-order statistics of MIMO channel capacity.

The book is organized as follows. In this chapter we briefly discuss models for signals used in this text. Chapter 2 describes a novel, four-parametric model of a SISO wireless channel and extends the concept to MIMO configuration. We also consider channels with a fluctuating number of scatterers and other deviations from Gaussian models. Chapter 3 expands on the modeling of MIMO channels based on scattering geometry and explores different geometry characteristics that effect the channel model. It also describes narrowband MIMO channel models while Chapter 4 is dedicated to wideband models. Chapter 5 treats topics related to the investigation of the capacity of the MIMO channel under different geometrical conditions, treats time variation of the capacity, and capacity of sparse channels. Chapter 6 deals with the methodology of MIMO channel prediction, while Chapter 7 deals with effects of errors on different aspects of communication system performance. Finally, Chapter 8 deals with the investigation of space-time code performance in generalized Gaussian MIMO channels.

Finally we would like to express our gratitude to a number of people and agencies that were instrumental in supporting the research that has resulted in most of the content of this book. First of all, we would like to express our admiration for our late teacher and colleague Prod. Daniil (Dani) Klovski. His investigation of diversity in the time-space communication channel in the late 1960s to the early 1980s laid a solid foundation of smart antenna techniques in the former USSR while also inspiring us to offer a generalized Gaussian model of MIMO channels.

We are also deeply indebted to our graduate students, now independent researchers themselves. Our deepest thanks to Drs. Vanja Subotic, Khaled Almustafa, Dan Dechene, and A. F. Ramos-Alarcón for their diligent work in developing most of the ideas presented in this manuscript. We would like to thank Drs. Tricia Willink and Karim Baddour from the Communications Research Centre Canada (CRC Canada) for numerous discussions, especially on topics related to MIMO channel estimation. Our research and graduate students have been financially supported through a number of research grants, provided by NSERC Canada, CONACYT Mexico, and CRC Canada. We are grateful to the University of Western Ontario and CINVESTAV-IPN, Mexico for creating excellent working conditions and the opportunity to spend a few months both in Mexico and Canada jointly working on the manuscript. We also would like to thank our colleagues at the University of Agder, Prof. Matthias Patzold, Drs. Gulzaib Rafiq, Dmitri Umanski, and others for a number of useful discussions and suggestions. And last, but not least, we would like to thank wonderful staff at Wiley for their patience and indispensable help in preparing the manuscript.

1.2 Signals, interference, and types of parallel channels

Here we consider the problem of transmission of digital information over a set of parallel channels. The discrete message is chosen from an alphabet $\mathcal{A} = \{x_k\}$ of size $m = \dim(\mathcal{A})$. Once a symbol, say x from the alphabet $\mathcal{A}$ is chosen it is encoded to a signal waveform $z_k(t)$. A unique waveform corresponds to each symbol of the alphabet. In general one would select n symbols from the alphabet at a single moment of time to transmit them over n channels simultaneously. This can be accomplished by using mn different signals $z_{rk}(t), r = 1, \cdots, m$, and $k = 1, \cdots, n$. We consider coherent systems where the duration of the symbols at every channel is fixed to be T and the start of the symbols in each channel coincide.

The received signals $z'_k(t)$ at the output of each of the parallel channels have a statistical relation to the transmitted signal $z_k(t)$, albeit one that does not coincide with it due to noise and interference. The received signals are processed by a decision-making block. We will focus on synchronous detection. This means that the decision-making algorithm observes a symbol over time period T and then decides which symbol has been transmitted.

Parallel channels are assumed to be linear and the signals $z_{rk}(t)$ narrowband. This means that most of the energy of the signals $z_{rk}(t)$ is concentrated in a frequency band F_{rk} that is much smaller than the carrier frequency f_{rk}. In this case the following representation is valid

$$z_{rk}(t) = E_{rk} \cos\left[2\pi f_{rk}t + \Phi_{rk}(t)\right], 0 \le t < T \tag{1.1}$$

In this case the envelope

$$E_{rk}(t) = \sqrt{z_{rk}^2 + \tilde{z}_{rk}^2} \tag{1.2}$$

and the initial phase

$$\Phi_{rk}(t) = \operatorname{atan} \frac{z_{rk}}{\tilde{z}_{rk}} - 2\pi f_{rk}t \tag{1.3}$$

are slowly varying factions with respect to $\cos(2\pi f_{rk}t)$. Here $\tilde{z}_{rk}$ is the Hilbert transform (Proakis 1997).

$$z_{rk}(t) = E_{rk} \sin\left[2\pi f_{rk}t + \Phi_{rk}(t)\right], \, 0 \le t < T \tag{1.4}$$

The low pass equivalent of this signal is thus

$$s_{rk} = E_{rk} \exp(j\Phi_{rk}) \tag{1.5}$$

The received signal can then be written as

$$z'_{rk} = \mu_k(t)z_{rk}\left[t - \tau_{rk}(t)\right] + \xi_k(t) \tag{1.6}$$

where $\mu_k(t)$ is a coefficient describing the attenuation of the signal transmitted through the k-th channel, τ_k is the delay associated with this channel, and $\xi_k(t)$ is the associated additive noise component. For the majority of realistic channels with variable parameters the time delay can be written as a sum

$$\tau_k(t) = \tau_{k \text{ av}} - \Delta\tau_k(t) \tag{1.7}$$

where $\tau_{k \text{ av}} = E\{\tau_k(t)\}$ is the average delay and $\Delta\tau_k(t)$ is the random fluctuation of the delay. The former can be associated with the overall propagation delay due to the finite distance between the transmit and the receive antennas, while the latter can be associated with variation in the channel and mutual position between the receiver and the transmitter. In order to have an effect on the performance of the system the fluctuating component should have a root-mean-square value comparable to the duration of the bit interval.

Let $\theta_k(t) = 2\pi f_{rk}\Delta\tau_k(t)$ be a random excessive phase associated with the k-th channel and r-th frequency. Then Equation (1.6) can be rewritten as

$$z'_{rk} = \mu_k(t)z_{rk}\left[t - \tau_{k \text{ av}}, \theta_k(t)\right] + \xi_k(t) \tag{1.8}$$

where $z(t - \tau, \theta)$ represents signal $z(t)$ delayed by τ and the phase of its carrier shifted by θ.

During a one-bit detection the receiver processes segments of the signal $z'_k(t)$ of duration T one by one. It is quite clear that without loss of generality a constant delay $\tau_{k \text{ av}}$ can be eliminated from consideration. Thus, the receiver observes a signal

$$z'_k(t) = \mu_k(t)z_{rk}[t, \theta_k(t)] + \xi_k(t), \, (l-1)T \le t < lT \tag{1.9}$$

where l indicates the order of the transmitted symbols.

For the narrowband process z'_k we can write its expansion in terms of in-phase and quadrature components (equation 2.44)

$$z'_k(t) = \mu_k(t)\cos\theta_k(t)z_{rk}(t) + \mu_k(t)\sin\theta_k(t)\tilde{z}_{rk}(t) + \xi_k(t), \, (l-1)T \le t < lT \tag{1.10}$$

Here $\mu_k(t)$ and $\theta_k(t)$ represent the magnitude and the phase of the transmission coefficient of the k-th channel.

In practice it is more efficient to represent fading either in terms of in-phase and quadrature components

$$\mu_{ck}(t) = \mu_k(t)\cos\theta_k(t) \tag{1.11}$$

$$\mu_{sk}(t) = \mu_k(t)\sin\theta_k(t) \tag{1.12}$$

or in its phasor (complex low-pass equivalent) form

$$\mu_k(t) = \mu_k(t)\exp[-j\theta_k(t)] \tag{1.13}$$

In this case

$$z'_k = \mu_{ck}(t)z_{rk}(t) + \mu_{sk}(t)\tilde{z}_{rk}(t) + \xi_k(t), \ (l-1)T \le t < lT \tag{1.14}$$

If $\mu_k(t)$ and $\theta_k(t)$ are random functions of time, they can be considered as multiplicative noise.

Without going into detail about the statistical properties of the random channel transmission coefficient it is important to provide a general classification of the digital communication systems utilizing such channels. The main purpose of a system with multiple channels is to increase the capacity of the channels. This can be achieved when the additive noise and the multiplicative noise are decorrelated.

Digital systems with multiple channels can be categorized as follows

- Channels with distinct media of propagation corresponding to different channels;

- Channels that share a media with the same physical properties.

The former category may include channels where the information is simultaneously transmitted over wired and wireless links. Such systems would mostly be used to provide redundancy and reliability of communication systems since the amount of information is the same regardless of number of channels used.

Systems belonging to the latter category are widely used in the diversity reception and recently in so-called MIMO systems. The improvement in the capacity of such systems is to a large degree defined by the number of channels used, the physical location of the antennas, and the signal processing techniques.

Depending on the method of forming multiple channels one can distinguish the following groups of parallel channels

1. systems with parallel channels formed on the transmitting side;

2. systems with parallel channels formed on the receiving side;

3. systems with parallel channels formed on both the receiving and transmitting side.

In the systems belonging to the first group the information about each bit/symbol is transmitted into the channel by means of multiple signals that are formed by the transmitter during the modulation stage. In this case the total number of different signals is mn where n is the number of parallel channels. Each bit/symbol can be transmitted simultaneously over all channels or sequentially over the same time interval T. Systems with frequency and time diversity are two well-known representatives of this class. In the case of frequency diversity the same signal is transmitted over different frequencies. The total signal in the receive antenna is thus

$$z'(t) = \sum_{k=1}^{n} \left[\mu_{ck}(t)z_{rk}(t) + \mu_{sk}(t)\tilde{z}_{rk}(t) \right] + \xi(t) \tag{1.15}$$

where $\mu_{ck}(t) + j * \mu_{sk}$ is a complex transmission coefficient for the signal $z_{rk}(t)$.

If the signals $z_{rk}(t)$ and $z_{rp}(t)$ are orthogonal for $k \neq p$ in the strong sense, then instead of a single received signal in Equation (1.15) one can consider a set of parallel channels with output signals defined by Equation (1.14) where $\mu_{ck}(t)$ and $\mu_{sk}(t)$ are in-phase and quadrature components of the transmission coefficient for the signal $z_{rk}(t)$ and $\xi_k(t)$ is the additive noise acting within the bandwidth allocated to the signal $z_{rk}(t)$. In general, quantities $\mu_{ck}(t)$, $\mu_{sk}(t)$, $\xi_k(t)$, and $\mu_{cp}(t)$, $\mu_{sp}(t)$, $\xi_p(t)$ are represented by correlated processes for all $k \neq p$. An optimal choice of signals $z_{rk}(t)$ will be such as to provide nearly complete decorrelation between the fading components and the additive noise of signals transmitted over different channels. For example, if all $z_{rk}(t)$ have non-overlapping spectra, than the additive noise $\xi_k(t)$ and $\xi_p(t)$ will be uncorrelated for $k \neq p$. Furthermore, if the propagation medium has a selectivity property,[1] one can choose carrier frequencies f_k in such a way that $\mu_{ck}(t)$, $\mu_{sk}(t)$ and $\mu_{cp}(t)$, $\mu_{sp}(t)$ are also almost completely decorrelated.

The simplest method of formation of two parallel channels is frequency shift keying (FSK). Indeed, FSK can be considered as two inverse amplitude modulations with different carriers.

---

[1] The majority of situations which involve the reflection of the transmitted signal from multiple obstacles will have this property.

2

Four-parametric model of a SISO channel

2.1 Multipath propagation

It has been recognized from the onset of communication theory that it is important to accurately represent communication channels, taking into account the specifics of electromagnetic wave propagation (Middleton 1960). Here we confine ourself to systems that use propagation in free (but perhaps non-homogeneous) media. In most situations such channels are random (stochastic) with randomly placed and varying in time inhomogeneities, such as reflecting surfaces, edges, absorbers, and so on. In this case an accurate representation of all inhomogeneities is rather difficult and even counterproductive due to the computational burden and lack of detailed information. Therefore, it seems only natural to deploy a phenomenological approach, that is to model only a selective set of (statistical) characteristics of channels, such as mean values, covariance functions, and so on, and disregard the specific mechanism of wave propagation assuming, in general, that the waves in question could be modeled well as a set of individual rays, which represent a high proportion of the power intersected by receive antennas. Each such ray may be the result of reflections from a relatively smooth surface,[1] diffracted ray, or others (Bertoni 2000). Such "hot spots" could be formed by a few volumes significantly separated in space, as shown in Figure 2.1. If the scattering region (volume) slightly changes in time, the "hot spots" start to move and change their intensity and new hot spots may appear. This is equivalent to variation of the phase and magnitude of each ray. Such a picture chimes well with the principles of physical optics,

---

[1] So called "hot spots".

Wireless Multi-Antenna Channels: Modeling and Simulation, First Edition.
Serguei L. Primak and Valeri Kontorovich.
© 2012 John Wiley & Sons, Ltd. Published 2012 by John Wiley & Sons, Ltd.

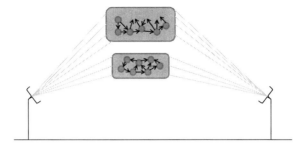

Figure 2.1   Physical optics model of scattering.

especially at the high propagation frequencies currently used in wireless systems. In the frame of this model each ray may be reflected a number of times before reaching the receive antenna. Often, the Born approximation is evoked to neglect all secondary reflections (see e.g., (Abdi and Kaveh 2002; Yu and Ottersten 2002)). In this case each ray can be considered as propagating in parallel to all others: we will call this mechanism a *parallel single-bounce scattering*. However, in many situations it is important to take into account multiple reflections (Middleton 1960; Salo et al. 2006). In this case we still have multiple rays propagating in parallel to each other, however each of them experience sequential reflections, attenuations, absorption, diffraction, and so on. This scenario will be referred to as multiple scattering or *sequential-parallel multiple scattering*.

   If a single bounce is assumed, the received signal $r(t, \mathbf{x}_r)$ can be readily written as a sum of the random numbers of independent components

$$r(t, \mathbf{x}_r) = \sum_{n=1}^{N(t)} r_n(t, \mathbf{x}_r, \mathbf{p}_n) \tag{2.1}$$

Here $\mathbf{x}_r$ is the vector position of the receive antenna, $\mathbf{p}_n$ is the vector of the parameters[2] associated with the n-th ray. This model will be considered in more details in Section 2.2. Often, the central limit theorem can be invoked to justify that the distribution of the received signal is Gaussian for a fixed number of scatterers.

   The case of multiple scattering often arises in situations where there is no line-of-sight and where there is a dense scattering environment between the receive and transmit sites (Bliss et al. 2002; Dohler et al. 2007; Gesbert et al. 2002; Oestges et al. 2003). The received signal can also be represented as a sum (2.1), however, all rays cannot be treated as independent. This could be explained by the fact that some of the components of the sum (2.1) are produced by the same "hot spots". General analysis of such a situation is quite complicated since the central limit theorem cannot be invoked without preconditions on the amount of correlation between the individual terms. The general structure of such channels can be described by a sequence of time-varying filters as shown in Figure 2.2. Among the many models with multiplicative scattering it is possible to select one in which rays are still

---

[2] Such as intensity, phase, delay, etc.

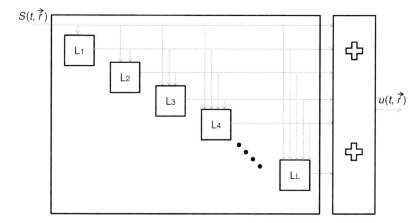

Figure 2.2 Mechanism of propagation with multiple scattering but independent rays.

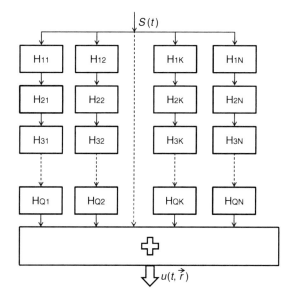

Figure 2.3 Mechanism of propagation with multiple scattering but independent rays.

independent but where each ray experiences multiple scattering. This situation is shown in Figure 2.3.

In this case, if the number of rays is large, the resulting received signal again converges to a Gaussian one, regardless of the distribution of components. However, if the number of rays is relatively small[3] the distribution of the signal in the receive antenna is dominated by the distribution of individual rays.

---

[3] $N < 10$ (Klovski 1982b).

Another important criteria of channel classification is related to the ability of the receiver to isolate different subrays (Rappaport 2002). This is defined by the aperture size on the receive and transmit sides as well as by the duration of the pulses used to transmit data symbols.

Let us assume that T is a duration of the transmitted symbol $s(t)$, $0 \leq t \leq T$ that occupies effective bandwidth Δf_s, due to the fact that the received signal is composed of multiple copies of the transmitted signal $s(t)$. The channel can be considered a single path channel if the maximum delay $\Delta \tau$ experienced by an individual ray is smaller than the resolution of the receiver

$$\Delta \tau \ll \frac{1}{\Delta f_s} \tag{2.2}$$

If the opposite is true, the channel should be considered multipath.

Even in the case of a single path channel, the resulting signal is a superposition of a large number of subpaths, mainly due to diffused scattering in the propagation environment.

In the case of a narrowband transmit signal on the interval $[iT : (i + 1)T]$

$$s_i(t) = S_i(t) \exp(j\omega_0 t) u(t, iT), \quad u(t, iT) = \begin{cases} 1 & \text{if } iT \leq t \leq (i+1)T \\ 0 & \text{otherwise} \end{cases} \tag{2.3}$$

the received signal $r_i(t)$ is the sum of delayed replicas of the radiated signal, that is

$$r_i(t) = \sum_{l=1}^{L} h_l S_i(t - \Delta \tau_l) \exp\left[j\omega_0(t - \Delta \tau_l - T_d)\right] u(t - \Delta \tau_l - T_d, iT) \tag{2.4}$$

or, for its based-band equivalent

$$R_i(t) = \sum_{l=1}^{L} h_l S_i(t - \Delta \tau_l - T_d) \exp(-j\omega_0 \Delta \tau_l) u(t - \Delta \tau_l - T_d) \tag{2.5}$$

Here L is the number of distinguishable rays[4] at the receiver, h_l is the complex channel coefficient for the l-th path, and T_d is the average propagation delay. Due to the assumption about the narrowband nature of the transmitted signal, the parameters h_l, $\Delta \tau_l$, and T_d are the same for all frequency components of the transmit signal.[5]

Due to changing environmental conditions and movement of the receiver (or the transmitter, or both) parameters h_l, $\Delta \tau_l$ could be treated as random functions. However, over the duration of a single symbol T of a transmitted signal these parameters can often be considered constants. These parameters, however, may be change from symbol to symbol (or from a block of symbols to a block of symbols). In Equation (2.5) the transient process due to symbol transitions is neglected because the signals considered are relatively narrowband.

---

[4] Assumed to be a constant for the moment.
[5] These conditions are only for frequency independence of the scattering parameters on frequency, and are not to be confused with the flat fading conditions.

As the next step let us assume that

$$t_l = \tau + \Delta t_l \tag{2.6}$$

where τ is the mean time of the ray propagation and Δt_l is the variation of propagation time from the mean value. Furthermore, let us assume that

$$|\Delta t_l| \ll \frac{1}{\Delta f_i} = \frac{T}{b_i} \tag{2.7}$$

$$|\Delta t_l| \ll T \tag{2.8}$$

In this case phase shifts of various frequency components of the signal are relatively small, the signal is not distorted in shape, and, therefore, the Equation (2.5) can be written as

$$r(t) = s(t - \tau) \sum_{l=1}^{L} \gamma_l \exp(-j\phi_l) = s(t - \tau) \sum_{l=1}^{L} h_l = h \cdot s(t - \tau) \tag{2.9}$$

Here $\phi_l = \omega_0 \Delta t_l$. The complex transmission coefficient $h = \gamma \exp(-j\phi)$, defined as

$$h = \gamma \exp(-j\phi) = \sum_{l=1}^{L} h_l = \sum_{l=1}^{L} \gamma_l \exp(-j\phi_l) = x + jy \tag{2.10}$$

is a complex random process with the in-phase and quadrature components x and y respectively:

$$x = \gamma \cos \phi = \sum_{l=1}^{L} x_l = \sum_{l=1}^{L} \gamma_l \cos \phi_l \tag{2.11}$$

$$y = \gamma \sin \phi = \sum_{l=1}^{L} x_l = \sum_{l=1}^{L} \gamma_l \sin \phi_l \tag{2.12}$$

where

$$\gamma = \sqrt{x^2 + y^2}, \quad \phi = \arctan \frac{y}{x} \tag{2.13}$$

$$\gamma_l = \sqrt{x_l^2 + y_l^2}, \quad \phi_l = \arctan \frac{y_l}{x_l} \tag{2.14}$$

When analyzing the channel, its transfer function

$$H(t, \omega_0) = \gamma \exp(-j\phi) \exp(j\omega_0 t) \tag{2.15}$$

must be treated as a non-stationary complex random process. However, over a short duration T_{st} the process could be treated as locally stationary and ergodic. The channel is considered to be locally stationary over time interval T_{st} in a frequency band F_{st}, if its time-frequency covariance function $C_{\omega_0, t}(\tau, \Omega)$ can be considered independent on t and ω_0, that is

$$C_{\omega_0, t}(\tau, \Omega) = C(\tau, \Omega) \tag{2.16}$$

Furthermore, if transmitted signals are narrowband the dependency on frequency in (2.16) vanishes

$$C_{\omega_0, t}(\tau, \Omega) = C(\tau) \tag{2.17}$$

If the parameters of distribution of all random variables γ, ϕ, and τ are the same for all positions of the transmitted symbols the channel is considered to be *symmetric*.

Let τ_k be the correlation interval of the parameters of the channel. If

$$T \ll \tau_k \tag{2.18}$$

that is, the channel coefficients vary slowly over the duration of many symbols, the channel is considered slow and the frequency flat. If the condition (2.18) is not satisfied, the fading is called time-selective, or fast.

It can be seen from Equation (2.11), that the in-phase and quadrature components are the result of the summation of a large number L of subrays with limited variance, such that

$$\lim_{l \to \infty} \frac{\sigma_{lx}^2}{\sigma_x^2} = 0, \quad \lim_{l \to \infty} \frac{\sigma_{ly}^2}{\sigma_y^2} = 0 \tag{2.19}$$

In this case, the central limit theorem could be applied, even for correlated subrays.

Simple calculations allow us to evaluate the mean and the variance of the in-phase and quadrature components based on the model (2.11). In this case

$$m_I = \sum_{l=1}^{L} \mathcal{E}\{\gamma_l\} \mathcal{E}\{\cos \phi_l\}, \quad m_Q = \sum_{l=1}^{L} \mathcal{E}\{\gamma_l\} \mathcal{E}\{\sin \phi_l\} \tag{2.20}$$

$$\sigma_I^2 = \sum_{l=1}^{L} \left[\mathcal{E}\{\gamma_l^2\} \mathcal{E}\{\cos^2 \phi_l\} - \mathcal{E}^2\{\gamma_l\} \mathcal{E}^2\{\cos \phi_l\} \right] \tag{2.21}$$

$$\sigma_Q^2 = \sum_{l=1}^{L} \left[\mathcal{E}\{\gamma_l^2\} \mathcal{E}\{\sin^2 \phi_l\} - \mathcal{E}^2\{\gamma_l\} \mathcal{E}^2\{\sin \phi_l\} \right] \tag{2.22}$$

$$R_{IQ} = \frac{1}{2} \sum_{l=1}^{L} \mathcal{E}\{\gamma_l^2\} \mathcal{E}\{\sin 2\phi_l\}$$

$$+ \sum_{k=1}^{L} \sum_{l \neq k}^{L} \mathcal{E}\{\gamma_l\} \mathcal{E}\{\gamma_k\} \{\sin \phi_l\} \{\sin \phi_k\} - m_Q m_Q \tag{2.23}$$

If all elementary scatterers have the same statistics, previous equations could be further simplified to produce

$$m_I = L \mathcal{E}\{\gamma\} \mathcal{E}\{\cos \phi\}, \quad m_Q = L \mathcal{E}\{\gamma\} \mathcal{E}\{\sin \phi\} \tag{2.24}$$

$$\sigma_I^2 = L \mathcal{E}\{\gamma^2\} \mathcal{E}\{\cos^2 \phi\} - \frac{m_I^2}{L} \tag{2.25}$$

$$\sigma_Q^2 = L\mathcal{E}\{\gamma^2\}\mathcal{E}\left\{\sin^2\phi\right\} - \frac{m_Q^2}{L} \tag{2.26}$$

$$R_{IQ} = \frac{L}{2}\mathcal{E}\{\gamma^2\}\mathcal{E}\{\sin 2\phi\} - \frac{m_I m_Q}{L} \tag{2.27}$$

Some general conclusions could be reached based on Equations (2.20)–(2.27):

1. If the distribution of the phases of each partial scatter are random and distributed uniformly on $[-\pi, \pi]$, then $m_I = m_Q = 0$, $\sigma_I^2 = \sigma_Q^2$, $r = 0$.

2. If fluctuations of the unwrapped phase are much wider than the interval $[-\pi, \pi]$, the wrapped distribution approaches the uniform on $[-\pi, \pi]$.

3. If the fluctuations of the phases are symmetrical with respect to their mean values, are equal to zero but with a variance that is significantly smaller than π, it can be easily seen that $m_Q = 0$, $m_I \neq 0$, and $\sigma_I^2 \neq \sigma_Q^2$, $r = 0$.

4. Finally, if the distribution of the phases is not symmetrical and has a small variance, one obtains that $m_Q \neq 0$, $m_I \neq 0$, and $\sigma_I^2 \neq \sigma_Q^2$, $r \neq 0$.

Therefore, the particular mechanism of scattering plays a significant role in the distribution of the resultant field. The distribution of the phases is more important in this case than particular distributions of magnitudes.

In the following we consider two cases of statistical description of the scattered field using representation (2.11): at first, we assume that L is large enough so that the distributions could be assumed to be Gaussian. This case is treated in Section 2.3 and is often used in communication applications. However, some deviations from Gaussian statistics are also found in experiments. Corresponding statistics are described in Section 2.2.1.

2.2   Random walk approach to modeling of scattering field

2.2.1   Random walk in two dimensions as a model for scattering field

Modeling the distribution of electromagnetic and optical fields in the presence of multiple scatterers has been of great interest since the earliest times of communications theory. While classical Rayleigh models can be traced back as far as the end of the 19th century, more sophisticated models were developed in the 1970s and 1980s (Barakat 1986, 1987; Jakeman and Pusey 1976; Jakeman and Tough n.d.). The content of this section follows mainly (Barakat 1986, 1987; Jakeman and Pusey 1976; Jakeman and Tough n.d.). First we consider the random walk in two dimensions as presented by R. Barakat in (Barakat 1986). In this case, the signal of interest $\xi(t)$ at a given moment of time t is considered to be a sum of complex phasors

$$\xi(t) = \sum_{n=1}^{N(t)} a_n(t) \exp(j\theta_n(t)) \qquad (2.28)$$

Magnitudes $a_n(t)$, phases $\theta_n(t)$, and number of scatterers $N(t)$ are independent random variables at any fixed moment of time. For simplicity it is assumed that the distribution of magnitudes and phases is identical for all scatterers, that is $p_{a,n}(A) = p_a(A)$, $p_{\phi,n}(\theta) = p_\phi(\theta)$ and the distribution of the number of scatterers is described by some known distribution $P(N)$.

2.2.2   Phase distribution and scattering strength

It is common to distinguish two scattering regimes (Barakat 1986): the *strong* scattering regime and the *weak* scattering regime. In the strong scattering regime rms scatterer size $\sigma > \lambda$ is larger than the wavelength λ associated with the carrier frequency of the signal. In this case, no phase value is given preference and the phase could be described as uniformly distributed

$$p_\phi(\theta) = \frac{1}{2\pi}, \ 0 \le \theta \le 2\pi \qquad (2.29)$$

In contrast, in the weak scattering regime, the size of scatterers is smaller than the wavelength $\sigma < \lambda$, and, therefore, a predominant value of the phases exists. A convenient model for such a non-uniform phase could be in the form of the von Mises–Tikhonov distribution (La Frieda and Lindsey 1973),[6] often used in communication applications (Abdi and Kaveh 2002; La Frieda and Lindsey 1973; Mardia and Jupp 2000)

$$p_\phi(\theta) = \frac{\exp[\kappa \cos(\theta - \mu)]}{2\pi I_0(\kappa)}, \ -\pi \le \theta \le \pi \qquad (2.30)$$

Here κ is the shape parameter, with $\kappa = 0$ corresponding to the uniformly distributed phase $p_\phi = 1/2\pi$. For large $\kappa > 10$, the von Mises-Tikhonov distribution is well approximated by a Gaussian distribution with mean μ and variance $\sigma^2 = 1/\kappa$ (Barakat 1986; Mardia and Jupp 2000). To emphasize the effect of phase distribution rather than the effects of the magnitude distribution, we assume, following (Barakat 1986), that the amplitudes of each ray are deterministic constant a.

2.2.3   Distribution of intensity

First let us assume that the number of rays $N(t)$ is deterministic and fixed but arbitrary: $N(t) = N$. In this case, the conditional density $p_\gamma(\gamma|N)$ of the process intensity $\gamma(t) = |\xi(t)|^2$ can be represented as (Barakat 1986)

$$p_\gamma(\gamma|N) = I_0^{-N}(\kappa) I_0 \left(\kappa \sqrt{\gamma}/a \right) \frac{1}{2} \int_0^\infty J_0^N(a\omega) J_0(\sqrt{\gamma}\omega)\omega d\omega \qquad (2.31)$$

---

[6] Also know as simply the von Mises distribution or Tikhonov distribution.

for $0 \leq \gamma N^2 a^2$. It was pointed out in (Barakat 1986), that the density (2.31) can be represented as a product of two terms: one, $I_0^{-N}(\kappa)I_0\left(\kappa\sqrt{\gamma}/a\right)$, which depends on the shape parameter κ of the phase distribution, and a term which is independent of the distribution shape, and therefore is the same as in the case of the uniform phase. Such factorization is rather due to a choice of phase PDF, rather than a general property. Integrals in (2.31) could be calculated explicitly for $N = 0, 1$ and $N = 2$ to produce

$$p_\gamma(\gamma|0) = I_0\left(\kappa\sqrt{\gamma}/a\right)\delta(\gamma) \tag{2.32}$$

$$p_\gamma(\gamma|1) = I_0^{-1}(\kappa)I_0\left(\kappa\sqrt{\gamma}/a\right)\delta(\gamma - a^2) \tag{2.33}$$

$$p_\gamma(\gamma|2) = I_0^{-2}(\kappa)I_0\left(\kappa\sqrt{\gamma}/a\right)\frac{1}{\pi\sqrt{\gamma}a\sqrt{1 - \gamma/4a^2}} \tag{2.34}$$

$$\gamma \ll 4a^2 \tag{2.35}$$

The next step is to utilize (2.31) to obtain statistics in the case of the randomly varying $N(t)$ by means of the Bayes theorem:

$$p_\gamma(\gamma) = \sum_{N=0}^{\infty} p_\gamma(\gamma|N)P(N)$$

$$= I_0\left(\kappa\sqrt{\gamma}/a\right)\frac{1}{2}\int_0^{\infty} J_0(\sqrt{\gamma}\omega)\omega d\omega \sum_{N=0}^{\infty} I_0^{-N}(\kappa)J_0^N(a\omega)P(N) \tag{2.36}$$

Various distributions have been adopted in the literature to model the number of rays in the received signal (Barakat 1986; Jakeman and Tough n.d.; Suzuki 1977; Zhao et al. 2002). The simplest candidate is of course the Poisson distribution

$$P(N) = \frac{\eta^N \exp(-\eta)}{N!} \tag{2.37}$$

where $\eta = < N >$ is the average number of rays and is the only parameter defining the distribution. It is well known that the Poisson distribution is a good model under the strict randomness condition. A more relaxed distribution is the so called negative binomial distribution, described by the following probability

$$P(N) = \binom{N + \alpha - 1}{N}\frac{(\eta/\alpha)^N}{(1 + \eta/\alpha)^{N+\alpha}}, \quad \alpha \geq 0 \tag{2.38}$$

It can be easily seen that the variance of the Poisson distribution is equal to its mean $\mathrm{var}(N) = \eta$ while for the negative binomial distribution

$$\mathrm{var}(N) = \eta\left(1 + \frac{\eta}{\alpha}\right) \geq \eta \tag{2.39}$$

that is, its variance always exceeds that of the Poisson distribution. In the limit $\alpha \to \infty$ the negative binomial distribution coincides with the Poisson distribution.

Using the extended binomial theorem[7] one obtains the following expression for the PDF of the intensity γ

$$
\begin{aligned}
p_\gamma(\gamma) &= \frac{1}{2} I_0\left(\frac{\kappa\sqrt{\gamma}}{a}\right) \int_0^\infty d\omega\, \omega J_0(\sqrt{\gamma}\omega) \\
&\quad \times \sum_{N=0}^\infty \binom{N+\alpha-1}{N} \left(\frac{\eta J_0(a\omega)}{\alpha I_0(\kappa)}\right)^N \left(1+\frac{\eta}{\alpha}\right)^{-N-\alpha} \\
&= \frac{1}{2} I_0\left(\frac{\kappa\sqrt{\gamma}}{a}\right) \int_0^\infty d\omega\, \omega J_0(\sqrt{\gamma}\omega) \Phi(\omega)
\end{aligned}
\tag{2.40}
$$

where

$$
\Phi(\omega) = \left\{ 1 + \frac{\eta}{\alpha}\left[1 - \frac{J_0(a\omega)}{I_0(\kappa)}\right] \right\}^{-\alpha}
\tag{2.41}
$$

In the limiting case of the Poisson distribution $\alpha \to \infty$, Equation (2.41) can be rewritten using the well known limit $(1 + x/\alpha)^{-\alpha} \to \exp(-x)$ to produce

$$
\Phi(\omega) = \exp\left\{ -\eta\left[1 - \frac{J_0(a\omega)}{I_0(\kappa)}\right]\right\}
\tag{2.42}
$$

Finally, if the scattering is strong $\kappa = 0$, $I_0(0) = 1$ and therefore

$$
\Phi(\omega) = \exp\left\{-\eta\left[1 - J_0(a\omega)\right]\right\}
\tag{2.43}
$$

In practice, the average number η of rays is significant, that is $\eta \gg 1$. However, the total power carrier by all rays is finite, therefore one can introduce the following normalized quantities which are finite even as η increases

$$
\bar{\kappa} = \sqrt{\eta}\kappa; \quad A = \sqrt{\eta}a
\tag{2.44}
$$

while $\kappa, a \to 0$. Under this condition, Equation (2.41) can be simplified to

$$
\Phi(\omega) \approx \left[\left(1 + \frac{\bar{\kappa}^2}{4\alpha}\right) + \frac{A^2}{4\alpha}\omega^2\right]^{-\alpha}
\tag{2.45}
$$

for small values of ω. Here we use the following approximation (Abramowitz and Stegun 1965) of the Bessel function series

$$
J_0(a\omega) \approx 1 - \frac{1}{4}a^2\omega^2, \quad I_0(\kappa) \approx 1 + \frac{\kappa^2}{4}
\tag{2.46}
$$

---

[7] The extended binomial theorem (Abramowitz and Stegun 1965) provides the following expansion for any arbitrary real positive number α:

$$
(1 + x)^\alpha = \sum_{N=0}^\infty \binom{\alpha}{N} x^\alpha
$$

where

$$
\binom{\alpha}{N} = \frac{(\alpha)(\alpha-1)(\alpha-2)\cdots(\alpha-N+1)}{N!}
$$

Finally, the expression for the probability density of the intensity γ can now be calculated using (2.45) in (2.40) (Barakat 1986)

$$
\begin{aligned}
p_\gamma(\gamma) &= \frac{1}{2} I_0\left(\frac{\kappa\sqrt{\gamma}}{a}\right) \int_0^\infty d\omega\omega J_0(\sqrt{\gamma}\omega)\Phi(\omega) \\
&\approx \frac{1}{2} I_0\left(\frac{\kappa\sqrt{\gamma}}{a}\right) \int_0^\infty d\omega\omega J_0(\sqrt{\gamma}\omega)\left[\left(1+\frac{\bar{\kappa}^2}{4\alpha}\right)+\frac{A^2}{4\alpha}\omega^2\right]^{-\alpha} \\
&= \frac{2\alpha}{\Gamma(\alpha)A^{\alpha+1}}\left(\frac{\alpha}{1+\frac{\bar{\kappa}^2}{4\alpha}}\right)\gamma^{(\alpha-1)/2} I_0\left(\frac{\bar{\kappa}}{A}\sqrt{\gamma}\right) K_{\alpha-1}\left\{\frac{2}{A}\left[\left(1+\frac{\bar{\kappa}}{4\alpha}\right)\alpha\gamma\right]^{1/2}\right\}
\end{aligned}
$$

$$(2.47)$$

This density describes the distribution of the intensity in the weak scattering regime with a random number of scatterers, distributed according to negative binomial distribution.

In the strong scattering limit, that is for $\bar{\kappa} = 0$, the expression (2.47) reduces to the well known K-distribution of Jakeman and Pusey (Jakeman and Tough n.d.)

$$
p_\gamma(\gamma) = \frac{2}{\Gamma(\alpha)A^{\alpha+1}}(\alpha\gamma)^{(\alpha-1)/2} K_{\alpha-1}\left(\frac{2}{A}\sqrt{\alpha\gamma}\right)
$$

$$(2.48)$$

At the another extreme, if the distribution of the number of scatterers indeed obeys the Poisson law, that is if $\alpha \to \infty$, one can show that the distribution of intensity approaches the well known von Laue–Rice intensity PDF (Barakat 1986), which corresponds to the Rice distribution of magnitudes:[8]

$$
p_\gamma(\gamma) = \frac{1}{A^2}\exp\left(-\bar{\kappa}^2/4\right)\exp\left(-\frac{\gamma}{A^2}\right) I_0\left(\frac{\bar{\kappa}\sqrt{\gamma}}{A}\right)
$$

$$(2.49)$$

2.2.4   Distribution of the random phase

The conditional joint distribution $p_{\gamma,\psi|N}$ of the phase and intensity for an arbitrary but deterministic N can be written as (Barakat 1987)

$$
p_{\gamma,\psi|N} = \left[2\pi I_0^N(\kappa)\right]^{-1}\exp\left[\kappa\sqrt{\frac{\gamma}{a^2}}\cos\psi\right]\frac{1}{2}\int_0^\infty [J_0(a\omega)]^N J_0\left(\omega\sqrt{\gamma}\right)\omega d\omega
$$

$$(2.50)$$

Unconditional density can be obtained through the use of Bayes theorem and the same manipulations as in the previous section (Barakat 1987). This results in the following joint PDF $p_{\gamma,\psi}$ of the phase and intensity

$$
p_{\gamma,\psi} = \frac{2^\alpha\alpha^\alpha \exp\left[\bar{\kappa}\sqrt{\gamma}/A\cos\psi\right]}{2\pi\,\Gamma(\alpha)A^2(4\alpha+\bar{\kappa}^2)^{(\alpha-1)/2}}\left(\frac{\gamma}{A^2}\right)^{(\alpha-1)/2} K_{\alpha-1}\left[\sqrt{\frac{4\alpha+\bar{\kappa}}{A^2}}\gamma\right]
$$

$$(2.51)$$

---

[8] It is interesting to note that in this case, the Rician distribution is obtained not due to the presence of a single coherent wave, known as line of sight (LoS), but rather due to the contribution of a random number of partially coherent rays.

The desired distribution of the phase $p_\psi(\psi)$ can now be calculated from (2.51) by integration over the intensity γ (Barakat 1987)

$$p_\psi(\psi) = \frac{2^\alpha \alpha^\alpha}{\pi \Gamma(\alpha)(4\alpha + \bar{\kappa}^2)^\alpha} \int_0^\infty \exp\left[ys \cos \psi\right] y^\alpha K_{\alpha-1}(y) dy \qquad (2.52)$$

where

$$s = \frac{\bar{\kappa}}{\sqrt{4\alpha + \bar{\kappa}^2}}, \quad 0 \le s \le 1 \qquad (2.53)$$

This integral cannot be evaluated analytically and its direct numerical evaluation is difficult due to the slow decay of the integrand. However, Barakat (Barakat 1987) has provided a convenient Fourier series for $p_\psi(\psi)$, first by expanding $\exp\left[ys \cos \psi\right]$ in terms of the modified Bessel functions of the first kind (Abramowitz and Stegun 1965)

$$\exp\left[ys \cos \psi\right] = \sum_{n=0}^\infty a_n I_n(sy) \cos n\psi, \quad a_0 = 1, \ a_n = 2 \text{ for } n \ge 1 \qquad (2.54)$$

and further integration which can be carrier analytical. As a result, the desired expression for $p_\psi(\psi)$ has the following form (Barakat 1987):

$$p_\psi(\psi) = \frac{1}{2\pi} + \sum_{n=1}^\infty f_n(\bar{\kappa}, \alpha) \cos n\psi \qquad (2.55)$$

where

$$f_n(\bar{\kappa}, \alpha) = \frac{s^n \left(\frac{n+\alpha/2}{2}\right) \Gamma\left(\frac{n+2}{2}\right)}{\pi \Gamma(\alpha) n!} {}_2F_1\left[\frac{n}{2}, \frac{n}{2} + 1 - \alpha, n+1; s^2\right] \qquad (2.56)$$

As could be expected, the PDF $p_\psi(\psi)$ does not depend on α in the two limiting cases of $\bar{\kappa} = 0$ and $\bar{\kappa} = \infty$. In the former it degenerates to the uniform density $p_\psi(\psi) = 1/2\pi$ while in the latter it is represented by the Dirac delta $p_\psi(\psi) = \delta(\psi)$, thus characterizing the extreme case of weak scattering, which can be attributed either to LoS conditions or two specular reflections.

Similarly, asymptotic results can be derived for the case of $\alpha \to \infty$.

2.3 Gaussian case

2.3.1 Four-parametric distribution family

If the number of independent scatterers N is large and the large number theorem is applied, both in-phase $h_I(t)$ and quadrature $h_Q(t)$ components can be considered as Gaussian random processes described by the mean values m_I, m_Q and variances σ_I^2 and σ_Q^2. In general, it cannot be assumed that $h_I(t)$ and quadrature $h_Q(t)$ have the same variance and are independent in the same moment of time. Therefore, $r = \mathcal{E}\{h_I(t)h_Q(t)\} \ne 0$. However, using a change of variables

$$x(t) + jy(t) = \exp(j\phi_0)\left(h_I(t) + jh_Q(t)\right) \qquad (2.57)$$

where[9]

$$\phi_0 = \frac{1}{2} \arctan 2r \frac{\sigma_I^2 \sigma_Q^2}{\sigma_I^2 - \sigma_2^2} \tag{2.58}$$

the problem can be converted to the consideration of two uncorrelated, and therefore independent, components. It is easy to see that the new mean and variance values are given by

$$m_x = \cos \phi_0 m_I - \sin \phi_0 m_Q, \quad m_x = \sin \phi_0 m_I + \cos \phi_0 m_Q \tag{2.59}$$

and

$$\sigma_{x,y}^2 = \frac{2\sigma_I^2 \sigma_Q^2 (1 - r^2)}{\sigma_I^2 + \sigma_Q^2} \left[1 \pm \sqrt{1 - (1 - r)^2 \frac{4\sigma_I^2 \sigma_Q^2}{\left(\sigma_I^2 + \sigma_Q^2\right)^2}} \right]^{-1} \tag{2.60}$$

where the sign "+" corresponds to the x component and the sign "−" corresponds to the y component. It is important to note that one obtains the spherically symmetric process $\sigma_x^2 = \sigma_y^2$ if and only if $\sigma_I^2 = \sigma_Q^2$ and $r = 0$. Thus, in general, one can consider independent in-phase and quadrature components, however the equality of the variances should not be considered in general.

The transformation (2.57) does not change the distribution of the magnitude $A(t) = |h(t)| = \sqrt{h_I^2(t) + h_Q^2(t)}$. However, the phase distribution is circularly rotated by the angle ϕ_0 defined by Equation (2.58). Thus, without loss of generality, one can assume that the in-phase and quadrature components are independent with mean m_I and m_Q and variances σ_I^2 and σ_Q^2.

The joint distribution of the in-phase and quadrature components is a Gaussian bivariate distribution, therefore

$$p_2(x, y) = \frac{1}{2\pi \sigma_I \sigma_Q} \exp\left[-\frac{(x - m_I)^2}{2\sigma_I^2} - \frac{(y - m_Q)^2}{2\sigma_Q^2} \right] \tag{2.61}$$

Change of variables $x = A \cos \phi$, $y = A \sin \phi$ with the Jacobian

$$J = \partial(x, y)/\partial(A, \phi) = A,$$

leads to the following joint distribution of the phase and magnitude

$$p_{A,\phi}(A, \phi) = \frac{A}{2\pi \sigma_I \sigma_Q} \exp\left[-\frac{(A \cos \phi - m_I)^2}{2\sigma_I^2} - \frac{(A \sin \phi - m_Q)^2}{2\sigma_Q^2} \right] \tag{2.62}$$

Integration over an auxiliary variable produces distributions $p_\phi(\phi)$ and $p_A(A)$ of the phase and the magnitude. Since these distributions depend on four parameters

---

[9] Note, that if initially $\sigma_I^2 = \sigma_Q^2$ the rotation angle is $\phi_0 = \pi/4$ regardless of the correlation coefficient r.

m_I, m_Q, σ_I, and σ_Q these distributions are known as four-parametric distributions (Klovski 1982b; Middleton 1960; Simon and Alouini 2000).

$$p_A(A) = \frac{A}{2\pi \sigma_I \sigma_Q} \int_0^{2\pi} \exp\left[-\frac{(A \cos\phi - m_I)^2}{2\sigma_I^2} - \frac{(A \sin\phi - m_Q)^2}{2\sigma_Q^2} \right] d\phi \quad (2.63)$$

$$p_\phi(\phi) = \frac{A}{2\pi \sigma_I \sigma_Q} \int_0^{\infty} \exp\left[-\frac{(A \cos\phi - m_I)^2}{2\sigma_I^2} - \frac{(A \sin\phi - m_Q)^2}{2\sigma_Q^2} \right] dA \quad (2.64)$$

2.3.2 Distribution of the magnitude

Let us focus for a moment on the four-parametric distribution of the magnitude. Since the expression cannot be calculated analytically, some alternative expressions can be derived to represent $p_A(A)$ as a series. One possibility is to expand the term

$$\exp\left[\frac{A m_Q \sin\phi}{\sigma_Q^2} - \frac{A^2}{2}\left(\frac{1}{\sigma_Q^2} - \frac{1}{\sigma_I^2} \right) \right]$$

into a power series with respect to A using the Newton binomial formula and then to integrate term-wise with respect to the phase ϕ. As a result one obtains

$$p_A(A) = \frac{A}{\sigma_Q \sigma_I} \exp\left[-\frac{A^2}{2\sigma_I^2} - \frac{m_I^2 \sigma_Q^2 + m_Q^2 \sigma_I^2}{2\sigma_I^2 \sigma_Q^2} \right]$$

$$\times \sum_{k=0}^{\infty} \sum_{s=0}^{\infty} \frac{(2k+2s-1)!}{k!(2s)!2^k} \cdot \frac{\left(\sigma_Q^2 - \sigma_I^2\right)^k m_Q^{2s\sigma_I^{2s}}}{\sigma_Q^{2k+4s} m_I^{k+s}} A^{k+s} I_{k+s}\left(\frac{m_I}{\sigma_I^2} A \right)$$

$$(2.65)$$

It is worth noting that the expression (2.65) does not change if the in-phase and quadrature components are interchanged.

While representation in terms of in-phase and quadrature components, means, and variances is convenient in terms of parameter estimation, it is often more revealing to consider a secondary set of four parameters which present a different view on the process under consideration.

The first parameter, q^2 is defined as the ratio between the power of the coherent component to the power of the fluctuating component

$$q^2 = \frac{m_I^2 + m_Q^2}{\sigma_I^2 + \sigma_Q^2} = \frac{m^2}{\sigma_I^2 + \sigma_Q^2} \quad (2.66)$$

The coherent component of the signal is connected to the presence of non-zero mean $m > 0$. The parameter q may vary from 0 (a channel without a coherent component) to infinity (a constant channel). In realistic channels the values of q^2 are limited to about 50 (Klovski 1982b).

The next parameter of interest is the ratio between the variances of the in-phase and quadrature components

$$\beta^2 = \frac{\sigma_I^2}{\sigma_Q^2} \tag{2.67}$$

This ratio characterizes asymmetry in the channel components. It is possible to restrict the values of β to the interval $0 \le \beta^2 \le 1$ since it is always possible to relabel I/Q components.[10]

The values of m_I and m_Q also define the phase angle ϕ_c of the coherent component

$$\phi_c = \arctan \frac{m_Q}{m_I} \tag{2.68}$$

$$m_I = m \cos \phi_c, \quad m_Q = m \sin \phi_c \tag{2.69}$$

Once again, by proper rotation and relabeling we can assume that the values of ϕ_c are confined to the first quadrant: $0 \le \phi_c \le \pi/2$.

Finally, the average power $\bar{P}$ of the received signal (assuming unit power applied to the input of the channel)

$$\bar{P} = m_I^2 + m_Q^2 + \sigma_I^2 + \sigma_Q^2 \tag{2.70}$$

accounts for both the power of the coherent and incoherent components.

The four parameters q^2, β^2, m, and $\bar{P}$ uniquely define the square of the mean values and variances of the in-phase and quadrature components through the following expressions

$$m_I^2 = \frac{q^2 \bar{P} \cos^2 \phi_c}{1 + q^2} \tag{2.71}$$

$$m_Q^2 = \frac{q^2 \bar{P} \sin^2 \phi_c}{1 + q^2} \tag{2.72}$$

$$\sigma_I^2 = \frac{\beta^2 \bar{P}}{(1 + \beta^2)(1 + q^2)} \tag{2.73}$$

$$\sigma_Q^2 = \frac{\bar{P}}{(1 + \beta^2)(1 + q^2)} \tag{2.74}$$

---

[10] It is important to mention that this will also exchange m_I and m_Q components. However, this has no effect on the parameter m or $\bar{P}$ as it is defined later.

The series (2.65) is not the only representation of the four-parametric PDF. Another useful approximation can be easily obtained from (2.63) (Middleton 1960)

$$
\begin{aligned}
p_A(A) = {} & \frac{A}{\sigma_I^2 \sigma_Q^2} \exp\left[-\frac{A^2}{4}\left(\frac{1}{\sigma_I^2} + \frac{1}{\sigma_Q^2}\right) - \frac{1}{2}\left(\frac{m_I^2}{\sigma_I^2} + \frac{m_Q^2}{\sigma_Q^2}\right) \right] \\
& \times \left\{ I_0\left[\frac{A^2}{4}\left(\frac{1}{\sigma_I^2} - \frac{1}{\sigma_Q^2}\right)\right] I_0\left[A\sqrt{\frac{m_I^2}{\sigma_I^4} + \frac{m_Q^2}{\sigma_Q^4}} \right] \right. \\
& + 2\sum_{k=1}^{\infty} I_k\left[\frac{A^2}{4}\left(\frac{1}{\sigma_I^2} - \frac{1}{\sigma_Q^2}\right)\right] I_{2k}\left[A\sqrt{\frac{m_I^2}{\sigma_I^4} + \frac{m_Q^2}{\sigma_Q^4}} \right] \\
& \left. \times \cos\left[2k \arctan\left(\frac{m_Q}{m_I} \cdot \frac{\sigma_I^2}{\sigma_Q^2}\right) \right] \right\}
\end{aligned}
\tag{2.75}
$$

However, Equation (2.65) is more convenient in applications to error probability analysis.

A number of particular results can be obtained from Equation (2.65) by restricting parameters m_I, m_Q, σ_I^2, and σ_Q^2 to a particular set. If, for example, $m_Q = 0$, $\phi_c = 0$, $m_I = m$, a series for so called three-parametric distribution (Middleton 1960) is obtained

$$
p_A(A) = \frac{A}{\sigma_I \sigma_Q} \exp\left(-\frac{A^2 + m^2}{2\sigma_I^2} \right) \sum_{k=0}^{\infty} \frac{(2k-1)!! \left(\sigma_I^2 - \sigma_Q^2\right)^k}{k! 2^k \sigma_Q^{2k} m^k} A^k I_k\left(\frac{mA}{\sigma_I^2}\right)
\tag{2.76}
$$

For a symmetric channel $\sigma_I^2 = \sigma_Q^2 = \sigma^2$ the Rician distribution with the Rician factor $K = q^2$:

$$
p_A(A) = \frac{A}{P} \exp\left(-\frac{A^2}{2\sigma^2} - q^2 \right) I_0\left(\sqrt{2}q\frac{A}{\sigma} \right)
\tag{2.77}
$$

The Rayleigh distribution is obtained by setting $m = q = 0$:

$$
p_A(A) = \frac{A}{\sigma^2} \exp\left(-\frac{A^2}{2\sigma^2} \right)
\tag{2.78}
$$

If $m_I = m_Q = m = q = 0$, and noting that

$$
\lim_{x \to 0} x^k I_k(x) = \frac{1}{2^k k!}
$$

then a two-parametric sub-Rayleigh distribution emerges

$$
p_A(A) = \frac{A}{\sigma_I \sigma_Q} \exp\left(-\frac{A^2}{2\sigma_I}\right) {}_1F_1\left[\frac{1}{2}, 1, \frac{A^2}{2}\left(\frac{1}{\sigma_I^2} - \frac{1}{\sigma_Q^2}\right)\right]
$$

$$
= \frac{A}{\sigma_I \sigma_Q} \exp\left[-\frac{A^2}{4}\left(\frac{1}{\sigma_I^2} + \frac{1}{\sigma_Q^2}\right)\right] I_0\left[\frac{A^2}{4}\left(\frac{1}{\sigma_I^2} - \frac{1}{\sigma_Q^2}\right)\right] \quad (2.79)
$$

There is significant evidence of sub-Rayleigh distribution (Barakat 1986).

If $\sigma_I^2 = \beta^2 = 0$ it can be seen from the expression (2.62) that the distribution of magnitudes reduces to

$$
p_A(A) = \frac{A}{\sigma_Q^2 \sqrt{2\pi}\sqrt{A^2 - m_I^2}} \left\{ \exp\left[-\frac{\left(\sqrt{A^2 - m_I^2} - m_Q^2\right)^2}{2\sigma_Q^2}\right] \right.
$$

$$
\left. + \exp\left[-\frac{\left(\sqrt{A^2 - m_I^2} + m_Q^2\right)^2}{2\sigma_Q^2}\right] \right\} \quad (2.80)
$$

A peculiarity of the distribution (2.80) is the fact that it is defined only for $A \geq m_I$ and it approaches infinity as $A \to m_Q$. This can be explained by the fact that one of the components (in-phase in this case) is constant and equal m_I.

If $m_I = m_Q = m = 0$ the distribution (2.80) collapses to a one-parametric one-sided Gaussian distribution

$$
p_A(A) = \frac{2}{\sqrt{2\pi\sigma_Q^2}} \exp\left(-\frac{A^2}{2\sigma_Q^2}\right) \quad (2.81)
$$

This distribution has the heaviest density around zero value of the magnitude (and therefore, the SNR) which results in the worst case performance estimate for a fading signal with a given average power.

An interesting feature of the the four-parametric distribution is the fact that it predicts the possibility of two local maxima in the distribution $p_A(A)$ of the magnitude for certain values of its parameters. This kind of phenomena has been discussed in (Alpert 1967) where bimodality has been explained by a Doppler shift of certain rays in the multipath propagation environment. However, the same phenomena could be explained without bringing the Doppler effect into consideration. It is sufficient to assume a certain relationship between the parameters m_I, m_Q, σ_I^2, and σ_Q^2.

To reveal an exact relationship between m_I, m_Q, σ_I^2, and σ_Q^2 that leads to a bimodality of the magnitude distribution is probably impossible. However, it is easy to provide the necessary conditions for such bimodality:

$$0 < \sigma_M^2 < M^2, \ 0 < \sigma_m^2 < m^2, \ m^2 > 0 \tag{2.82}$$

where the following notations are used

$$M^2 = \max(m_I^2, m_Q^2), \ m^2 = \min(m_I^2, m_Q^2)$$

σ_M^2 is the variance of the component with the mean squared equal to M^2 and σ_m^2 is the variance of the component with the mean squared equal to m^2.

In many publications a two-parametric distribution

$$p_A(A; m, \Omega) = \frac{2}{\Gamma(m)} \frac{m^m}{\Omega^m} A^{2m-1} \exp\left(-\frac{m}{\Omega} A^2\right), \ m \geq 0.5 \tag{2.83}$$

known as the Nakagami distribution (Simon and Alouini 2000) is used. Here m is the ratio between the mean power of the received signal and the variance of the power

$$m = \frac{\Omega^2}{\mathcal{E}\left\{(A^2 - \Omega)^2\right\}} \tag{2.84}$$

where $\Omega = \mathcal{E}\{A^2\}$. Therefore, the parameter m is a convenient measure of the fading depth. In particular, the inverse quantity $AF = 1/m$ is known as the amount of fading (Simon and Alouini 2000). The maximum value of $AF = 2$ corresponds to the value $m = 0.5$, that is, for the case of one-sided normal distribution. Value $AF = 1$, $m = 1$ corresponds to the Rayleigh fading. Finally, a constant channel corresponds to $AF = 0$, $m = \infty$.

Originally, distribution (2.83) was derived as an approximation to the problem of distribution of non-negative functions of many random arguments. Since then this distribution has been supported by a large amount of experimental data. This can also be explained by the fact that the Nakagami distribution provides a good fit to the four-parametric distribution, which in itself has a very clear physical backing.

It follows from the definition (2.84) that the Nakagami parameter m should be defined in terms of the parameters of the four-parametric distribution as

$$
\begin{aligned}
m &= \frac{1}{2} \frac{\left(m_I^2 + m_Q^2 + \sigma_I^2 + \sigma_Q^2\right)^2}{\sigma_I^4 + \sigma_Q^4 + 2\sigma_I^2 m_I^2 + 2\sigma_Q^2 m_Q^2} \\
&= \frac{1}{2} \frac{\left(1 + \beta^2 + q^2\right)^2}{1 + \beta^2 + 2q^2(1 + \beta^2)(\beta^2 \cos^2 \phi_c + \sin^2 \phi_c)}
\end{aligned} \tag{2.85}
$$

Thus, in order to compare the four-parametric distribution with the Nakagami distribution one has to assume constant average power $\bar{P} = \Omega$ and, given β, q, and ϕ_c calculate the corresponding value of m using Equation (2.85).

For the three-parametric distribution (2.76) ($m_Q = m = \phi_c = 0$, $m_Q = m$), Equation (2.85) produces

$$m_3 = \frac{1}{2} \frac{(1 + \beta^2)(1 + q^2)^2}{1 + \beta^4 + 2q^2\beta^2(1 + \beta^2)} \tag{2.86}$$

The same equation, specialized to the case of the Rice ($\beta^2 = 1$) and sub-Rayleigh distribution ($q = 0$) produces

$$m_{\text{Rice}} = 1 + \frac{q^4}{1 + 2q^2} = \frac{(1 + K)^2}{1 + 2K} \tag{2.87}$$

and

$$m_{\text{subR}} = \frac{(1 + \beta^2)^2}{1 + \beta^4} \tag{2.88}$$

It can be seen from Equations (2.85)–(2.88) that the same value of m can correspond to different combinations of parameters m_1, m_Q, σ_I^2, and σ_Q^2, and this leads to miscalculations in the probability of errors in some modulation schemes. Only for three values of m: $m = 1/2$, $m = 1$, and $m = \infty$ is the correspondence unique. This should not come as a surprise, since one would not expect that a two-parametric distribution would represent completely the fine structure of a four-parametric distribution.

By careful analysis one can conclude that both distributions show better agreement in the areas of high values of signal magnitude. In addition, discrepancies increase as the value of the parameter β increases. Use of the Nakagami PDF significantly reduces the complexity of calculations. However, such analysis allows us to see only trends rather the a fine structure of the obtained results.

In general, the four-parametric distribution of magnitude can be thought of as a distribution of the radius vector

$$\varrho = \sqrt{\varepsilon_1^2 + \varepsilon_2^2}$$

with independent Gaussian components which have means m_1 and m_2 and variances σ_1^2 and σ_2^2 respectively.

The cumulative distribution function $P(r) = \text{Prob}\{\varrho < r\}$ can be written as a function of four parameters A, B, C, and D as follows

$$P_A(r) = \int_0^r p_A(r)dr = \sqrt{\frac{\sigma_1^2 + \sigma_2^2}{2}} \int_0^r p_A(x)dx$$

$$= \frac{1 + B^2}{4B\pi} \int_0^r x \left\{ \int_0^{2\pi} \exp\left[-\frac{1}{2}\left(\sqrt{\frac{1 + B^2}{2}} x \cos\phi - \frac{m_1}{\sigma_1} \right) \right. \right.$$

$$\left. \left. -\frac{1}{2}\sqrt{\frac{1 + B^2}{2}} x \sin\phi - \frac{m_2}{\sigma_2} \right) \right] d\phi \right\} dx = F(A, B, C, D) \tag{2.89}$$

where

$$A = \frac{r\sqrt{2}}{\sqrt{\sigma_1^2 + \sigma_2^2}}, \quad B = \frac{\sigma_1}{\sigma_2}, \quad C = \sqrt{\frac{m_1^2 + m_2^2}{\sigma_1^2 + \sigma_2^2}}, \quad D = \arctan\frac{m_2}{m_1}$$

and

$$\frac{m_1}{\sigma_1} = \frac{C\sqrt{1+B^2}\sin D}{B}, \quad \frac{m_2}{\sigma_2} = C\sqrt{1+B^2}\cos D$$

The range of parameters A, B, C, and D can be chosen such that

$$A \geq 0, \quad 0 \leq B \leq 1, \quad C \geq 0, \quad 0 \leq D \leq \frac{\pi}{2} \tag{2.90}$$

If $B = 1$, that is $\sigma_1 = \sigma_2 = \sigma$, the value of $F(A, 1, C, D)$ is reduced to the CDF of the Rice distribution, which can be expressed in terms of the Marcum Q function (Cantrell and Ojha 1987; Helstrom 1960).

The integral function $F(A, B, C, D)$ for large values of A can be approximated by

$$F(A, 1, C) = \frac{1}{2}\left\{1 + \Phi\left[A - C\sqrt{2} - O\left(\frac{1}{C}\right)\right]\right\} \tag{2.91}$$

where

$$C = \sqrt{\frac{m_1^2 + m_2^2}{2\sigma^2}} \gg 1 \tag{2.92}$$

and

$$\Phi(x) = \frac{2}{\sqrt{2\pi}} \int_0^x \exp\left(-\frac{t^2}{2}\right) dt$$

is the Kramp function (Abramowitz and Stegun 1965).

It is possible to derive similar approximations for the four-parametric CDF as long as

$$C = \sqrt{\frac{m_1^2 + m_2^2}{\sigma_1^2 + \sigma_2^2}} \gg 1 \tag{2.93}$$

If $B = 0$, then taking (2.80) into account one obtains

$$2F(A, 0, C, D) = \left\{\Phi\left[\sqrt{\frac{A^2}{2} - C^2\cos^2 D} + C\sin D\right] \right.$$
$$\left. + \Phi\left[\sqrt{\frac{A^2}{2} - C^2\cos^2 D} - C\sin D\right]\right\} \tag{2.94}$$

If $A \leq C\sqrt{2}\cos D$ then $F(A, 0, C, D) = 0$.

2.3.3   Distribution of the phase

The required distribution of the phase $\phi(t) = \arctan h_Q(t)/h_I(t)$ can be obtained from the joint distribution (2.62) by integrating over the magnitude A. This leads to the following expression for $p_\phi(\phi)$

$$p_\phi(\phi) = \frac{1}{2\pi\sigma_I\sigma_Q} \exp\left(-\frac{m_I^2}{2\sigma_I^2} - \frac{m_Q^2}{2\sigma_Q^2}\right) \int_0^\infty \gamma \exp\left[-\frac{\gamma^2}{2}\left(\frac{\cos^2\phi}{\sigma_I^2} + \frac{\sin^2\phi}{\sigma_Q^2}\right)\right.$$

$$\left. + \gamma\left(\frac{\cos\phi_c m_I}{\sigma_I^2} + \frac{\sin\phi_c m_Q}{\sigma_Q^2}\right)\right] d\gamma \tag{2.95}$$

After integration this reduces to

$$p_\phi(\phi) = \frac{\sigma_I\sigma_Q}{2\pi} \exp\left(-\frac{m_I^2}{2\sigma_I^2} - \frac{m_Q^2}{2\sigma_Q^2}\right) \frac{1 + k\sqrt{\pi}\exp(k^2)\left[1 + \Phi(\sqrt{2}k)\right]}{\sigma_Q^2\cos^2\phi + \sigma_I^2\sin^2\phi} \tag{2.96}$$

where

$$k = \frac{m_I\sigma_Q^2\cos\phi + m_Q\sigma_I^2\sin\phi}{\sigma_I\sigma_Q\sqrt{\sigma_Q^2\cos^2\phi + \sigma_I^2\sin^2\phi}}$$

Analysis of (2.96) shows that the distribution $p_\phi(\phi)$ is an asymmetric distribution on the interval $[-\pi, \pi]$ and has no more than six extrema at points

$$\phi_1 = 0, \ \phi_{2,3} = \pm\frac{\pi}{2}, \ \phi_4 = \pm\pi, \ \phi_5 = \arctan\frac{m_Q}{m_I}, \ \phi_6 = \arctan\frac{m_Q}{M_i} - \pi \ \text{if } \phi_c > 0 \tag{2.97}$$

In the case of three-parametric distribution $\phi_c = m_Q = 0$ the distribution of the phase is symmetric with respect to the origin of the coordinate system and has no more than four extrema at ϕ_1–ϕ_4. In this case the location of the maximum depends on the ratio of variances β^2.

If $\sigma_I^2 = \sigma_Q^2 = \sigma^2$, that is in the case of a Rician channel, the distribution of the phase is symmetric, has a maximum at $\phi_1 = \arctan\phi_c$, and a minimum at $\phi_1 = \arctan\phi_c - \pi$.

Let us emphasize that, in general, the magnitude and the phase of the vectors distributed according to the four-parametric distribution are dependent, except for the case of the Rayleigh process. In the case of modeling using the Nakagami-m distribution of magnitude the question of the phase distribution remains open. Often, it is assumed that such distribution is uniform, that is $p_\phi(\phi) = 1/2\pi$ and it is independent on the distribution of the magnitude:

$$p_2(A, \phi) = \frac{1}{2\pi} p_A(A) \tag{2.98}$$

and, therefore, the joint distribution of the in-phase and quadrature components is given by

$$p_2(x, y) = \frac{1}{2\pi} \frac{p_A \left(\sqrt{x^2 + y^2} \right)}{\sqrt{x^2 + y^2}} = \frac{m^m \left(x^2 + y^2 \right)^{m-1}}{\pi \Gamma(m) \Omega^m} \exp \left[-\frac{m}{\Omega} \left(x^2 + y^2 \right) \right]$$

(2.99)

Integration over one of the variables, say y, leads to the following distribution of the in-phase (or quadrature) components

$$p(x) = \frac{m^{\frac{2m-1}{4}} \exp \left(-\frac{mx}{2\bar{P}} \right) x^{m-3/2}}{\sqrt{\pi} \Gamma(m) \bar{P}^{\frac{2m-1}{4}}} W_{\frac{2m-1}{4}, \frac{2m-1}{4}} \left(\frac{mx^2}{\bar{P}} \right)$$

(2.100)

where $W_{\alpha,\beta}(x)$ is the Whittaker function (Abramowitz and Stegun 1965). It can be seen from Equation (2.100) that the assumption of the uniform but independent phase leads to the non-Gaussian distribution of the components if $m \neq 1$.

It is also worth mentioning that different distributions of the in-phase and quadrature components may give rise to the same distribution of the envelope. It can be seen that distribution of the magnitude of

$$A = \sqrt{(x + A_0 \cos \theta)^2 + (y + A_0 \sin \theta)^2}$$

of a complex random variable $x + jy$ with in independent zero-mean equal variance σ^2 Gaussian components and a complex random variable $A_0 \exp(j\theta)$ with a constant magnitude A_0 and arbitrary distributed phase θ still results in the Rice distribution of A

$$p(A) = \frac{A}{\sigma^2} \exp \left(-\frac{A^2 + A_0^2}{2\sigma^2} \right) I_0 \left(\frac{A_0 A}{\sigma^2} \right)$$

(2.101)

while the distribution of in-phase and quadrature components $x + A_0 \cos \theta$ and $y + A_0 \sin \theta$ may have a distribution that is significantly different from Gaussian if the phase θ is not uniform.

Therefore, it is important to emphasize that the statistics of the channel are more accurate if they are estimated from the in-phase and quadrature components. It is also important to remove the correlation between the components at the same moment of time.

In general, once the correlation between I and Q components is removed by rotations, the variances of these components are not equal: $\sigma_I^2 \neq \sigma_Q^2$. However, it is possible to describe situations when equalization of variances could take place.

If scatterers are distributed in a volume in such a way that the spatial dimension of the volume in one direction is much larger than in another, then in this case the large number of scatterers will appear independent.

A different scenario can be observed when the signal is reflected from a number of clusters (layers). In this case elementary scatterers could be correlated within each cluster and the resulting variances of in-phase and quadrature components will not materialize, unless the number of clusters is large. Let the received signal be written in the form

$$r_i(t) = s_i(t - \tau) \sum_{k=1}^{K} s_{ik} = s_i(t - \tau) \sum_{k=1}^{K} h_{ik} = s_i(t - \tau)h(t) \qquad (2.102)$$

2.3.4 Moment generating function, moments and cumulants of four-parametric distribution

The following expression for the moment generating function of the instantaneous SNR $\gamma = A^2 E_S/N_0$ is given in (Simon and Alouini 2000)

$$M_\gamma(s) = \frac{1}{\sqrt{\left(1 - 2\sigma_I^2 \frac{E_s}{N_0} s\right)\left(1 - 2\sigma_Q^2 \frac{E_s}{N_0} s\right)}} \exp\left[\frac{m_I^2 \frac{E_s}{N_0} s}{1 - 2\sigma_I^2 \frac{E_s}{N_0} s} + \frac{m_Q^2 \frac{E_s}{N_0} s}{1 - 2\sigma_Q^2 \frac{E_s}{N_0} s}\right]$$

$$(2.103)$$

or, equivalently

$$M_\gamma(s) = \frac{(1 + \beta^2)(1 + q^2)}{\sqrt{\left[(1 + \beta^2)(1 + q^2) - 2\beta^2 \bar{P}s\right]\left[(1 + \beta^2)(1 + q^2) - 2\bar{P}s\right]}}$$

$$\times \exp\left[\frac{q^2 \frac{r^2}{1+r^2}(1 + \beta^2)\bar{P}s}{(1 + \beta^2)(1 + q^2) - 2\beta^2 \bar{P}s} + \frac{q^2 \frac{r^2}{1+r^2}(1 + \beta^2)\bar{P}s}{(1 + \beta^2)(1 + q^2) - 2\bar{P}s}\right]$$

$$(2.104)$$

where $r^2 = m_I^2/m_Q^2$. These equations can be obtained by noting that the moment generating function of a sum of two independent variables is the product of the corresponding moment generating functions of the summands.

2.3.5 Some aspects of multiple scattering propagation

Let us consider an individual path, experiencing multiple reflections, as shown in Figure 2.3. At each hope of such propagation the signal experiences random attenuation (Barakat 1986; Bertoni 2000).

$$H_l = A_l \exp(j\phi_l) = \exp(\ln \chi_l + j\phi_l) \qquad (2.105)$$

Therefore, total attenuation H over the whole path could be represented as

$$H = \prod_{l=1}^{L} H_l = \exp\left(\sum_{l=1}^{L} \chi_l + j \sum_{l=1}^{L} \phi_l\right)$$

$$= \exp\left(\sum_{l=1}^{L} \chi_l\right) \exp\left(j \sum_{l=1}^{L} \phi_l\right) = A \exp(j\phi) \qquad (2.106)$$

If the number of hops L is large, each individual term χ_l and ϕ_l approaches zero, and therefore the central limit theorem can be applied. In other words

$$\chi = \sum_{l=1}^{L} \chi_l, \quad \phi = \sum_{l=1}^{L} \phi_l \tag{2.107}$$

could be considered Gaussian and statistically independent on each other. If $\chi = \ln A$ is distributed according to a Gaussian low with the (log)-mean m_χ and the (log)-variance σ_χ^2, the distribution of the magnitudes is described as a log-normal distribution (Simon and Alouini 2000)

$$p_A(A) = \frac{1}{\sqrt{2\pi \sigma_{chi}^2}\, A} \exp\left[-\frac{(\ln A - m_\chi)^2}{2\sigma_\chi^2}\right] \tag{2.108}$$

The log-normal distribution of the magnitude can be approximated by the Nakagami distribution discussed above by equating the amount of fading produced by both distributions. This results in a simple relation

$$\sigma_c hi^2 = \frac{1}{2}\ln\left(1 + \frac{1}{m}\right), \quad m^{-1} = \exp(4\sigma_\chi^2) - 1 \tag{2.109}$$

between the parameter m and σ_χ^2. For small log-variance $\sigma_\chi^2 \leq 0.5$ the log-normal distribution is approximated by the Nakagmi distribution with $m \geq 3$ and therefore approaches the Rice distribution. However, for much larger values of the log-variance $\sigma_\chi^2 \geq 0.5$ such approximation is not appropriate.

As has been mentioned earlier, the distribution of the total (unwrapped) phase also approaches Gaussian with the mean value m_ϕ and the variance σ_ϕ^2. However, this distribution must be wrapped onto the interval $[-\pi, \pi]$ since only this interval of phases can be uniquely measured. In general, if the unwrapped phase is described by distribution $p_\phi(\phi)$ with the characteristic function $\Theta_\phi(s)$ the distribution of the wrapped phase could be described by the following series (Mardia and Jupp 2000)

$$p_{w\phi} = \frac{1}{2\pi}\left[1 + 2\sum_{s=1}^{\infty} |\Theta_\phi(r)| \cos s\phi\right] \tag{2.110}$$

In the case of Gaussian distribution of the phase, the characteristic function $\Theta_\phi(s)$ is also Gaussian, which leads to the following expression for the wrapped distribution (Klovski 1982b; Middleton 1960)

$$p_{w\phi} = \frac{1}{2\pi}\left[1 + 2\sum_{s=1}^{\infty} \exp\left(-\frac{r^2\sigma_\phi^2}{2}\right)\cos s\phi\right] = \frac{1}{2\pi}\vartheta_3\left(\frac{\phi}{2}, \exp\left[-\frac{\sigma_\phi^2}{2}\right]\right) \tag{2.111}$$

where ϑ_3 is Jacobi's theta function (Abramowitz and Stegun 1965).

In many practical cases, the variance $\sigma_\phi^2 \gg 1$ and the wrapped distribution quickly approach the uniform distribution of the phase.

$$p_{w\phi} = \frac{1}{\pi} \qquad (2.112)$$

For smaller variances the wrapped distribution could be quite accurately approximated by the von Mises-Tikhonov distribution (Mardia and Jupp 2000)

$$p_{w\phi} \approx \frac{\exp[\kappa \cos(\phi - m_\phi)]}{2\pi I_0(\kappa)}, \quad \kappa \geq 0 \qquad (2.113)$$

where $\kappa = 1/\sigma_\phi^2$.

If the number of hops is not significant, more complicated distributions could be used to describe statistics of the magnitude A. For example, in a situation known as a "keyhole," the distribution of the magnitudes could be described as a distribution of a product of two Rayleigh variables, the so called double Rayleigh distribution (Paulraj et al. 2003). This leads to a form of K distribution considered above in Section 2.2.3 with $\alpha = 0$. Generalized K distribution could be used as an approximation for the case of multiple reflections which deviate from Gaussian/Rician statistics.

3

Models of MIMO channels

3.1 General classification of MIMO channel models

There are a number of review papers dealing with MIMO channel modeling
(Almers et al. 2007; Yu and Ottersten 2002) and many others (Correia, Ed. 2007).
We base our description on the paper (Almers et al. 2007) as the most current and advanced review. The authors classify models into physical, based on
a particular mechanism of propagation of radio waves, and analytical,[1] based on
reproducing some, mostly statistical, properties of the channels. We would adhere
to such a classification with the terms "analytical" and "phenomenological" used
interchangeably.

3.2 Physical models

Physical channel models characterize an environment on the basis of electromagnetic wave propagation by describing the double-directional multipath propagation
between the transmit and the receive arrays (Almers et al. 2007). Such models
directly model physical propagation parameters such as the complex amplitude,
DoD, DoA, polarization and time variation, reflective surfaces, scatterers, and so on.
Physical models are independent of antenna configurations (antenna pattern, number
of antennas, array geometry, polarization, mutual coupling) and system bandwidth.
However, in order to achieve accuracy, some models may imply significant complexity. In addition, physical models are usually applied to a particular scenario
and have to be changed as the location of the transmit–receive pair changes.

A very detailed overview of propagation modeling with a focus on MIMO
channel modeling is presented in (Correia, Ed. 2007). The book presents a

---

[1] Such models are often referred to as phenomenological (Primak et al. 2004).

Wireless Multi-Antenna Channels: Modeling and Simulation, First Edition.
Serguei L. Primak and Valeri Kontorovich.
© 2012 John Wiley & Sons, Ltd. Published 2012 by John Wiley & Sons, Ltd.

comprehensive summary of concepts, models, measurements, parameterization, and validation results based on the developments within the COST 273 framework.[2]

In order to emphasize the nature of the parameters modeled, physical MIMO channel models can be further classified (Almers et al. 2007) as deterministic models, geometry-based stochastic models, and non-geometric stochastic models. In this terminology deterministic models (DM), such as ray tracing and physical optics models and their refinements, characterize the physical propagation parameters in a completely deterministic manner. Such methods are applicable in very static environments, such as indoor environments. Geometry-based stochastic channel models (GSCM) can be considered as DM applied to the propagation environment randomly drawn from certain distributions. One and two ring models are good examples of geometric models (Abdi and Kaveh 2002). In contrast, non-geometric stochastic models (NGSM) impose statistical distribution on the physical parameters being modeled (DoD, DoA, delay, and so on) by their PDF without considering a particular underlying geometry. Prominent examples are the extensions of the Saleh–Valenzuela model (Wallace and Jensen 2002) and Zwik's model (Zwick et al. 2002).

3.2.1   Deterministic models

"Deterministic" physical models aim to predict the actual physical radio propagation process for a given environment. Often environmental parameters can be stored in a central database, easily retrieved, and used to predict the propagation process. While deterministic models are considered to be physically meaningful and potentially accurate, high accuracy requires substantial computational efforts and the preparation of a detailed database. It is also believed that such models can eventually augment and even substitute measurement data.

Due to substantial distance (in wavelength) between the transmitter and the receiver the method of choice for numerical simulations is ray tracing (Bertoni 2000). In short, the received signal is modeled as a sum of rays, propagating according to the laws of optics between the received and the transmitter sites and described by a layered tree structure with each node describing an obstacle and each leaf describing an unobstructed (line of sight-like) propagation scenario. The model factors in such factors as free space path loss, diffraction, distance decay law, and so on (Bertoni 2000; Lee 1997).

In order to translate the calculated EM signal into the output of the antenna elements one has to take into account antenna parameters such as the directivity of array elements, polarization, mutual coupling, and so on. Such models account for specular reflections and diffusions. However, additional effort is needed to account for diffusive reflections due to reflective surface roughness. Such phenomenon can be interpreted either as an increased number of beams (Degli-Esposti et al. 2004) or can be accounted for through considering the ray as having a finite non-zero beamwidth (Primak and Kontorovich 2009).

---

[2] See www.lx.it.pt/cost273 for details.

3.2.2   Geometry-based stochastic models

A geometry-based model is defined by locations of the scatterer that contribute to the signal in the receive antenna. In deterministic geometrical approaches, such as ray tracing, the scatterer locations are prescribed based on a particular propagation scenario. In contrast, geometry-based stochastic channel models (GSCM) choose the scatterer locations in a stochastic (random) fashion according to a certain probability distribution. A sample of the channel impulse response is then found by a simplified ray tracing-like procedure.

3.2.2.1   Single-bounce scattering

The origins of the GSCM can be traced to simulation in systems with receiver diversity. One of the first such models (Lee 1997) deterministically placed scatterers around the receive antenna and assumed that propagation was affected only by reflection from such a scatterer. The technique was refined by augmenting this single-scattering model with randomly placed scatterers (Blanz and Jung 1998; Norklit and Andersen 1998; Petrus et al. 2002). While it is claimed that random placement reflects physical reality much better,[3] it is the single-scattering assumption that makes ray tracing extremely simple: all paths, excluded from the LoS, consist of two subpaths connecting the scatterer to the transmitter and the receiver. These are described by the DoD, DoA, attenuation, and polarization rotation. In the wideband case geometry also defines the propagation time or relative delay. The scatterer interaction itself can be taken into account via an additional random phase shift and a possible change in polarization. A GSCM has a number of important advantages (Almers et al. 2007):

1. It has an immediate relation to physical reality; important parameters (like scatterer locations) can often be determined via simple geometrical considerations.

2. Many effects are implicitly reproduced: small-scale fading is created by the superposition of waves from individual scatterers; DoA and delay drifts caused by MS movement are implicitly included.

3. All information is inherent to the distribution of the scatterers; therefore, dependencies of power delay profile (PDP) and angular power spectrum (APS) do not lead to a complication of the model.

4. Tx/Rx and scatterer movement as well as shadowing and the (dis)appearance of propagation paths (e.g., due to blocking by obstacles) can be easily implemented; this allows one to include long-term channel correlations in a straightforward manner.

---

[3] The random positioning of scatterers has a randomizing effect on the phase of the received signal but may not contribute significantly to attenuation of the signal if scatterers are concentrated close to the receiver or transmitter. Therefore, as far as narrowband modeling is concerned, there is no difference between placing scatterers in a random manner or in a fixed geometrical pattern, such as rings or a disk. However, the situation is significantly different since the location of the scatterer defines the propagation delay as well and cannot be accounted for by a random phase distribution.

Three types of scatterers should be considered in GSCM models. One group is located around the transmitter. This group is seen as a very narrow angle cone from the transmitter side. However, multiple rays arrive with well randomized phase and are independent. They create a similar Doppler shift in the received signal and, as a result, have narrow PSD. Such scatterers create strong correlation between the received signals in different antenna elements, while preserving the spherical symmetry of such processes. The second group of scatterers is located around the receiver itself. As a result, such scatterers create often loosely correlated signals in different antenna elements and a wide PSD to significant diversity in AoA. Finally, a group of scatterers that is located far from both the transmitter and the receiver sides could be seen as a near point scatterer cluster. The signal reflected from such clusters has a partially randomized phase but a very narrow AoA spread. This situation is treated in (Primak and Kontorovich 2009).

3.2.2.2 Multiple-bounce scattering

One of the characteristic features of the single-bounce model considered in the previous section is that it imposes a rigid dependance between the AoA and AoD.[4] In other words

$$p(\phi_T, \phi_R) = p_T(\phi_T)\delta(\phi_R - f[\phi_R]) \tag{3.1}$$

where $f(x)$ is some deterministic function. This could be appropriate picocells, when propagation is dominated by LoS and a single-bounce reflection. However, in a microcell environment with a low placed base station, or even to some degree in macrocells, the propagation mostly involves propagation along guiding structures, such as street canyons. Such propagation could be treated as propagation with multiple bounces (subject to reflection and diffraction) from the walls of the guiding structure (Kuchar et al. 2000; Suzuki 1977; Toeltsch et al. 2002).

In the narrowband case this could be easily circumvented by randomizing the phase of the received signal. Indeed, in this case the delay spread of the signal caused by the difference in propagation path cannot be resolved, thus all rays arriving at the receiver with a given AoA can be attributed a single scattering cluster. Rigid dependence of the arrival phase on the distance traveled, and thus sensitivity to the cluster location, is removed by phase randomization. This is why single-bounce models of narrowband channels are popular and representative. However, in the case of wideband channels, additional measures are required, as described in Chapter 4.

3.3 Analytical models

An alternative approach to physical models, known as analytical models, ignores any physical propagation mechanisms and reproduces the statistical properties of

---

[4] In addition, in the wideband case it also imposes a rigid dependency between AoA, AoD, and ToA.

the MIMO matrix in the corresponding domain. Analytical models are often useful when the operation of communication system design is considered on the algorithm selection, evaluation, and verification level. It allows one to avoid the computationally costly reproduction of a channel's physical properties, which are the same for the different algorithms under test. It is assumed at this stage that how to translate the physical model into a proper analytical model is known, either by means of simulation, measurements, or theoretical argument (Debbah and Muller 2005).

Two main classes of analytical models can be clearly identified: propagation-motivated models and correlation-based models (Almers et al. 2007). The former, which includes the finite scatterer model (Burr 2003; Zhang et al. 2006), the maximum entropy model (Debbah and Muller 2005), and the virtual channel representation (Sayeed 2002), represents the channel matrix via propagation parameters; the latter, such as IID (Paulraj et al. 2003), Kronecker (Kermoal et al. 2002), and Weichselberger (Weichselberger et al. 2006), characterizes the MIMO channel matrix statistically in terms of the correlations between the matrix entries.

Most of the narrowband analytical models are based on a multivariate complex Gaussian distribution of the MIMO channel coefficients. In the case of the proper (or symmetric) complex Gaussian process (Schreier and Scharf 2003a) this leads to the well known Rayleigh or Rician fading models, almost universally accepted in research development (Paulraj et al. 2003). In the more complicated case of improper complex Gaussian a few different approaches have been undertaken as explained below in Section 3.3.1.

3.3.1   Channel matrix model

Let us consider a MIMO channel that is formed by N_T transmit and N_R receive antennas. The $N_R \times N_T$ channel matrix

$$\mathbf{H} = \mathbf{H}_{LoS} + \mathbf{H}_{diff} + \mathbf{H}_{sp} \tag{3.2}$$

can be decomposed into three components. The line of sight component $\mathbf{H}_{LoS}$ could be represented as

$$\mathbf{H}_{LoS} = \sqrt{\frac{P_{LoS}}{N_T N_R}} \mathbf{a}_L \mathbf{b}_L^H \exp(j\phi_{LoS}) \tag{3.3}$$

Here P_{LoS} is the power carried by the LoS component, $\mathbf{a}_L$ and $\mathbf{b}_L$ are receive and transmit antenna manifolds (van Trees 2002), and ϕ_{LoS} is a deterministic constant phase. Elements of both manifold vectors have unity amplitudes and describe a phase shift in each antenna with respect to some reference point.[5] Elements of the matrix $\mathbf{H}_{diff}$ are assumed to be drawn from proper (spherically-symmetric) complex Gaussian random variables with zero-mean and correlation between elements, imposed by the joint distribution of angles of arrival and departure (Almers et al. 2007). This is due to the assumption that the diffusion component is composed of

---

[5] This is not true when the elements of the antenna arrays are not identical or different polarizations are used.

a large number of waves with independent and uniformly distributed phases due to large and rough scattering surfaces. Both the LoS and diffusive components are well studied in the literature. A combination of the two leads to the well known Rice model of MIMO channels (Almers et al. 2007).

Proper statistical interpretation of the specular component $\mathbf{H}_{sp}$ is much less developed in MIMO literature, despite its wide application in optics and random surface scattering (Beckmann and Spizzichino 1963). Specular components represent an intermediate case between an LoS and a purely diffusive component. Formation of such a component is often caused by mild roughness, therefore the phases of different partial waves have either strongly correlated phases or non-uniform phases.

In order to model the contribution of specular components to the MIMO channel transfer function we consider first a contribution from a single specular component. Such a contribution can be easily written in the following form

$$\mathbf{H}_{sp} = \sqrt{\frac{P_{sp}}{N_T N_R}} \, [\mathbf{a} \odot \mathbf{w}_a] \, [\mathbf{b} \odot \mathbf{w}_b]^H \, \xi \tag{3.4}$$

Here P_{sp} is the power of the specular component, and $\xi = \xi_R + j\xi_I$ is a random variable drawn according to Equation (2.53) from a complex Gaussian distribution with parameters $m_I + jm_Q$, σ_I^2, σ_Q^2, and independent in-phase and quadrature components. Since specular reflection from a moderately rough or very rough surface allows reflected rays to be radiated from the first Fresnel zone it appears as a signal with some angular spread. This is reflected by the window terms $\mathbf{w}_a$ and $\mathbf{w}_b$ (Primak and Sejdic 2008; van Trees 2002). It is shown in (Primak and Sejdic 2008) that it could be well approximated by so called discrete prolate spheroidal sequences (DPSS) (Percival and Walden 1993) or by a Kaiser window (Percival and Walden 1993; van Trees 2002). If there are multiple specular components, formed by different reflective rough surfaces such as those in an urban canyon scenario 3.1, the resulting specular component is a weighted sum of terms, similar to those defined by (3.4). Each such term is defined for different angles of arrival and departure:

$$\mathbf{H}_{sp} = \sum_{k=1} \sqrt{\frac{P_{sp,k}}{N_T N_R}} \, [\mathbf{a}_k \odot \mathbf{w}_{a,k}] \, [\mathbf{b}_k \odot \mathbf{w}_{b,k}]^H \, \xi_k \tag{3.5}$$

It is important to mention that in the mixture (3.5), unlike the LoS component, the absolute value of the mean term is not the same for different elements of the matrix $\mathbf{H}_{sp}$. Therefore, it is not possible to model them as identically distributed random variables. Their parameters (mean values) also have to be estimated individually. However, if the angular spread of each specular component is very narrow, the windows $\mathbf{w}_{a,k}$ and $\mathbf{w}_{b,k}$ could be assumed to have only unity elements. In this case variances of the in-phase and quadrature components of all elements of the matrix $\mathbf{H}_{sp}$ are the same.

The main focus of analytical models is on representing the diffusive component $\mathbf{H}_{diff}$ of the channel matrix. Usually, its deterministic part is absorbed into the LoS or specular component, therefore it is possible to assume that elements of $\mathbf{H}_{diff}$ are proper zero-mean complex Gaussian processes. A complete statistical description of such a matrix can be achieved by specifying properties of $N_T N_R \times N_T N_R$ the covariance matrix

$$\mathbf{R_h} = \mathcal{E}\left\{\mathbf{h}_{diff}\, \mathbf{h}_{diff}^H\right\} \tag{3.6}$$

of the vector $\mathbf{h}_{diff} = \text{vec}\,\mathbf{H}_{diff}$. The matrix $\mathbf{R_h}$ is known as the full covariance matrix, has a high dimension, and is very difficult to obtain from measurements (Paulraj et al. 2003). Therefore, less general but more practically achievable descriptions are required.

In the rare cases when the matrix $\mathbf{R_h}$ is known, either from measurements or through theoretical assumptions,[6] there is a simple method of generating independent samples of the channel realization. Indeed, let $\mathbf{R_h} = \mathbf{U}\mathbf{\Lambda}\mathbf{U}$ be the eigenvalue decomposition of $\mathbf{R_h}$. Then, if one defines $\mathbf{A} = \mathbf{R_h}^{1/2} = \mathbf{U}\mathbf{\Lambda}^{1/2}\mathbf{U}$, $\mathbf{R_h} = \mathbf{A}\mathbf{A}^H$, the vector $\mathbf{h}$ can be generated from i.i.d. zero-mean unit variance complex Gaussian vector $\mathbf{w}$ of size $N_T N_R \times 1$, $\mathcal{E}\left\{\mathbf{w}\mathbf{w}^H\right\} = \mathbf{I}$ as

$$\mathbf{h} = \mathbf{A}\mathbf{w} = \mathbf{R_h}^{1/2}\mathbf{w}, \quad \mathbf{H} = \text{unvec}\,\mathbf{h} \tag{3.7}$$

There are a number of simplified models and their hierarchy is considered in (Weichselberger 2004). In this formulation, elements of the matrix $\mathbf{R_h}$ are considered as 2×2 tensors. Indeed, the description in the form of the covariance matrix $\mathbf{R_h}$ hides the directional nature of the MIMO link.[7]

A more nuanced approach is to consider elements

$$R_{mp}^{nq} = \mathcal{E}\left\{h_m^n h_p^{q*}\right\} \tag{3.8}$$

of the correlation matrix $\mathbf{R_h}$ as tensors of order 4, twice covariant and twice contravariant. The tensor of order 4 can be mapped onto a matrix (tensor of order 2) in two different ways. One, defined by Equation (3.6) defines a Hermiatian symmetric correlation tensor, which represents correlations between individual links.[8] Alternatively, link end oriented arrangements can be achieved through the following, generally asymmetric $N_R^2 \times N_T^2$ "link-end-oriented" matrix (Weichselberger 2004)

$$\mathbf{Q_H} = \mathcal{E}\left\{\mathbf{H}^* \otimes \mathbf{H}\right\} \tag{3.9}$$

This matrix provides more structure since it clearly distinguishes between the ends of the link.

While tensor $\mathcal{R} = R_{mp}^{nq}$, correlation matrices $\mathbf{R_h}$, and $\mathbf{Q_H}$ provide a complete statistical description of a MIMO channel, it is their orthogonal decompositions that are particularly revealing of the channel properties, since they indicate degrees of freedom and variation imbedded into the channel. Such decompositions can be achieved in a number of ways as described below.

---

[6] A good example of such assumptions are numerous ring models (Abdi and Kaveh 2002).

[7] In other words, it is impossible to identify which element corresponds to the transmit and which to the receive antennas, with the first element being an exception.

[8] Such a definition ignores the role of the transmit and receive side since one antenna from each side participates in each link.

3.3.1.1 Eigenmodes

All random realizations of a MIMO chanel $\mathbf{H}$ belong to a space, which is spanned by no more than $N_T N_R$ eigenmodes $\mathbf{U}_k$, or, equivalently by order 2 tensors[9] $U_m^n[k]$. Since R_{mp}^{nq} is an order 4 tensor, it can be represented as a sum of the product of two tensors of order 2 (matrices):

$$\mathcal{R} = R_{mp}^{nq} = \sum_{k=1}^{N_T N_R} \sigma_k^2 U_m^n[k] U_p^q[k] \qquad (3.10)$$

This is a tensor representation of the well known SVD decomposition (Golub and van Loan 1996)

$$\mathbf{R_h} = \sum_{k=1}^{N_T N_R} \sigma_k^2 \cdot \operatorname{vec} \mathbf{U}_k \operatorname{vec} \mathbf{U}_k^H = \sum_{k=1}^{N_T N_R} \sigma_k^2 \cdot \mathbf{u}_k \mathbf{u}_k^H \qquad (3.11)$$

where $\mathbf{u}_k = \operatorname{vec} \mathbf{U}_k$. The eigenmodes $U_m^n[k]$ must satisfy the invariance equation

$$R_{mp}^{nq} U_r^s[k] \delta^{rp} \delta sq = \sigma_k^2 U_m^n[k] \qquad (3.12)$$

or, in its equivalent vector notation

$$\mathbf{R_h} \mathbf{u}_k = \sigma_k^2 \mathbf{u}_k \qquad (3.13)$$

Equation (3.13) provides a convenient computational form for finding eigenmodes.

Since $\mathbf{R_h}$ is a positive semidefinite Hermiatian matrix, all the eigenvalues σ_k^2 are non-negative. Furthermore, using identity (A117) in (van Trees 2002) one obtains

$$\operatorname{tr} \mathbf{U}_l^H \mathbf{U}_k = \operatorname{vec}^H \mathbf{U}_l \operatorname{vec} \mathbf{U}_k = \delta_{lk} \qquad (3.14)$$

that is, the eigenmodes are orthonormal to each other. The sum of eigenvalues

$$\sum_{k=1}^{N_T N_R} \sigma_k^2 = \mathcal{E}\{\|\mathbf{H}\|_F^2\} \qquad (3.15)$$

is the average energy/power of the transfer matrix. Finally, Equation (3.7) can be rewritten in terms of eigenmodes as

$$\mathbf{H} = \sum_{k=1}^{N_T N_R} \xi_k \sigma_k \mathbf{U}_k \qquad (3.16)$$

where ξ_k are mutually independent zero-mean proper complex random processes with unit variance.

Equation (3.16) can be considered a generalization of the Karhunen–Loéve expansion of an ordinary random variable. In fact, this is the only expansion of $\mathbf{H}$ in terms of independent components (Weichselberger 2004). The number of

---

[9] Since upper and lower case indices in tensor notation are used to indicate the coordinates of the tensor, we would use a number in square brackets to indicate an order of a tensor in a set of tensors. For example, $U_m^n[k]$ will be the k-th tensor of order 2 corresponding to a matrix $\mathbf{U}_k$.

significant[10] non-zero eigenvalues σ_k^2 and their power specify the amount of spatial diversity offered by the channel.

Since each eigenmode $\mathbf{U}_k$ is itself a tensor of order 2, it can be further represented as a sum of rank 1 modes through the corresponding SVD decomposition

$$\mathbf{U}_k = \sum_{l=1}^{N_{min}} \mu_{l,k}^2 \mathbf{u}_l[k]\mathbf{w}_l^H[k] \tag{3.17}$$

where $N_{min} = \min(N_T, N_R)$. Rank 1 eigenmodes, or *fibers*, $\mathbf{u}_l[k]$ and $\mathbf{w}_l[k]$ are associated with the receive and transmit side of each eigenmode $\mathbf{U}_k$. Fibers $\mathbf{u}_{l_1}[k_1]$ and $\mathbf{u}_{l_2}[k_2]$ belonging to different eigenmodes $\mathbf{U}_k$ are not orthogonal in general. The number of non-zero singular values $\mu_{k,l}^2$ greater than one is either the result of multipath components which induce a non-orthogonal array response in the corresponding antenna array, or is caused by highly correlated or extended scattering clusters (Weichselberger 2004).

The existence of higher order fibers with non-zero power indicates that, at least theoretically, the spatial multiplexing and spatial diversity are completely decoupled. Indeed, any full rank eigenmode has full spatial multiplexing potential but no diversity. At the same time, the directivity of a MIMO channel limits the excitation opportunities to a subset of orthogonal modes, thereby limiting the amount of multiplexing that can be realistically achieved.

As has been mentioned earlier, it is not realistic to expect that the full correlation matrix will be accurately known, especially when it is estimated from measurements in real time. Therefore it is important to consider simplified models, which are based on a smaller amount of information about the correlation matrix. In particular, it is important to understand and incorporate so-called transmit

$$\mathbf{R_T} = \mathcal{E}\left\{\mathbf{H}_{diff}^H \, \mathbf{H}_{diff}\right\} = \mathbf{U_T}\mathbf{\Lambda}_T\mathbf{U}_T^H \tag{3.18}$$

and receive

$$\mathbf{R_R} = \mathcal{E}\left\{\mathbf{H}_{diff} \, \mathbf{H}_{diff}^H\right\} = \mathbf{U_R}\mathbf{\Lambda}_R\mathbf{U}_R^H \tag{3.19}$$

correlation matrices, since they are easier to estimate from the data. At the same time, these matrices are related to the end properties of the scattering environment on each side of the link.

3.3.1.2   IID model

The simplest analytical MIMO model is the so-called i.i.d. model (sometimes referred to as the canonical model). In this case

$$\mathbf{R_h} = \sigma_h^2 \, \mathbf{I}_{N_T N_R \times N_T N_R} \tag{3.20}$$

where σ_h^2 is the variance of an individual link. Such a situation corresponds to a rich scattering environment with far spaced antennas. The model has only a single

---

[10] Proper characterization of the diversity allotted by a MIMO channel can only be defined by a complete spectrum of eigenvalues σ_k^2.

parameter σ_h^2 and is somewhat unrealistic, especially when a significant number of antennas is used. However, it is a very simple and useful first approximation in problems related to the evaluation of capacity as a function of the number of antenna elements. From an information-theoretical point of view, the i.i.d. model (3.20) arises as the best approximation of the full covariance matrix $\mathbf{R_h}$ when only its average energy $\sigma_h^2 = ||\mathbf{R_h}||_F^2$ is known.

3.3.1.3 Kronecker product

Let us consider the SVD decomposition of the matrix $\mathbf{Q}_H$, defined by Equation (3.9)

$$\mathbf{Q_H} = \mathcal{E}\left\{\mathbf{H}^* \otimes \mathbf{H}\right\} = \sum_{k=1}^{N_{min}} \mu_k \, \mathbf{v}_k \mathbf{w}_k^H \tag{3.21}$$

where the eigenmodes $\mathbf{v}_k$ and $\mathbf{w}_k$ can be thought of as the vectorization of some matrices $\mathbf{V}_k$ and $\mathbf{W}_k$ respectively

$$\mathbf{v}_k = \text{vec}\,\mathbf{V}_k, \quad \mathbf{w}_k = \text{vec}\,\mathbf{W}_k \tag{3.22}$$

Let us note that $\mathbf{V}_k$ is a size $N_R \times N_R$ and $\mathbf{W}_k$ is a size $N_T \times N_T$. In other words, each of these matrices describes properties of the channel from the receive and transmit sides respectively.

It is easy to see that

$$\mathbf{Q_H}\mathbf{w}_k = \mu_k \mathbf{v}_k \tag{3.23}$$

$$\mathbf{v}_k^H \mathbf{Q_H} = \mu_k \mathbf{w}_k \tag{3.24}$$

as well as

$$\text{tr}\,\mathbf{V}_l^H \mathbf{V}_k = \delta_{kl}, \; \text{tr}\,\mathbf{W}_l^H \mathbf{W}_k = \delta_{kl} \tag{3.25}$$

Furthermore, the following expansion of the full covariance matrix can be easily obtained by careful inspection of the elements of $\mathbf{Q}_H$ and $\mathbf{R_h}$ (Pitsianis and van Loan 1993)

$$\mathbf{R_h} = \sum_{k=1}^{N_{min}} \mu_k \mathbf{W}_k \otimes \mathbf{V}_k \tag{3.26}$$

where the operator $\otimes$ stands for the Kronecker product of two matrices (tensors) (Golub and van Loan 1996).

It is proven in (Pitsianis and van Loan 1993) and rediscovered in (Weichselberger 2004) that if the eigenvalues μ_k are ordered in descending order, then $\mu_1 > 0$, $\mathbf{W}_1 = \mathbf{W}_1^H$, and $\mathbf{V}_1 = \mathbf{V}_1^H$ are semipositive Hermian matrices, that is, they correspond to some correlation functions. Furthermore, $\mathbf{R}_1 = \mu_1 \mathbf{W}_1 \otimes \mathbf{V}_1$ is a solution which minimizes $||\mathbf{R_h} - \mathbf{R}_1||_F$ such that $\mathbf{R}_1$ is a Kronecker product of two matrices.[11] Unfortunately, only $\mu_1 > 0$ is assured in such a

---

[11] In fact, $\mathbf{R}_K = \sum_{k=1}^K \mu_k \mathbf{W}_k \otimes \mathbf{V}_k$ is the best approximation of $\mathbf{R_h}$ by the sum of $K < N_{min}$ Kronecker products. However, when a partial sum $\mathbf{R}_K$ is always positive definite, the approximation error $\mathbf{R_h} - \mathbf{R}_K$ is, in general, indefinite (Pitsianis and van Loan 1993; Weichselberger 2004).

decomposition, therefore only μ_1 can be treated as an energy/power of the corresponding modes.

Thus, the simplest model that accounts for some correlation on each side of the link can be found in the form

$$\mathbf{R_h} = \mu_1 \mathbf{W}_1 \otimes \mathbf{V}_1 \tag{3.27}$$

In this case, representation (3.7) can be written as

$$\mathbf{h} = \mathbf{R_h}^{1/2}\boldsymbol{\xi} = \sqrt{\mu_1}\,\mathbf{W}_1^{1/2} \otimes \mathbf{V}_1^{1/2}\boldsymbol{\xi} \tag{3.28}$$

Equivalently, one obtains

$$\mathbf{H} = \sqrt{\mu_1}\,\mathbf{V}_1^{1/2}\mathbf{H}_w\mathbf{W}_1^{1/2} \tag{3.29}$$

where $\boldsymbol{\xi} = \text{vec}\,\mathbf{H}_w$ is a vector of unit variance zero-mean independent complex Gaussian random variables.

In order to clarify the meaning of $\mathbf{V}$ and $\mathbf{W}$ in this model one can calculate the transmit and the receive correlation matrices as defined by (3.18) and (3.19). On one side

$$\mathbf{R}_T = \mathcal{E}\left\{\mathbf{H}^H\mathbf{H}\right\} = \mu_1\mathbf{W}_1^{1/2}\mathcal{E}\left\{\mathbf{H}_w^H\mathbf{H}_w\right\}\mathbf{W}_1^{1/2} = \mu_1 N_R\mathbf{W}_1 \tag{3.30}$$

Similarly

$$\mathbf{R}_R = \mu_1 N_T\mathbf{V}_1 \tag{3.31}$$

Thus, Equation (3.29) can be rewritten as

$$\mathbf{H} = \frac{1}{\sqrt{N_T N_R}}\,\mathbf{R}_T^{1/2}\mathbf{H}_w\mathbf{R}_R^{1/2} \tag{3.32}$$

This form was first postulated in (Kermoal et al. 2002) without explicit reference to Pitsianis work.

It follows from representation (3.32) that the receive and the transmit side of the channel link could be changed independently. This translates into the fact that the distribution of AoA and AoD are statistically independent. The distributions of AoA are often available from numerous measurements conducted for SIMO systems and range from a uniform distribution, to the Laplace, Gaussian, and von-Mises, and their weighted sums (Abdi and Kaveh 2002). Depending on the propagation scenario, the distribution of AoD could also be chosen from these models. For example, for a high elevation of BS antenna it is possible to assume that AoD form a narrow cone around the LoS direction. In contrast, the low placed antenna of a mobile, used as a transmitter, should be associated with a wide range of AoD.

The Kronecker model can be easily expanded to accommodate more complicated scenarios. This leads to the following representation

$$\mathbf{H} = \sum_{k=1}^{K}\alpha_k\mathbf{V}_k^{1/2}\mathbf{H}_{wk}\mathbf{W}_k^{1/2} \tag{3.33}$$

where $\mathbf{H}_{wk}$ are $N_R \times N_T$ matrices of complex i.i.d. Gaussian processes with unit variances. However, due to the fact that some of the coefficients α_k could be negative, the generalization (3.33) does not allow for any useful physical interpretation.

Further modification of the Kronecker model can be advanced by the observation that the covariance matrices $\mathbf{R}_T$ and $\mathbf{R}_R$ contain two sorts of information, encoded into their SVD decompostion (3.18) and (3.19). The eigenvectors of such a decomposition describe the "virtual" beamforming directions, which produce orthogonal transmit or receive modes. There are always as many such directions as the size of the corresponding antenna array. The eigenvalues determine the amount of total energy allocated/collected from a particular eigendirection.

In this case, representation (3.32) can be rewritten as

$$\mathbf{H} = \frac{1}{\sqrt{N_T N_R}} \mathbf{U}_T^H \mathbf{\Lambda}_T^{1/2} \mathbf{U}_T \mathbf{H}_w \mathbf{U}_R^H \mathbf{\Lambda}_R^{1/2} \mathbf{U}_R^H = \frac{1}{\sqrt{N_T N_R}} \mathbf{U}_T^H \mathbf{\Lambda}_T^{1/2} \mathbf{H}_{w1} \mathbf{\Lambda}_R^{1/2} \mathbf{U}_R^H \tag{3.34}$$

where $\mathbf{H}_{w1} = \mathbf{U}_T \mathbf{H}_w \mathbf{U}_R^H$ is still a matrix of i.i.d. unit variance complex Gaussian variables, since unitary transforms $\mathbf{U}_T$ and $\mathbf{U}_R$ do not change the statistical nature of the matrix $\mathbf{H}_w$.

Now, the transfer matrix between receiver and the transmitter can be thought of as a sequence of the following steps:

- Transmit mode formation using $\mathbf{U}_T$.

- Transmit mode loading using $\mathbf{\Lambda}_T^{1/2}$.

- Transfer among transmit and received modes using $\mathbf{H}_{w1}$.

- Receive power loading using $\mathbf{\Lambda}_R^{1/2}$.

- Received mode mixing using $\mathbf{U}_R$.

While modes are not associated with any physical feature (in other words they are virtual modes), such a representation is very convenient from at least two perspectives: a) it allows us to better understand the mechanism of transfer of power from one side of the link to another in the presence of link correlation; b) it allows us to construct a number of more general models. In fact, representation (3.34) indicates that a channel with a Kroneker structure provides the same amount of diversity as an i.i.d. channel, since the covariance matrices on each side will normally have a full rank.[12] The loss of capacity could then be attributed to uneven mode loading, which could be partially compensated by water-filling in systems with a feedback.

Before proceeding to considerations of possible generalizations of the Kronecker model, let us investigate power transfer between virtual modes. This could be described by the so-called mode-coupling matrix $\mathbf{G} = \{g_{kl}\}$ (or coupling vector $\mathbf{g} = \mathrm{vec}\, \mathbf{G}$) (Sayeed 2002) which represents the average power of the virtual path that corresponds to the k-th transmit and l-th receive mode. Mathematically

---

[12] An important exception is the scenario with a small number of small scattering centers.

it can be achieved by calculating the diagonal of the covariance matrix of

$$\mathbf{h}_{w1} = \text{vec} \sqrt{N_T N_R} \mathbf{U}_T \mathbf{H} \mathbf{U}_R^H = \sqrt{N_T N_R} \, \mathbf{U}_R^* \otimes \mathbf{U}_T \, \text{vec} \, \mathbf{h} \tag{3.35}$$

$$\mathbf{g} = N_T N_R \, \text{diag} \, \mathcal{E} \left\{ \mathbf{h}_{w1} \mathbf{h}_{w1}^H \right\} = N_T N_R \mathbf{U}_R^* \otimes \mathbf{U}_T \mathbf{R_h} \mathbf{U}_T^H \otimes \mathbf{U}_R^T \tag{3.36}$$

It must be noted that Equation (3.36) also gives a practical way of estimating $\mathbf{g}$ from experimental data. Furthermore, if $\mathbf{G}_2$ is such a matrix that[13] $\mathbf{G} = \mathbf{G}_2 \odot \mathbf{G}_2$, Equation (3.34) can be finally rewritten in a form most suitable for generalization

$$\mathbf{H} = \mathbf{U}_T^H (\mathbf{G}_2 \odot \mathbf{H}_{w1}) \mathbf{U}_R \tag{3.37}$$

In the case of the Kronecker channel, it is easy to see that the matrix $\mathbf{G}$ is a rank 1 matrix formed by the eigenvalues of the transmit and receive correlation matrices:

$$\mathbf{G} = \boldsymbol{\lambda}_R \boldsymbol{\lambda}_T^H, \quad \boldsymbol{\lambda}_R = \text{diag} \, \boldsymbol{\Lambda}_R, \quad \boldsymbol{\lambda}_T = \text{diag} \, \boldsymbol{\Lambda}_T \tag{3.38}$$

There are a number of ways one can choose to generalize the model (3.38). One choice is to enforce the structure of the modes on both sides. If such modes are given by a DFT matrix, so-called virtual channel representation (VCR) is obtained (Sayeed 2002); if such modes are given by discrete prolate spheroidal sequences, a model described in (Kontorovich et al. 2008) can be recovered. The maximum entropy case is treated in Section 3.3.1. If matrix $\mathbf{G}$ can be allowed to have a rank higher than 1, the Weichselberger or UIU model is obtained (Tulino et al. 2005a; Weichselberger et al. 2006). If one allows non-square matrices $\mathbf{U}_R$ and $\mathbf{U}_T$ with orthonormal columns $\mathbf{U}_R^H \mathbf{U}_R = \mathbf{I}_{N_R}, \mathbf{U}_T^H \mathbf{U}_T = \mathbf{I}_{N_T}$ one can decompose radiated and received signals in a plane or spherical waves as in (Lamahewa et al. 2006; Teal et al. 2002). Finally, the double-bouncing model could be suggested as in (Oestges Clerckx 2007) to account for such effects as keyhole.

3.3.1.4 Separable maximum entropy model

The ME principle can be invoked when only limited information is available about the channel. In this section we assume that the receive and transmit correlation matrices $\mathbf{R}_R$ and $\mathbf{R}_T$ are known and that no other information about the channel is available. Following (Wallace and Maharaj 2007) we will look at such a distribution $p(\mathbf{H})$ of $\mathbf{H}$ that maximizes the entropy

$$\mathcal{H} = \int p(\mathbf{H}) \ln p(\mathbf{H}) d\mathbf{H} \tag{3.39}$$

given the following obvious constraints

$$p(\mathbf{H}) \geq 0, \quad \int p(\mathbf{H}) d\mathbf{H} = 1, \quad \sum_{k=1}^{N_T} H_{ik} H_{jk}^* = \mathbf{R}_R[i, j], \quad \sum_{k=1}^{N_R} H_{ki} H_{kj}^* = \mathbf{R}_T[i, j] \tag{3.40}$$

---

[13] The matrix $\mathbf{G}_2$ is element wise square root of the matrix $\mathbf{G}$.

Setting the Lagrange function $\mathcal{L}(\mathbf{H})$

$$\mathcal{L}(\mathbf{H}) = \int p(\mathbf{H}) \ln p(\mathbf{H}) d\mathbf{H} - \sum_{k=1}^{N_R} \sum_{l=1}^{N_R} a_{kl} \left(\sum_{n=1}^{N_R} H_{kn} H_{ln}^* - \mathbf{R}_T[k,l] \right)$$

$$- \sum_{k=1}^{N_R} \sum_{l=1}^{N_R} b_{kl} \left(\sum_{n=1}^{N_T} H_{kn} H_{ln}^* - \mathbf{R}_R[k,l] \right) \tag{3.41}$$

and differentiating with respect to elements H_{ij} of $\mathbf{H}$ one obtains that the joint distribution is Gaussian

$$p(\mathbf{H}) = C_0 \exp\left(-\mathbf{h}^H \mathbf{R}_{ME}^{-1} \mathbf{h} \right) \tag{3.42}$$

where $\mathbf{h} = \mathrm{vec}\, \mathbf{H}$ and

$$\mathbf{R}_{ME} = (\mathbf{I}_T \otimes \boldsymbol{\mu}_R + \boldsymbol{\mu}_T \otimes \mathbf{I}_R)^{-1} \tag{3.43}$$

where $\boldsymbol{\mu}_T = \{a_{kl}\}$ and $\boldsymbol{\mu}_R = \{b_{kl}\}$ are matrices of Lagrangian multipliers. It is clear from Equation (3.43) that the ME solution is not the same as the corresponding Kronecker model (Wallace and Maharaj 2007). To gain a deeper insight into the form of (3.43) let us consider an SVD decomposition of $\boldsymbol{\mu}_T = \{a_{kl}\}$ and $\boldsymbol{\mu}_R = \{b_{kl}\}$:

$$\boldsymbol{\mu}_T = \mathbf{U}_{\mu T} \boldsymbol{\Lambda}_{\mu T} \mathbf{U}_{\mu T}^H, \quad \boldsymbol{\mu}_R = \mathbf{U}_{\mu R} \boldsymbol{\Lambda}_{\mu R} \mathbf{U}_{\mu R}^H \tag{3.44}$$

Using these decompositions, one can rewrite (3.43) as

$$\mathbf{R}_{ME} = \mathbf{U}_\mu \left(\mathbf{I}_T \otimes \boldsymbol{\Lambda}_{\mu R} + \boldsymbol{\Lambda}_{\mu T} \otimes \mathbf{I}_R \right)^{-1} \mathbf{U}_\mu^H = \mathbf{U}_\mu \boldsymbol{\Xi}_\mu^{-1} \mathbf{U}_\mu^H \tag{3.45}$$

where

$$\mathbf{U}_\mu = \mathbf{U}_{\mu T} \otimes \mathbf{U}_{\mu R}, \quad \text{and} \quad \boldsymbol{\Xi}_\mu = \mathbf{I}_T \otimes \boldsymbol{\Lambda}_{\mu R} + \boldsymbol{\Lambda}_{\mu T} \otimes \mathbf{I}_R \tag{3.46}$$

It is easy to see that the matrix $\boldsymbol{\Xi}_\mu$ is a diagonal matrix with elements

$$\xi_{ll} = \lambda_{\mu,T}[l] + \lambda_{\mu,R}[l],$$

where $\lambda_{\mu,T}[l]$ and $\lambda_{\mu,R}[l]$ are the corresponding elements of $\boldsymbol{\Lambda}_{\mu T}$ and $\boldsymbol{\Lambda}_{\mu R}$ respectively. It is shown in (Wallace and Maharaj 2007) that the eigenvectors of $\mathbf{R}_{ME}$ are just Kronecker products of the eigenvectors of the transmit and receive correlation matrices, that is $\mathbf{U}_\mu = \mathbf{U}_T \otimes \mathbf{U}_R$. In addition, the eigenvalues $\lambda_{\mu,T}[l]$ and $\lambda_{\mu,R}[l]$ are related to those of transmit and receive correlation matrices

$$\lambda_T[nn] = \sum_{m=1}^{N_R} \frac{1}{\lambda_{\mu,R}[m] + \lambda_{\mu,T}[n]} \tag{3.47}$$

$$\lambda_R[mm] - \sum_{n=1}^{N_T} \frac{1}{\lambda_{\mu,R}[m] + \lambda_{\mu,T}[n]} \tag{3.48}$$

Solution of Equations (3.47)–(3.48) is impossible analytically and some numerical procedure must be devised.

3.3.1.5 UIU model

The generalization of the Kronecker model (3.32), first considered in (Weichselberger et al. 2006) and applied for the calculation of capacity in (Tulino et al. 2005a) removes a restriction that the matrix $\mathbf{G}$ must have rank 1. In general, the Weichselberger model can be written as

$$\mathbf{H} = \mathbf{U}_T^H \left(\mathbf{G}_2 \odot \mathbf{H}_{w1} \right) \mathbf{U}_R \qquad (3.49)$$

where unitary matrices $\mathbf{U}_T$ and $\mathbf{U}_R$ are obtained from SVD of the transmit and receive correlation matrices. More generally, the UIU model used in (Tulino et al. 2005a) treats $\mathbf{U}_T$ and $\mathbf{U}_R$ as arbitrary unitary matrices of appropriate size.

The elements of $\mathbf{G}_2$ can be found as an expectation of the linear transformation of the channel matrix $\mathbf{H}$ as (Weichselberger et al. 2006)

$$\mathbf{G}_2[m, n] = \mathcal{E}_\mathbf{H} \left\{ \left| \mathbf{u}_R^H[m] \, \mathbf{H} \, \mathbf{u}_T[m] \right|^2 \right\} \qquad (3.50)$$

where $\mathbf{u}[m]$ is the m-th column of $\mathbf{U}$.

As has been observed in (Sayeed 2002), the structure of the matrix $\mathbf{G}_2$ reflects the spatial arrangements of scatterers and defines diversity and multiplexing properties of the corresponding MIMO channels. In the frame of the UIU model each eigenvector $\mathbf{u}_T[m]$ defines a certain transmit array pattern and each eigenvector $\mathbf{u}_R[n]$ defines a receive array pattern. The value of the coefficient $\mathbf{G}_2[m, n]$ defines how much energy is transmitted through a particular pair $m - n$ of modes.

3.4 Geometrical phenomenological models

The propagation of electromagnetic waves in the urban environment is a very complicated phenomenon (Beckmann and Spizzichino 1963; Bertoni 2000; Rappaport 2002). The received waves are usually a combination of line of site (LoS) and a number of specular and diffusive components. Mobile-to-mobile communications (Sen and Matolak 2008) introduce a new geometry of radio wave propagation, especially in urban settings. In this case antennas are located on a level well below rooftops or even treetops. Therefore propagation is dominated by rays reflected and diffracted from buildings, trees, other cars, and so on as in Figure 3.1. Grazing

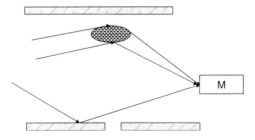

Figure 3.1 Mobile-to-mobile propagation scenario. In addition to the LoS and diffusive components (not shown) there are specular reflections from rough surfaces such as building facades and trees.

angles are also often very small in such scenarios, therefore reflective surfaces cannot always be treated as smooth, resulting in an LoS-like specular component, nor as very rough, resulting in a purely diffusive component (Beckmann and Spizzichino 1963). It is also common to assume that the resulting superposition of multiple reflected and diffused rays results in a spherically symmetric random process (Schreier and Scharf 2003b). However, it is shown in (Beckmann and Spizzichino 1963), that there is an intermediate case that results in a partially coherent reflection and, as such, to an improper complex Gaussian process, representing the channel transfer function. When the number of reflected waves is not sufficient, the resulting distribution is highly non-Gaussian (Barakat 1986). We defer investigation of such cases to future work and existing literature (Barakat 1986; Jakeman and Tough n.d.). Here we focus on the origin of the four-parametric distribution (Klovski 1982a) and estimation of its parameters.

3.4.1   Scattering from rough surfaces

Let us consider a rough surface of extent $2L \gg \lambda$ of the first Fresnel zone which is illuminated by a plane wave at the incidence angle $\theta_i = \pi/2 - \gamma$ as shown in Figure 3.2.

We assume a simple Gaussian model of the surface roughness (Beckmann and Spizzichino 1963), which is described by a random deviation $\zeta(x)$ from the mean level. The process $\zeta(x)$ has variance σ_r^2, the spatial covariance function $C(\Delta x) = \sigma_r^2 c(\Delta x)$, and the correlation length X (Beckmann and Spizzichino 1963).

Let us consider first the reflection from a surface portion of length X equal to the correlation length of the roughness. In the specular direction phases of elementary waves have a random component

$$\eta_\phi = \frac{4\pi}{\lambda} \zeta(x) \cos \theta_i \tag{3.51}$$

Thus, the variance σ_ϕ^2 of the random phase deviation could be evaluated as

$$\sigma_\phi^2 = 16\pi^2 \frac{\sigma_r^2}{\lambda^2} \cos^2 \theta_i \tag{3.52}$$

If $\sigma_\phi^2 \gg 4\pi^2$, that is

$$g = 2\frac{\sigma_r}{\lambda} \cos \theta_i \gg 1 \tag{3.53}$$

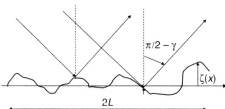

Figure 3.2   Rough surface geometry. Size of the patch 2L corresponds to the size of the first Fresnel zone. The rough surface is described by a random process, $\zeta(x)$.

then the variation of phase is significantly larger than 2π, the distribution of the wrapped phase (Mardia and Jupp 2000) is approximately uniform, and the resulting wave could be considered as a purely diffusive component. However, in the opposite case of $0 < g \ll 1$ the variation of phase is significantly less than 2π and cannot be considered uniform. For a perfectly smooth surface $g = 0$ the phase is deterministic, similar to LoS.

If the first Fresnel zone has extent $2L$, then there is approximately $N = 2L/X$ independent sections of the rough surface patches which contribute independently to the resultant field. Therefore, one can assume the following model for the reflected field/signal in the specular direction

$$\xi = \sum_{n=1}^{N} A_n \exp(j\phi_n) \tag{3.54}$$

where ϕ_n is a randomly distributed phase with the variance given by Equation (3.52). If $\sigma_\phi^2 \gg 4\pi^2$ the model reduces to the well accepted spherically symmetric diffusion component model; if $\sigma_\phi^2 = 0$ LoS-like conditions for the specular component are observed with the rest of the values spanning an intermediate scenario.

Detailed investigation of the statistical properties of model (3.54) can be found in (Beckmann and Spizzichino 1963) and some consequent publications, especially in the field of optics (Barakat 1986; Jakeman and Tough n.d.). Assuming that the central limit theorem holds, as in (Beckmann and Spizzichino 1963), one comes to the conclusion that $\xi = \xi_I + j\xi_Q$ is a Gaussian process with zero-mean and *unequal* variances σ_I^2 and σ_Q^2 of the real and imaginary parts. Therefore ξ is an improper random process (Schreier and Scharf 2003b). Coupled with a constant term $m = m_I + jm_Q$ from the LoS-type components, the model (2.53) gives rise to a large number of different distributions of the channel magnitude, including Rayleigh ($m = 0$, $\sigma_I = \sigma_Q$), Rice ($m \neq 0$, $\sigma_I = \sigma_Q$), Hoyt ($m \neq 0$, $\sigma_I > 0$ $\sigma_Q = 0$) m, and many others (Klovski 1982a; Simon and Alouini 2000). Following (Klovski 1982a) we will refer to the general case as a four-parametric distribution, defined by the following parameters

$$m = \sqrt{m_I^2 + m_Q^2}, \quad \phi = \arctan \frac{m_Q}{m_I} \tag{3.55}$$

$$q^2 = \frac{m_I^2 + m_Q^2}{\sigma_I^2 + \sigma_Q^2}, \quad \beta = \frac{\sigma_Q^2}{\sigma_I^2} \tag{3.56}$$

Two parameters q^2 and β are the most fundamental since they describe the power ratio between the deterministic and stochastic components (q^2) and the asymmetry of the components β. Our further study is focused on these two parameters.

3.5   On the role of trigonometric polynomials in analysis and simulation of MIMO channels

The importance of joint distribution was indicated in (Xu et al. 2004). Weichselberger (Weichselberger 2004) suggested that the correlation matrix could be decomposed into the sum of Kronecker products but has failed to assign any particular meaning to eigenmodes higher than one.

3.5.1   Measures of dependency

The correlation coefficient

$$\rho = \frac{\text{cov}(\xi, \eta)}{\sqrt{\text{var}(\xi)\text{var}(\eta)}} \tag{3.57}$$

provides a useful insight into dependence properties of random variables, especially in the joint Gaussian case where the correlation coefficients completely describe such dependency. It also represents linear dependence, which can be thought of as the strongest type of dependency (Hutchinson and Lai 1990). However, when describing highly non-Gaussian processes such as joint distribution of AoA and AoD, the correlation coefficient is not the best quantity to describe such a dependency. A great number of alternatives have been suggested in literature (Hutchinson and Lai 1990; Stewart and Sun 1990) and others, all of them must satisfy requirements suggested by Renyi (Hutchinson and Lai 1990; Rényi 1970). An important measure is the maximal correlation (sup correlation) studied by Sarmanov and Lancaster (Lancaster 1958; Sarmanov 1958a,b).

$$\rho' = \sup \rho\,[F_1(\xi)F_2(\eta)] \tag{3.58}$$

where F_1 and F_2 are arbitrary non-linear functions. The importance of this correlation rests with the fact that it coincides (Hutchinson and Lai 1990) with the first canonical coefficient ρ_1 and is relevant to the sum of Kronecker product expansions of the full covariance matrix, as discussed later.

If the bivariate distribution $p_2(x, y)$ is consistent with the marginals $p_{1x}(x)$ and $p_{1y}(y)$, which in turn induce an orthonormal basis $\{p_n(x)\}$ and $\{q_m(y)\}$ respectively, that is

$$\int_{-\infty}^{\infty} p_{1x}(x)p_n(x)p_m(x)dx = \delta_{nm}, \quad p_0(x) = 1 \tag{3.59}$$

$$\int_{-\infty}^{\infty} p_{1y}(y)q_n(y)q_m(y)dy = \delta_{nm}, \quad q_0(x) = 1 \tag{3.60}$$

then the matrix $\mathbf{R} = \{\rho_{ij}\}$ where $\rho_{ij} = \mathcal{E}\{p_i(x)\}$ also defines the degree of correlation between two random variables. The scalar

$$\phi^2 = \text{trace } \mathbf{R}\mathbf{R}^H = \sum_{i=0}^{\infty}\sum_{j=0}^{\infty} \rho_{ij} \tag{3.61}$$

known as the mean square contingency also serves as a measure of dependence. If $\phi^2 < \infty$ the distribution $p_2(x, y)$ is called ϕ^2 bounded (Hutchinson and Lai 1990).

It was shown by Lancaster (Lancaster 1958) that sets $\{p_n(x)\}$ and $\{q_m(y)\}$ can be transformed into two new orthonormal sets $\{P_n(x)\}$ and $\{Q_m(y)\}$ such that

$$\bar{\rho}_{kl} = \mathcal{E}\{P_n(x)Q_m(y)\} = \delta_{kl}\rho_k, \quad 1 \geq \rho_1^2 \geq \rho_2^2 \cdots \tag{3.62}$$

and therefore

$$p_2(x, y) = p_{1x}(x)p_{1y}(y)\left[1 + \sum_{k=1}^{\infty} \rho_k P_k(x) Q_k(y)\right] \qquad (3.63)$$

Functions $P_n(x)$ and $Q_n(y)$ are known as canonical variables, ρ_i are called canonical correlations and (3.63) is known as a canonical expansion of the bivariate distribution (Hutchinson and Lai 1990).

There many known canonical expansions where canonical variables are polynomials (Hutchinson and Lai 1990; Ord 1972). It was shown that it is not always the case and that the necessary and sufficient conditions are

$$\mathcal{E}\{x^n|y\} = \text{a polynomial in } y \text{ of degree } \leq n \qquad (3.64)$$

$$\mathcal{E}\{y^n|x\} = \text{a polynomial in } x \text{ of degree } \leq n \qquad (3.65)$$

3.5.2 Non-negative trigonometric polynomials and their use in estimation of AoD and AoA distribution

Non-negative trigonometric polynomials $\varrho(\theta)$ are widely used in approximation theory (Szegö 1939). It was shown by Fejér (Szegö 1939) that any non-negative trigonometric polynomial of order N is uniquely defined by some regular polynomial

$$h(z) = \sum_{n=0}^{N} c_n z^n \qquad (3.66)$$

of order N through a simple, but unfortunately non-linear equation:

$$\varrho(\theta) = \left|h\left(e^{j\theta}\right)\right|^2 = \sum_{n=-N}^{N} a_n e^{j\theta} = \alpha_0 + \sum_{k=1}^{N} (\alpha_k \cos k\theta + \beta_k \sin k\theta) \qquad (3.67)$$

where

$$\alpha_0 = a_0 = \sum_{k=0}^{n} |c_k|^2 \qquad (3.68)$$

and

$$2a_k = \alpha_k - j\beta_k = 2\sum_{l=0}^{n-k} c_{l+k}c_l^*, \ k = 1, 2, \cdots, n \qquad (3.69)$$

3.5.3 Approximation of marginal PDF using non-negative polynomials

Since the distribution of AoD and AoA is limited to a range $[-\pi, \pi]$ it could be properly approximated by a non-negative trigonometric polynomial of the form (3.67) with the additional condition that it should integrate to unity, which is equivalent to the following constraint on the coefficients of the generating polynomial $h(z)$

$$\alpha_0 = a_0 = \sum_{k=0}^{n} |c_k|^2 = \frac{1}{2\pi} \tag{3.70}$$

The rest of the coefficients could be found by proper constrained optimization as in (Fernández-Durán 2004). Instead of fitting angular data, which are not directly available, we note that an expression for the covariance matrix in the antenna array can be obtained using the series (3.67). Indeed

$$\begin{aligned}
\rho(d) &= \int_{-\pi}^{\pi} \exp\left(j2\pi \frac{d}{\lambda} \sin\theta \right) p(\theta) \\
&= \sum_{n=-N}^{N} a_n \int_{-\pi}^{\pi} \exp\left(j2\pi \frac{d}{\lambda} \sin\theta \right) \exp(jn\theta) \\
&= \sum_{n=-N}^{N} 2\pi a_n J_n\left(2\pi \frac{d}{\lambda} \right)
\end{aligned} \tag{3.71}$$

Therefore, the set of coefficients c_n can be optimized to minimize the difference between a given covariance matrix $\mathbf{R}_m$ and the one obtained by sampling of Equation (3.71).

3.6   Canonical expansions of bivariate distributions and the structure MIMO channel covariance matrix

3.6.1   Canonical variables and expansion

Let $p_2(\theta_T, \theta_R)$ be a bivariate PDF which describes the joint distribution of AoA and AoD and $p_T(\theta_T)$ and $p_R(\theta_R)$ are corresponding marginal distributions, that is

$$p_T(\theta_T) = \int_{-\pi}^{\pi} p_2(\theta_T, \theta_R) d\theta_R, \quad p_R(\theta_R) = \int_{-\pi}^{\pi} p_2(\theta_T, \theta_R) d\theta_T \tag{3.72}$$

Therefore, the bivariate density can be represented as

$$p_2(\theta_T, \theta_R) = p_T(\theta_T) p_R(\theta_R) \Omega(\theta_T, \theta_R) \tag{3.73}$$

where $\Omega(\theta_T, \theta_R) \geq 0$ is a non-negative function, often referred to as the Radon–Nykodim derivative of the bivariate density with respect to its marginals (Shilov and Gurevich 1978). The function $\Omega(\theta_T, \theta_R)$ can be approximated by its Fourier series

$$\begin{aligned}
\Omega(\theta_T, \theta_R) &= \sum_{k=-L_T}^{L_T} \sum_{l=-L_R}^{L_R} c_{kl} \exp(-jk\theta_T) \exp(jl\theta_R) \\
&= \mathbf{f}_T^H(\theta_T) \mathbf{C} \mathbf{f}_R(\theta_R)
\end{aligned} \tag{3.74}$$

where

$$\mathbf{f}_T(\theta_T) = \left[\exp(-jL_T\theta_T), \cdots, 1, \cdots, \exp(jL_T\theta_T)\right]^T$$

$$\mathbf{f}_R(\theta_R) = \left[\exp(-jL_R\theta_R), \cdots, 1, \cdots, \exp(jL_R\theta_R)\right]^T \qquad (3.75)$$

and the $L_R \times L_T$ matrix $\mathbf{C} = \{c_{ij}\}$ contains elements of the Fourier series. Functions $\exp(-jk\theta_T)$ and $\exp(jl\theta_R)$ are linearly independent and therefore can be converted to orthonormal sets $\mathbf{g}_T(\theta_T)$ and $\mathbf{g}_R(\theta_R)$ through a linear transformation

$$\mathbf{g}_T(\theta_T) = \mathbf{D}_T\mathbf{f}_T(\theta_T), \quad \mathbf{g}_R(\theta_R) = \mathbf{D}_R\mathbf{f}_R(\theta_R) \qquad (3.76)$$

where $\mathbf{D}_T$ and $\mathbf{D}_R$ are $L_T \times L_T$ and $L_R \times L_R$ transformation matrices. Therefore, Equation (3.74) can be rewritten as

$$\Omega(\theta_T, \theta_R) = \mathbf{g}_T^H(\theta_T)\mathbf{D}_T^{-1\,H}\mathbf{C}\mathbf{D}_T^{-1}\mathbf{g}_R(\theta_R) \qquad (3.77)$$

Finally, using SVD decomposition (Franklin 2000) of the matrix $\mathbf{D}_T^{-1\,H}\mathbf{C}\mathbf{D}_T^{-1} = \mathbf{U}\boldsymbol{\Sigma}\mathbf{V}^H$ and using transformations $\mathbf{p}(\theta_T) = \mathbf{U}^H\mathbf{g}_T(\theta_T)$ and $\mathbf{q}(\theta_R) = \mathbf{V}^H\mathbf{g}_R(\theta_R)$ the bivariate density $p_2(\theta_T, \theta_R)$ can be represented in the form

$$p_2(\theta_T, \theta_R) = p_T(\theta_T)p_R(\theta_R)\mathbf{p}(\theta_T)^H\boldsymbol{\Sigma}\mathbf{q}(\theta_R)$$

$$= p_T(\theta_T)p_R(\theta_R)\left[\sum_{l=1}^{L}\sigma_l P_l(\theta_T)Q_l(\theta_R)\right] \qquad (3.78)$$

which is clearly the canonical representation of the bivariate density $p_2(\theta_T, \theta_R)$. Here L is the number of non-zero singular values of $\mathbf{D}_T^{-1\,H}\mathbf{C}\mathbf{D}_T^{-1}$. Since matrices $\mathbf{D}_T$ and $\mathbf{D}_R$ have full rank the value of L is defined by the rank of the matrix $\mathbf{C}$. If the bivariate density is estimated from the data obtained by using $N_T \times N_R$ MIMO system $L \leq \max(N_T, N_R)$.

The canonical representation (3.78) also defines a canonical structure of the full correlation matrix of a MIMO channel. Indeed, the correlation coefficient between any two pairs of the transmit–receive antennas is defined as

$$\rho(d_T, d_R) = \int_{-\pi}^{\pi}\int_{-\pi}^{\pi}\exp\left[j\frac{2\pi}{\lambda}(d_T\sin\theta_T - d_R\sin\theta_R)\right]$$

$$\times p_2(\theta_T, \theta_R)d\theta_T d\theta_R = \sum_{l=1}^{L}\sigma_l\rho_l(d_T)\rho_l^*(d_R) \qquad (3.79)$$

where

$$\rho_l(d_T) = \int_{-\pi}^{\pi}\exp\left[j\frac{2\pi}{\lambda}d_T\sin\theta_T\right]P_l(\theta_T)d\theta_T \qquad (3.80)$$

$$\rho_l(d_R) = \int_{-\pi}^{\pi}\exp\left[j\frac{2\pi}{\lambda}d_R\sin\theta_R\right]Q_l(\theta_R)d\theta_R \qquad (3.81)$$

3.6.2   General structure of the full covariance matrix

The representation (3.79) implies that a sampled covariance matrix $\mathbf{R}_h$ of a MIMO channel in general can be represented as

$$\mathbf{R_h} = \sum_{l=1}^{L} \sigma_l \mathbf{R}_{Tl} \otimes \mathbf{R}_{Rl} \tag{3.82}$$

This representation is not new in itself. For example, Weichselberger (Weichselberger 2004) considers such a representation based on the Pitsianis–Golub (Golub and van Loan 1996) decomposition of $\mathbf{R}_h$, or equivalently, based on the SVD decomposition of the permuted covariance matrix $\mathbf{Q}_h = \mathcal{E}\{\mathbf{H} \otimes \mathbf{H}^H\}$. However, no interpretation of each term has been given, since the decomposition (3.82) cannot be interpreted in terms of energy, except for the first term. This stems from the fact that functions $\rho_l(d_T)$ and $\rho_l(d_T)$ are not proper covariance functions for $l > 1$.

However, our analysis shows that a meaningful statistical interpretation can be given to such representation. Indeed, while each of the functions $\rho_l(d_T)$ and $\rho_l(d_T)$ itself is not a proper covariance function, a sum of any first $L_1 \leq L$ of them is. Therefore a partial sum

$$\mathbf{R_h} = \sum_{l=1}^{L_1} \sigma_l \mathbf{R}_{Tl} \otimes \mathbf{R}_{Rl} \tag{3.83}$$

can be considered as an approximation of the full covariance matrix, an approximation which takes into account $L_1 - 1$ canonical correlations. In the degenerate case of $L_1 = 1$ no correlation at all is taken into account between the transmit and the receive side. This corresponds to a Kronecker product model widely accepted in the literature (Kermoal et al. 2000). The first non-trivial case takes into consideration both effects of the uncorrelated scattering and the strongest canonical correlation coefficient $\rho_1 = \sigma_2$. The correlation coefficient itself describes how strong the interaction is between the sides; however, the nature of such an interaction is encoded in the form of the canonical variables $\mathbf{g}_T(\theta_T)$ and $\mathbf{g}_R(\theta_R)$, and therefore into the quasi-correlation matrices $\mathbf{R}_{T2}$ and $\mathbf{R}_{R2}$.

3.6.3   Relationship to other models

It is interesting to compare models based on the joint distribution of AoA and AoD with other techniques used in the literature. Among these, one and two ring geometrical models constitute a significant proportion (Abdi and Kaveh 2002; Yu and Ottersten 2002). In general, geometrical models assume that a ray, radiated from an transmit array is reflected by a point scatterer before reaching the receiver array. This, therefore, imposes a condition that the AoA and the AoD are deterministically linked through some, possibly non-linear, dependance $\phi_T = h(\phi_R)$ or $\phi_R = g(\phi_T)$. As a result, the conditional densities are δ-like functions

$$p(\phi_T|\phi_R) = \delta\left(\phi_T - h(\phi_R)\right) \left| \frac{d}{d\phi_R} h(\phi_R) \right| \tag{3.84}$$

$$p(\phi_R|\phi_T) = \delta\,(\phi_R - g(\phi_T))\left|\frac{d}{d\phi_T}g(\phi_T)\right| \qquad (3.85)$$

The geometric models are idealistic since real scatterers are rather large bodies (compared to the wavelength) distributed in space rather than concentrated in a single point. Therefore, if the transmitting array illuminates the field in a certain fixed direction ϕ_T it is reasonable to expect that a number of scatterers will be illuminated and the received field is perceived as arriving from a certain angular spread rather than from a single direction. Such a picture could be visualized through the conditional PDF

$$p_{R|T}(\phi_R|\phi_T) = \frac{p_2(\phi_T, \phi_R)}{p_T(\phi_T)} \qquad (3.86)$$

In the case of point scatterers, the conditional PDF (3.86) is a weighted sum of δ functions each corresponding to a single point scatterer with magnitudes proportional to their cross-section. On the other extreme, independent AoA and AoD will produce a conditional PDF $p_{R|T}(\phi_R|\phi_T) = p_R(\phi_R)$ which does not depend on the illuminating direction. This is possible when scatterers are dispersed over a certain region without much concentration in single points.

3.6.3.1 One ring models

Let us for simplicity consider a single ring model with a ring of scatterers surrounding transmitter, under the usual assumptions that the ring radius R is much smaller than the distance $D \gg R$ between the antennas. If the ring is situated around the transmitter antenna then

$$\sin\phi_R \approx \phi_R \approx \frac{R}{D}\sin\phi_T = \Delta_T\phi_T \qquad (3.87)$$

where $2\Delta_T \ll 1$ describes the angular spread on the receiving side. A similar relation

$$\sin\phi_T \approx \phi_T \approx \frac{R}{D}\sin\phi_R = \Delta_R\phi_R, \;\; \Delta_R \ll 1 \qquad (3.88)$$

can be derived for the ring situated around the receiver. Therefore, the joint distribution of AoA and AoD can be written as

$$p_{2T}(\phi_T, \phi_R) = p_T(\phi_T)\delta\left(\phi_R - \frac{R}{D}\sin\phi_T\right)\frac{R}{D}\cos\phi_T \qquad (3.89)$$

$$p_{2R}(\phi_T, \phi_R) = p_R(\phi_R)\delta\left(\phi_T - \frac{R}{D}\sin\phi_R\right)\frac{R}{D}\cos\phi_R \qquad (3.90)$$

Using these equations for the joint distribution and choosing von-Mises PDF for AoA/AoD on the proper side one can derive the expressions for the spatial covariance function identical to that obtained in (Abdi and Kaveh 2002).

3.6.3.2   Two ring model

If scattering geometry is such that scattering centers are present around both sides of the link the following mixture model can be suggested

$$p_2(\phi_T, \phi_R) = \eta p_{2T}(\phi_T, \phi_R) + (1 - \eta) p_{2R}(\phi_T, \phi_R) \qquad (3.91)$$

This model has been suggested and analyzed in (Abdi and Kaveh 2002; Zhang et al. 2005).

3.6.3.3   A disk of scatterers

In this model scatterers are assumed to be uniformly spread within two concentric rings of radii $L \ll R_1 < R_2 \ll D$ where L is the size of the antenna and D is the distance between sides. Without loss of generality we can assume that the scattering ring is around the receiver site. In this case to every fixed AoA ϕ_R there is a certain spread of AoD, which can be approximated by the following conditional distribution

$$p_{D|A}(\phi_T|\phi_R) \approx \frac{D}{R_2 - R_1} \sin \phi_R \Pi \left(\phi_T - \frac{R_1 + R_2}{2D} \sin \phi_R, \frac{R_2 - R_1}{2D} \sin \phi_R \right)$$
$$(3.92)$$

where

$$\Pi(s - s_0, \alpha) = \begin{cases} 1 & \text{if} \qquad s_0 - \alpha \leq s \leq s_0 + \alpha \\ 0 & \text{otherwise} \end{cases}$$

Therefore, the joint density is

$$p_2(\phi_T, \phi_R) \approx \frac{D \sin \phi_R}{2\pi (R_2 - R_1)} \Pi \left(\phi_T - \frac{R_1 + R_2}{2D} \sin \phi_R, \frac{R_2 - R_1}{2D} \sin \phi_R \right) \quad (3.93)$$

3.7   Bivariate von Mises distribution with correlated transmit and receive sides

While there are infinitely many bivariate distributions with von Mises marginals[14] we construct a class of bivariate distributions which is flexible enough to describe a wide class of distributions of AoA/AoD and at the same time allows for some analytical investigation.

3.7.1   Single cluster scenario

Let us consider a scenario represented in Figure 3.3. A single cluster is illuminated by the transmit antenna array with AoD distributed according to the von Mises PDF with parameters κ_T and μ_T. The reradiated field reaches the receive antenna

---

[14] None of them are given in (Mardia and Jupp 2000) for unknown reasons. There are some discussions in (La Frieda and Lindsey 1973). Ince's three-parametric equation, after transformation to self-adjoint form, is known as the Whittacker–Hill Equation (Arscott 1964).

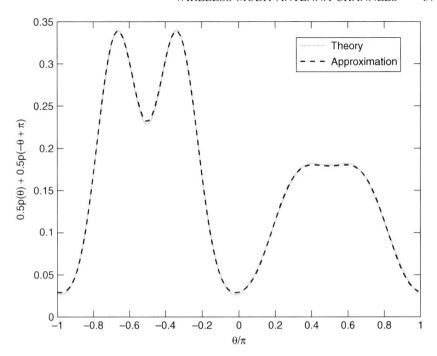

Figure 3.3 Single cluster scattering scenario (approximation by bivariate von Mises PDF).

array with the angular spread also described by the von Mises PDF but with parameters κ_R and μ_R. In general, it is quite possible that AoD and AoA are not independent since a fraction of power can be described by the theory of physical optics (Bertoni 2000), and is therefore strongly correlated in direction. We will model such a scenario by a bivariate distribution based on the Lancaster canonical form of bivariate PDF

$$p(\theta_T, \theta_R) = p_T(\theta_T) p_R(\theta_R) \left[1 + \sum_{k=1}^{K} \rho_k P_k(\theta_T) Q_k(\theta_R) \right] \qquad (3.94)$$

where $p_T(\theta_T)$ and $p_T(\theta_R)$ are von Mises marginal PDF and $P_k(\theta_T)$ and $Q_k(\theta_R)$ are orthonormal non-negative sine polynomials with weights $p_T(\theta_T)$ and $p_T(\theta_R)$ respectively.

The simplest such distribution is

$$p_2(\phi_1, \phi_2; \kappa_1, \kappa_2, \mu_1, \mu_2, \rho, m_1, m_2)$$
$$= \frac{\exp\left[\kappa_1 \cos(\phi_1 - \mu_1)\right]}{2\pi I_0(\kappa_1)} \frac{\exp\left[\kappa_2 \cos(\phi_2 - \mu_2)\right]}{2\pi I_0(\kappa_2)}$$
$$\times \left[1 + \rho \sin m_1(\phi_1 - \mu_1) \sin m_2(\phi_2 - \mu_2)\right] \qquad (3.95)$$

It is easy to see that the marginals of this PDF are von Mises ones with parameters (μ_1, κ_1) and (μ_2, κ_2) respectively. The parameter $|\rho| \leq 1$ defines the (non-linear) correlation between two components. If $\rho = 0$ one obtains a case of independent scattering between the receiver and the transmitter side, and as a result the correlation between any two transmit–receive links factorizes in the product of the two spatial correlation functions, each representing one side of the link:

$$C_0(d_T, d_R) = \frac{1}{I_0(\kappa_1)I_0(\kappa_2)} \times I_0 \left(\sqrt{\kappa_1^2 - 4\pi^2 \frac{d_T^2}{\lambda^2} + j4\pi\kappa_1 \frac{d_T}{\lambda} \cos \mu_1} \right)$$

$$\times I_0 \left(\sqrt{\kappa_2^2 - 4\pi^2 \frac{d_R^2}{\lambda^2} + j4\pi\kappa_2 \frac{d_R}{\lambda} \cos \mu_2} \right) \tag{3.96}$$

In the case of finite ρ, the spatial covariance factorized in the sum of the two products

$$C(d_T, d_R, \rho) = C_0(d_T, d_R) + \rho C_{1T}(d_T)C_{1R}(d_R) \tag{3.97}$$

where

$$C_1(d) = \int_{-\pi}^{\pi} \frac{\exp[\kappa \cos(\phi - \mu)]}{I_0(\kappa)} \sin m(\phi - \mu) d\phi$$

$$= \int_{-\pi}^{\pi} \frac{\exp[\kappa \cos(\phi - \mu)]}{I_0(\kappa)} \frac{\exp[jm(\phi - \mu)] - \exp[-jm(\phi - \mu)]}{2j} d\phi$$

$$\tag{3.98}$$

3.7.2 Multiple clusters scenario

Let us now consider a scenario represented in Figure 3.4. We will assume that clusters are illuminated independently and therefore there is no statistical dependence between AoA/AoD corresponding to a different cluster. Therefore, the bivariate distribution of AoA and AoD can be described in terms of mixture models

$$p(\theta_T, \theta_R) = \sum_{k=1}^{K} P_k p_k(\theta_T, \theta_R), \quad P_k \geq 0, \quad \sum_{k=1}^{K} P_k = 1 \tag{3.99}$$

where PDF $p_k(\theta_T, \theta_R)$ is given by equation (3.94).

It can be seen that the marginal distributions corresponding to the mixture model (3.99) are mixtures of the von Mises PDFs on each side of the link.

3.8 Bivariate uniform distributions

3.8.1 Harmonic coupling

In this section we derive a representation of the spatial covariance function related to the bivariate uniform distribution. Of course, such a distribution is not unique and much effort has been spent on finding different classes of such distributions (Hutchinson and Lai 1990). In particular we will focus on representation using

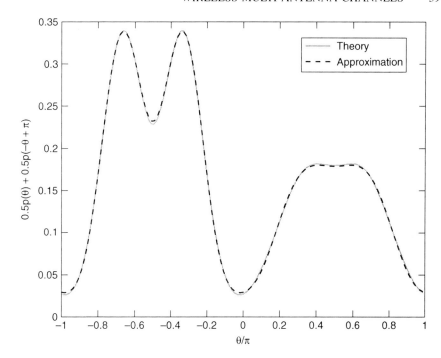

Figure 3.4 Multiple clusters scattering scenario (approximation by uniform distribution).

trigonometrical polynomials. The simplest such PDF is given in the form

$$p_2(\phi_T, \phi_R; \Delta_1, \Delta_2, \alpha, m, n) = \frac{1}{\Delta_1 \Delta_2} \left[1 + \alpha\, f_1(\phi_T) f_2(\phi_R) \right] \tag{3.100}$$

where

$$f_1(\phi_T) = \sqrt{2} \sin\left(2\pi m \frac{\phi_T}{\Delta_T} \right), \quad f_2(\phi_R) = \sqrt{2} \sin\left(2\pi n \frac{\phi_R}{\Delta_R} \right) \tag{3.101}$$

Parameters Δ_T and Δ_R describe the angle spread on each side, the parameter $|\alpha| < 1/2$ governs the correlation between the sides, while the integers m and n control the non-linear character of the correlation. It is easy to check by direct calculations that the marginal PDF corresponding to the bivariate density (3.100) is indeed the uniform PDF with angular spread Δ_1 and Δ_2.

It is easy to see by direct calculations that sets $\{1, f_1(\phi_T)\}$ and $\{1, f_2(\phi_R)\}$ are indeed orthonormal sets with respect to the marginal PDFs $p_T(\phi_T) = 1/\Delta_T$ and $p_R(\phi_R) = 1/\Delta_R$ respectively. Therefore, PDF (3.100) is written in its canonical form and α is the maximum correlation coefficient.

3.8.1.1   Narrow angular spread

In this section we assume that scattering centers are far removed from the antennas, therefore the angular spread on each side is small, that is $\Delta_{R,T}/2\pi \ll 1$. In this case the following approximation of the complex exponent

$$\exp\left(j2\pi\frac{d}{\lambda}\sin(\phi+\phi_0)\right) \approx \exp\left(j2\pi\frac{d}{\lambda}\cos\phi_0\right)\exp\left(j2\pi\frac{d}{\lambda}\phi\sin\phi_0\right) \quad (3.102)$$

can be effective when used for calculation of the integrals. The correlation function corresponding to the case $\alpha = 0$, that is to independent receive and transmit sides with a single cluster seen at the angles ϕ_{T0} and ϕ_{R0} and narrow angle spreads Δ_1 and Δ_2, can be obtained in the same way as described in (Zhao and Loyka 2004b):

$$\rho_0(d_T, d_R; \Delta_T, \Delta_R) = \rho_0(d_T, \Delta_T, \phi_{T0})\rho_0(d_R, \Delta_R, \phi_{R0}) \quad (3.103)$$

where

$$\rho(d, \Delta) = \frac{1}{\Delta}\int_{-\pi}^{\pi}\exp\left[j2\pi\frac{d}{\lambda}\sin\phi\right]d\phi$$

$$\approx \exp\left[j2\pi\frac{d}{\lambda}\sin\phi_0\right]\text{sinc}\left(\Delta\frac{d}{\lambda}\cos\phi_0\right) \quad (3.104)$$

In the presence of correlation, that is for $\alpha \neq 0$ the covariance function $\rho(d_T, d_R, \Delta_T, \Delta_R)$ has an additional term

$$\alpha\rho_1(d_T, d_R; \Delta_T, \Delta_R, m, n) = 2\alpha\rho_1(d_T, \Delta_T, m)\rho_1(d_R, \Delta_R, n) \quad (3.105)$$

where

$$\rho_1(d, \Delta, m) = \frac{1}{\Delta}\int_{-\Delta/2}^{\Delta/2}\exp\left[j2\pi\frac{d}{\lambda}\sin(\phi+\phi_0)\right]\sin m\phi\, d\phi$$

$$\approx \frac{\exp\left[j2\pi\frac{d}{\lambda}\sin(\phi_0)\right]}{\Delta}\int_{-\Delta/2}^{\Delta/2}\exp\left[j2\pi\frac{d}{\lambda}\phi\right]\sin m\phi\, d\phi$$

$$= (-1)^m j\exp\left[j2\pi\frac{d}{\lambda}\sin(\phi_0)\right]\frac{m\sin\left(\Delta\pi\cos\phi_0\, d/\lambda\right)}{\pi\left[\Delta^2\cos^2\phi_0\, d^2/\lambda^2 - m^2\right]} \quad (3.106)$$

3.8.1.2 Isotropic scattering

In this case $\Delta_T = \Delta_R = 2\pi$ and the correlation on each side is described by the Clark's formula, that is for $\alpha = 0$

$$\rho_0(d_T, d_R; 2\pi, 2\pi) = J_0\left(2\pi\frac{d_T}{\lambda}\right)J_0\left(2\pi\frac{d_R}{\lambda}\right) \quad (3.107)$$

If the correlation between the transmit and receive sides is not zero, then the covariance function is given in the form

$$\rho(d_T, d_R; 2\pi, 2\pi) = \rho_0(d_T, d_R; 2\pi, 2\pi) + \alpha\rho_1(d_T)\rho_1(d_R) \quad (3.108)$$

where

$$
\rho_d(d) = \frac{1}{2\pi} \int_{-\pi}^{\pi} \exp\left(j2\pi \frac{d}{\lambda} \sin\theta\right) \sin m\theta \, d\theta
$$

$$
= \frac{1}{2\pi} \int_{-\pi}^{\pi} \exp\left(j2\pi \frac{d}{\lambda} \sin\theta\right) \frac{\exp(jm\theta) - \exp(-jm\theta)}{2j} d\theta
$$

$$
= \frac{j(1 - (-1)^m)}{2} J_m\left(2\pi \frac{d}{\lambda}\right) \tag{3.109}
$$

Interestingly enough, for the case of odd $m = 2k$ the additional term in the covariance function vanishes, thus resulting in the same covariance function as in the uncorrelated case.

3.8.2 Markov-type bivariate density

It can be seen from the section above that the linear correlation, corresponding to the simplest models of the form (3.100–3.101) is relatively small even for the extreme values of non-linear correlation. In this section we consider the generalization of the bivariate density based on the stationary Markov phase process as considered in (La Frieda and Lindsey 1973) and other sources in applications to PLL performance. In this case the bivariate PDF becomes

$$
p_2(\phi_T, \phi_R) = \frac{1}{\Delta_T \Delta_R} \left[1 + 2 \sum_{n=1}^{\infty} \rho^{n^2} \cos n \left(\frac{2\pi}{\Delta_T} \phi_T - \frac{2\pi}{\Delta_R} \phi_R\right)\right] \tag{3.110}
$$

Of course, this density is also easily represented in terms of trigonometric polynomials. An important property of the bivariate density (3.110) is the fact that it describes a situation where both sides are deterministically related. Indeed, if $\rho = 1$ in (3.110), one obtains

$$
p_2(\phi_T, \phi_R) = \frac{1}{\Delta_R} \delta \left(\phi_T - \frac{\Delta_T}{\Delta_R} \phi_R\right) \tag{3.111}
$$

On the other hand, when correlation is relatively small, only a single term in the expansion (3.110) needs to be retained, therefore producing

$$
p_2(\phi_T, \phi_R) = \frac{1}{\Delta_T \Delta_R} \left[1 + 2\rho \cos \left(\frac{2\pi}{\Delta_T} \phi_T - \frac{2\pi}{\Delta_R} \phi_R\right)\right] \tag{3.112}
$$

For narrow angular spreads on both sides the following expression for the correlation between two links becomes

$$
\rho(d_T, d_R) = \exp\left[j \frac{2\pi}{\lambda} (d_T \sin\phi_{0T} - d_R \sin\phi_{0R})\right]
$$

$$
\times \operatorname{sinc}\left(d_T \cos\phi_{0T} - d_R \frac{\Delta_R}{\Delta_T} \cos\phi_{0R}\right) \tag{3.113}
$$

3.9   Analytical expression for the diversity measure of an antenna array

Diversity measure

$$\Psi(\mathbf{R}) = \frac{\text{trace}^2(\mathbf{R})}{||\mathbf{R}||_F^2} = \frac{\left(\sum_{n=1}^{N}\lambda_n\right)^2}{\sum_{n=1}^{N}\lambda_n^2} \qquad (3.114)$$

has been recently suggested in (Ivrlač and Nossek 2005) and successfully used to describe a degree of diversity and correlation provided by an antenna array and MIMO link (Herdin et al. 2005; Özcelik et al. 2005). Here λ_n is the n-th eigenvalue of the matrix $\mathbf{R}$. However, the authors resort either to numerical calculations or to a very specific case of the covariance matrix with all off-diagonal elements equal. Here we provide an analytical expression for the diversity quantity for the case of uniform and von-Mises distribution of the angle of arrival. Such an expression will allow us to evaluate the dependence of $\Psi(\mathbf{R})$ on the beamwidth and the average angle of arrival.

3.9.1   Relation of the shape of the spatial covariance function to trigonometric moments

It is interesting to relate, if possible, the behavior of the spatial covariance function with parameters of the angular distribution such as trigonometric moments. Since ULA cannot distinguish between AoA θ and $\pi - \theta$ we can express the spatial covariance function in terms of distribution

$$\hat{p}(\theta) = \begin{cases} p(\theta) + p(\pi - \theta) & \text{if } \theta \in [-\pi/2; \pi/2] \\ 0 & \text{otherwise} \end{cases} \qquad (3.115)$$

Furthermore, using substitution $\omega = \sin\theta$, one can rewrite the expression for the spatial covariance function as

$$\begin{aligned}
\rho(d) &= \int_{-\pi/2}^{\pi/2} \exp\left(-j2\pi\frac{d}{\lambda}\sin\theta\right) \hat{p}(\theta)d\theta \\
&= \int_{-1}^{1} \exp\left(-j2\pi\frac{d}{\lambda}\omega\right) \frac{\hat{p}(\arcsin\omega) + \hat{p}(\pi - \arcsin\omega)}{\sqrt{1-\omega^2}}d\omega \\
&= \int_{-1}^{1} \exp\left(-j2\pi\frac{d}{\lambda}\omega\right) w(\omega)d\omega = \exp\left[\Phi(2\pi d/\lambda)\right]
\end{aligned} \qquad (3.116)$$

Here $w(\omega)$ can be recognized as the PDF of the sine of the angle of arrival and $\Phi(2\pi d/\lambda)$ is the corresponding cumulant generating function (Papoulis 1991; Primak et al. 2004):

$$\Phi\left(2\pi\frac{d}{\lambda}\right) = \sum_{k=1}^{\infty} j^k \frac{\kappa_k}{k!}\left(2\pi\frac{d}{\lambda}\right)^k \qquad (3.117)$$

The representation in terms of the cumulant generating function is more beneficial than that of the moment generating function used in (Salz and Winters 1994) as it converges faster since higher order cumulants approach zero. Using four first cumulants of $w(\omega)$ one can obtain so-called excess approximation (Primak et al. 2004) of the covariance function and its magnitude squared

$$\rho(z) \approx \exp\left(j\kappa_1 z - \frac{\kappa_2}{2} z^2 - j\frac{\kappa_3}{6} z^3 + \frac{\kappa_4}{24} z^4 \right) \qquad (3.118)$$

$$|\rho(z)|^2 \approx \exp\left(-\kappa_2 z^2 + \frac{\kappa_4}{12} z^4 \right) \qquad (3.119)$$

Here $z = 2\pi d/\lambda$. It is transparent from representation (3.118) that the oscillating nature of the covariance function is dominated by the first and the third cumulants, while the behavior of the diversity factor is dominated by the second and the fourth cumulants.

Since the first cumulant κ_1 coincides with the mean value of the random variable we obtain

$$\kappa_1 = \int_{-\pi/2}^{\pi/2} \omega \hat{p}(\omega) d\omega = \int_{-\pi}^{\pi} \sin\theta p(\theta) d\theta = \frac{m_{-1} + m_1}{2j} \qquad (3.120)$$

where

$$m_k = \int_{-\pi}^{\pi} \exp(jk\theta) p(\theta) d\theta \qquad (3.121)$$

is the circular moment of the distribution $p(\theta)$ (Mardia and Jupp 2000). A similar relation can be found for other cumulants. Most importantly, the variance of the $\sin\theta$ is given by

$$\kappa_2 = \int_{-\pi}^{\pi} \sin^2\theta p(\theta) d\theta - \kappa_1^2 = \frac{1}{2} - \frac{m_{-2} + m_2}{4} - \kappa_1^2 \qquad (3.122)$$

While Equation (3.118) provides a very good approximation for small values of d it is difficult to trace its asymptotic behavior.

$p(\theta)$	κ_1
$C \cos^n(\theta - \theta_0) \Pi(\Delta)$	C
$\frac{\exp[\kappa(\theta - \theta_0)]}{2\pi I_0(\kappa)}$	$\sin\theta_0 \frac{I_1(\kappa)}{I_0(\kappa)}$

It can be seen from (3.118) that behavior at small distances z is dominated by the first two cumulants. However, for larger z the series (3.118) may not provide a good approximation, especially in terms of asymptotic behavior. What we suggest is to represent the covariance function in the form

$$\rho(z) = \exp(j\kappa_1 z)\mathrm{sinc}\left(z\sqrt{\frac{3\kappa_2}{\pi^2}} \right)[1 + g(z)] \qquad (3.123)$$

where $g(0) = 0$. Function $g(z)$ must be chosen in such a way that it provides a proper asymptotic behavior of $\rho(z)$. Such behavior is defined by how smooth the function $w(\omega)$ is, rather than being defined by such characteristics as the mean or the variance. In order to be consistent with results obtained by using the von Mises PDF, one has to chose $g(z) \sim \sqrt{z}$, while the uniform distribution in a given sector requires $g(z) = 0$. It is believed at this stage that

$$g(z) = \alpha z^{\beta} \tag{3.124}$$

will provide a good choice, since it covers cases of von Mises $\beta = 1/2$, uniform $\beta = 0$, and discrete (fixed angle) $\beta = 1$ AoA.

The following integral, if it exists, will be needed in the evaluation of the diversity measure for large matrices

$$\int_{-L}^{L} |\rho(x)|^2 \left(1 - \frac{x}{L}\right) dx \tag{3.125}$$

Fortunately, this integral can be calculated analytically if approximation (3.123)–(3.124) is used.

3.9.2   Approximation of the diversity measure for a large number of antennas

Since the diversity measure (3.114) is defined solely in terms of eigenvalues, we can use the approximation of the Toeplitz matrix $\mathbf{R}$ by a circulant matrix $\mathbf{C(R)}$ which has a similar set of eigenvalues. Such an approximation is well developed by Szegö and Grenander (Grenander and Szegö 1984). Two main results follow from the Szegö theorem: a) for large $N = 2M + 1$ eigenvalues of $\mathbf{R}$ and

$$\mathbf{C(R)} = \text{circ}\,\{1, r_1, r_2, \cdots, r_M, r_{M-1}, r_{M-2}, \cdots, r_1\} \tag{3.126}$$

are close, and b) if $N \to \infty$ the eigenvalues of $\mathbf{R}$ approach the power spectral density $S(\omega)$ corresponding to the covariance function $\rho(\tau)$ giving rise to the covariance matrix $\mathbf{R}$.

The approximation of eigenvalues is based on the following theorem of Szegö (Grenander and Szegö 1984): for properly defined continuous function $F(\lambda)$

$$\lim_{n \to \infty} \frac{F\left(\lambda_1^{(n)}\right) + F\left(\lambda_1^{(n)}\right) + \cdots + F\left(\lambda_1^{(n)}\right)}{n + 1} = \frac{1}{2\pi} \int_{-\pi}^{\pi} F\left[f(x)\right] dx \tag{3.127}$$

where $\lambda_i^{(n)}$, $i = 1, \cdots, n + 1$ are the eigenvalues of a Toeplitz matrix $\mathbf{R}$ of size $(n + 1) \times (n + 1)$ and

$$f(x) = \frac{1}{N} \sum_{k=-n}^{n} r_n \exp(jnx) \tag{3.128}$$

By setting $F(\lambda) = \lambda^s$, $s = 1, 2$ one can easily obtain the following asymptotic expressions for the sum and sum of squares of the eigenvalues

$$\sum_{k=1}^{n} \lambda_k^{(n-1)} \approx n\frac{1}{2\pi} \int_{-\pi}^{\pi} f(x) dx \tag{3.129}$$

$$\sum_{k=1}^{n-1} \left(\lambda_k^{(n-1)}\right)^2 \approx n \frac{1}{2\pi} \int_{-\pi}^{\pi} f^2(x)dx \tag{3.130}$$

Therefore, the diversity measure $\Psi(\mathbf{R})$ can be approximated by

$$\Psi(\mathbf{R}) \approx \frac{N}{2\pi} \frac{\left[\int_{-\pi}^{\pi} f(x)dx\right]^2}{\int_{-\pi}^{\pi} f^2(x)dx} \tag{3.131}$$

Furthermore, taking (3.128) into account, integrals in (3.131) can be evaluated as

$$\int_{-\pi}^{\pi} f(x)dx = \frac{2\pi}{N} r_0, \quad \int_{-\pi}^{\pi} f^2(x)dx = \frac{1}{N^2} \sum_{k=-N}^{N} |r_n|^2 \tag{3.132}$$

and, therefore

$$\Psi(\mathbf{R}) \approx \frac{N}{\sum_{k=-N}^{N} |r_n/r_0|^2} = \frac{N}{\sum_{k=-N}^{N} |\rho_n|^2} \tag{3.133}$$

where ρ_n are elements of a normalized covariance matrix. It is interesting to note the opposite trend defined by Equation (3.133) as the number of antennas increases. On one side the diversity measure tends to increase in proportion to the number of antennas. However, the opposite trend of increasing the sum in the denominator may render this increase slower. Only if the correlation decays quickly can we observe a linear increase in diversity measure.

If d_T is a distance between two antennas of a ULA, then $L = N d_T$ represents its length and equation (3.133) can be modified to produce

$$\Psi(\mathbf{R}) = \frac{L}{\sum_{k=-N}^{N} |\rho_n|^2 d_T} \approx \frac{L}{\int_{-L}^{L} |\rho(s)|^2 ds} \tag{3.134}$$

However, all calculations can be made much simpler if the Frobenius norm is calculated directly by grouping elements of the corresponding sub-diagonals. In this case

$$\|\mathbf{R}\|_F^2 = \sum_{k=1}^{N} \sum_{l=1}^{N} |r_{kl}|^2 = \sum_{n=-N}^{N} (N - |n|) |r_n|^2$$

$$= N \sum_{n=-N}^{N} \left(1 - \frac{|n|}{N}\right) |r_n|^2 \tag{3.135}$$

and, therefore,

$$\Psi(\mathbf{R}) = \frac{N}{\sum_{n=-N}^{N} (1 - |n|/N) |r_n|^2} \approx \frac{(1 + 1/N)L}{\int_{-L}^{L} |\rho(s)|^2 (1 - s/L) ds}$$

$$\approx \frac{L}{\int_{-L}^{L} |\rho(s)|^2 (1 - |s|/L) ds} \tag{3.136}$$

Furthermore, if the amplitude of the covariance function $\rho(s)$ decays quickly over the aperture of the antenna, the term $1 - |s|/L \approx 1$ for significant values of $|\rho(s)|$, therefore Equation (3.134) could be recovered. While Equation (3.136) is more accurate, it may produce more complicated results.

3.9.3   Examples

A few interesting examples can be worked out analytically and compared to the results of numerical simulations.

3.9.3.1   Nossek example (Ivrlač and Nossek 2003)

The original definition of the diversity of the correlation matrix has been based on the matrix with constant off-diagonal terms: $\mathbf{R}_{ii} = 1$ and $\mathbf{R}_{ij} = \rho$, $j > i$. In this case approximation to the diversity measure becomes

$$\Psi(\mathbf{R}) \approx \frac{N}{1 + (N-1)|\rho|^2} \tag{3.137}$$

This is an exact result which can be obtained by the direct calculation of the trace and the Frobenius norm of the covariance matrix. Interestingly, for large $N|\rho|^2 \gg 1$ the diversity index becomes constant and depends only on ρ:

$$\Psi(\mathbf{R}) \approx \frac{1}{|\rho|^2} \tag{3.138}$$

3.9.3.2   Exponential distribution

In this case elements of the normalized covariance matrix are given by $\mathbf{R}_{ij} = \rho^{|i-j|}$. Therefore the approximation of the diversity order becomes

$$\Psi(\mathbf{R}) \approx \frac{N(1 - |\rho|^2)}{1 + |\rho|^2 - 2|\rho|^{2N}} \approx \frac{N(1 - |\rho|^2)}{1 + |\rho|^2} \tag{3.139}$$

It can be seen that in this case the diversity order increases linearly with the number of antennas, however its slope is factored by $(1 - |\rho|^2)/((1 + |\rho|^2)) < 1$. Results of numerical simulation are shown in Figure 3.5.

3.9.3.3   Narrow uniformly spread beam

For example, let us consider a uniform linear array (ULA) with N elements separated by distance d_A. If the AoA is uniformly distributed with a small angular spread $\pm \Delta/2$ around the average AoA θ_0, the following approximation for the covariance function can be easily derived (Salz and Winters 1994; Zhao and Loyka 2004b)

$$\rho(d) = \exp\left(j2\pi \frac{d}{\lambda}\sin\theta_0\right)\operatorname{sinc}\left(\Delta\frac{d}{\lambda}\cos\theta_0\right) \tag{3.140}$$

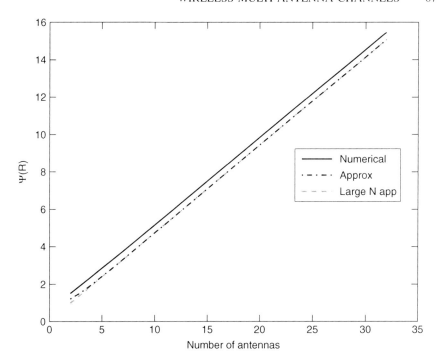

Figure 3.5   Approximation of diversity index $\Psi(\mathbf{R})$ *using Equation (3.139). Number of antennas* $N = 8$; *distance between the antenna elements* $d_A = 0.5\lambda$.

In this case

$$|\rho(d)|^2 = \text{sinc}^2\left(\Delta\frac{d}{\lambda}\cos\theta_0\right) \tag{3.141}$$

and therefore

$$\Psi(\mathbf{R}) = \frac{(1 + 1/N)L^2b^2\pi^2}{\text{Ci}(2\pi Lb) - 2\sin^2(\pi Lb) + 2\pi Lb\text{Si}(2\pi Lb) - \ln(2\pi Lb) - \gamma} \tag{3.142}$$

Here

$$b = \frac{\Delta}{\lambda}\cos(\phi_0)$$

$\gamma = 0.5772\ldots$ is the Euler constant (Abramowitz and Stegun 1965), $\text{Si}(x)$ and $\text{Ci}(x)$ are the integral sine and integral cosine functions respectively (Abramowitz and Stegun 1965).

This expression does not give a clear indication of how the diversity measure $\Psi(\mathbf{R})$ depends on the parameters Δ and θ_0 and a proper asymptotic must be developed. The simplest of such asymptotics results in a simple expression

$$\psi(\mathbf{R}) \approx \Delta\frac{L}{\lambda}\cos\phi_0 \tag{3.143}$$

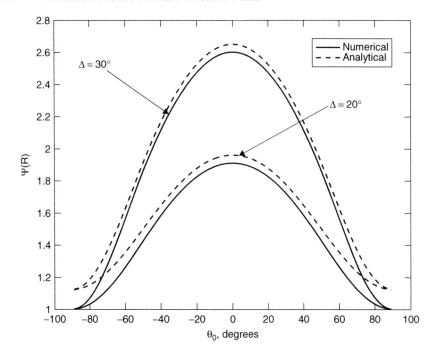

Figure 3.6   Approximation of diversity index $\Psi(\mathbf{R})$ using equation (3.142). Number of antennas $N = 8$; distance between the antenna elements $d_A = 0.5\lambda$.

while an additional term produces

$$\psi(\mathbf{R}) \approx \Delta \frac{L}{\lambda} \cos \phi_0 + \frac{1 - \ln(2\pi \Delta \cos \phi_0 L/\lambda)}{\pi^2} \tag{3.144}$$

Interestingly, this coincides with the Shannon formula[15] (Slepian et al. 1961). Equations (3.143) and (3.144) can be used only for $\Delta L/\lambda \gg 1$ The quality of approximation can be seen from Figure 3.6.

The case of narrow-beam uniform AoA/AoD can be investigated much further. Indeed, since the spatial covariance function (3.140) has the form

$$\rho(d) = \exp(j\omega)\operatorname{sinc}(ad) \tag{3.145}$$

its eigenvalues μ_n and eigenfunctions $\tilde{\psi}_n(d)$ can be expressed (Willink et al. 2006) in terms of the prolate spherodial wave functions (PSWF) $\psi(x, a)$ and associated eigenvalues $\lambda_n(a)$ (Slepian et al. 1961), (Percival and Walden 1993) as

$$\tilde{\psi}_n(d) = \exp\left(-j2\pi \frac{d}{\lambda}\right) \psi_n\left(d; \frac{L}{\lambda} \Delta \cos \theta_0\right) \tag{3.146}$$

$$\mu_n = \lambda_n(\Delta \cos \theta_0) \tag{3.147}$$

---

[15] This formula tells us that the signal of duration T concentrated in bandwidth F has $2FT + c_1 \ln(2FT) + c_2$ degrees of freedom, where c_1 and c_2 are some constants. Of course, only the first part of this equation became famous.

If a ULA with N elements and spacing d_T between elements is considered, the covariance matrix $\mathbf{R}$ can be represented as

$$\mathbf{R} = \mathbf{U}(N, \Delta, \theta_0) \Lambda(N, \Delta, \theta_0) \mathbf{U}^H(N, \Delta, \theta_0) \qquad (3.148)$$

where

$$\Lambda(N, \Delta, \theta_0) = diag\{\lambda_n(N\Delta\theta_0)\}, \ n = 1, \cdots, N \qquad (3.149)$$

$$\mathbf{U}_{k,n} = u_n(k; N, N\Delta\theta_0) \qquad (3.150)$$

is a diagonal matrix of eigenvalues corresponding to the discrete prolate spheroidal sequence $u_n(k; N, N\Delta\theta_0)$, $n, k = 1, \cdots, N$ (Percival and Walden 1993; Slepian et al. 1961). The first $N\Delta \cos\theta_0$ eigenvalues are extremely close to the unity, while the others decay rapidly (see (Zemen and Mecklenbräuker 2005) for a nice plot). Therefore, the approximation of the diversity measure is the same as (3.143). The representation of type (3.148) has been used in (Alcocer et al. 2006), albeit only for $\theta_0 = 0$.

3.9.3.4   Two uniformly-spread beams

In practical situations, when omnidirectional antennas are used, it is possible that the received/transmit field is originated from a number of angularly small but separated scatterers. For simplicity let us consider a case of two such scatterers seen at the azimuths $\theta_{10} < \theta_{20}$ and the beamwidths Δ_1 and Δ_2 respectively. Therefore, the probability density can be thought of as a weighted sum of two uniform densities:

$$p(\theta) = \alpha p_1(\theta, \theta_{10}, \Delta_1) + \beta p_2(\theta, \theta_{20}, \Delta_2) \qquad (3.151)$$

where $\alpha + \beta = 1, \alpha > 0, \ \beta > 0$. Therefore, the spatial covariance is also a weighted sum of covariances corresponding to single clusters, that is

$$\rho(d) = \alpha\rho_1(d) + \beta\rho_2(d) = \alpha \exp\left(j2\pi \frac{d}{\lambda} \sin\theta_{10} \right) sinc\left(\Delta_1 \frac{d}{\lambda} \cos\theta_{10} \right)$$

$$+ \beta \exp\left(j2\pi \frac{d}{\lambda} \sin\theta_{10} \right) sinc\left(\Delta_1 \frac{d}{\lambda} \cos\theta_{10} \right) \qquad (3.152)$$

However, the expression for $|\rho(d)|^2$, needed for the calculation of the diversity measure is not linear and contains three terms

$$|\rho(d)|^2 = \alpha^2 |\rho_1(d)|^2 + \beta^2 |\rho_2(d)|^2 + 2\alpha\beta\Re\rho_1(d)\rho_2^*(d)$$

$$= \alpha^2 sinc^2\left(\Delta_1 \frac{d}{\lambda} \cos\theta_{10} \right) + \beta^2 sinc^2\left(\Delta_2 \frac{d}{\lambda} \cos\theta_{20} \right)$$

$$+ 2\alpha\beta \cos\left[2\pi \frac{d}{\lambda} (\sin\theta_{10} - \sin\theta_{20}) \right]$$

$$\times sinc\left(\Delta_1 \frac{d}{\lambda} \cos\theta_{10} \right) sinc\left(\Delta_2 \frac{d}{\lambda} \cos\theta_{20} \right) \qquad (3.153)$$

Evaluation of the integral (3.134) for the first two terms in Equation (3.153) is identical to that of a single cluster case. Calculation of the last term is somewhat more complicated and can be accomplished as follows. First let us note that the product of a cosine term and two sinc terms can be written a sum of two terms, each of which is a product of two sinc terms. Specifically

$$
\cos\left[2\pi \frac{d}{\lambda}(\sin\theta_{10} - \sin\theta_{20})\right] \mathrm{sinc}\left(\Delta_1 \frac{d}{\lambda}\cos\theta_{10}\right)
$$
$$
= \frac{1}{2}\frac{a+b}{b}\mathrm{sinc}\left[\frac{d}{\lambda}(2\sin\theta_{10} - 2\sin\theta_{20} + \Delta_1\cos\theta_{10})\right] \tag{3.154}
$$

3.9.3.5 Isotropic scattering

In the case of isotropic scattering the spatial correlation function is represented by $\rho(d) = J_0(2\pi d/\lambda$. In this case

$$
\int_{-L}^{L}\left(1 - \frac{|s|}{L}\right) J_0^2\left(2\pi \frac{s}{\lambda}\right) ds
$$
$$
= L \times \left[2\,_2F_3\left(\frac{1}{2},\frac{1}{2}; 1, 1, \frac{3}{2}; -4L^2\pi^2/\lambda^2\right) - J_0\left(2\pi\frac{L}{\lambda}\right) - J_1\left(2\pi\frac{L}{\lambda}\right)\right] \tag{3.155}
$$

and, therefore

$$
\Psi(\mathbf{R}) = \left(1 + \frac{1}{N}\right)\left[2\,_2F_3\left(\frac{1}{2},\frac{1}{2}; 1, 1, \frac{3}{2}; -4L^2\pi^2/\lambda^2\right)\right.
$$
$$
\left. - J_0\left(2\pi\frac{L}{\lambda}\right) - J_1\left(2\pi\frac{L}{\lambda}\right)\right]^{-1} \tag{3.156}
$$

The results compare well with the numerical simulation as can be seen from Figure 3.7.

It can be seen that as the number of antennas increases the amount of diversity per antenna $\Psi(\mathbf{R})/N$ approaches zero, although $\Psi(\mathbf{R}) \to \infty$ as $N \to \infty$ due to the slow divergence of the integral (3.155). For realistic antenna sizes, that is $N < 32$, the value of diversity per antenna can be calculated numerically for a given value of distance between antennas.

3.9.4   Leading term analysis of degrees of freedom

In order to consider asymptotic expansions of integrals of type (3.197) and (3.198) let us use the following generic way of calculating the diversity index (Bender and Orszag 1999). As we have seen the index depends on the value of the following sum

$$
S_N(\rho_1, \rho_2) = \sum_{-N+1}^{N+1} \rho_1(nd_T)\rho_2^*(nd_T)\left(1 - \frac{|n|}{N}\right)
$$

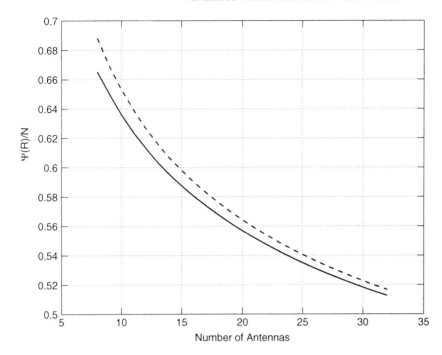

*Figure 3.7 Approximation of diversity measure $\Psi(\mathbf{R})$ per antenna using Equation
(3.156). $d_A = 0.3\lambda$.*

$$= N \sum_{-N+1}^{N+1} \rho_1\left(\frac{n}{N} N d_T\right) \rho_2^*\left(\frac{n}{N} N d_T\right)\left(1 - \frac{|n|}{N}\right)\frac{1}{N}$$

$$\times N \int_{-1}^{1} \rho_1\left(N d_T x\right) \rho_2^*\left(N d_T x\right)\left(1 - |x|\right) dx$$

$$= 2N \int_{0}^{1} \rho_1\left(Lx\right) \rho_2^*\left(Lx\right)\left(1 - x\right) dx \qquad (3.157)$$

We would like to find an asymptotic formula of Shannon type

$$\Psi(\mathbf{R}) \sim \alpha \frac{L}{\lambda} + \beta \ln \frac{L}{\lambda} + \delta \qquad (3.158)$$

For simplicity we will assume that $\lambda = 1$ in Equation (3.158). Dividing both sides
by L and taking the limit as $L \to \infty$ one can easily obtain

$$\alpha = \lim_{L\to\infty} \frac{\Psi(\mathbf{R})}{L} = \lim_{L\to\infty} \frac{2}{d_T}\left[\int_{0}^{1} |\rho(Lx)|^2(1-x)dx\right]^{-1}$$

$$= \lim_{L\to\infty} \frac{2}{d_T}\left[\int_{0}^{1} |\rho(Lx)|^2 dx\right]^{-1} = \int_{-\infty}^{\infty} |\rho(x)|^2 dx \qquad (3.159)$$

This readily results in $\alpha = \Delta \cos \theta_0$ for the uniform narrow beam, $\alpha = -d_T / \ln \rho$ for exponential correlation and $\alpha = 0$ for the uniform distribution of phase and the constant correlation. This shows that the degree of freedom grows more slowly than the number of antennas. Equation (3.159) also reveals the conditions of having linear growth in the diversity with the rate of decay of the covariance function: it is necessary that the covariance decays at least as fast as $1/d$.

3.10   Effect of AoA/AoD dependency on the SDoF

If the receive and transmit sites are independent, the corresponding covariance function factorizes $\rho(d_T, d_R) = \rho_T(d_T)\rho_R(d_R)$ which is equivalent to Kronecker factorization of the full correlation matrix $\mathbf{K_h} = \mathbf{R}_R \odot \mathbf{R_T}^T$. It is easy to check that in this case $\mathrm{trace}\mathbf{K_h} = \mathrm{trace}\mathbf{K}_R \mathrm{trace}\mathbf{K}_T$ and $||\mathbf{R}_h||_F = ||\mathbf{R}_T||_F ||\mathbf{R}_R||_F$, and therefore, $\Psi(\mathbf{R_h}) = \Psi(\mathbf{R}_T)\Psi(\mathbf{R}_R)$.

We focus on the normalized covariance matrix, in other words disregard the power imbalance for now (Zhao and Loyka 2004b).

3.11   Space-time covariance function

3.11.1   Basic equation

First let us consider the space-time covariance function on one side of the link: without loss of generality we will assume the receiver side. The expression will be easily expanded on using canonical factorization 3.63. Let γ be an angle between the broadside direction of the ULA (or two elements of a more generic array) and the direction of the movement of the array. If the incoming wave forms an angle θ with the broad-side of the array, then it experiences the Doppler shift $f_D \cos(\theta - \gamma)$ where $f_D = f_0 v/c$ is the maximum Doppler shift provided that the carrier frequency is f_0 and the velocity of the vehicular is v. In this case the expression for the normalized space-time covariance function becomes

$$\rho(d, \tau) = \int_{-\pi}^{\pi} e^{j2\pi\left[\frac{d}{\lambda}\sin\theta + f_D\tau\cos(\theta-\gamma)\tau\right]} p(\theta)d\theta \tag{3.160}$$

After simple manipulations, Equation (3.160) can be written in the form

$$\rho(d, \tau) = \int_{-\pi}^{\pi} \exp\left[j2\pi x(d, \tau)\sin(\theta + \alpha(d, \tau))\right] p(\theta)d\theta \tag{3.161}$$

where

$$x^2(d, \tau) = \frac{d^2}{\lambda^2} + 2\frac{d}{\lambda} f_D\tau \sin\gamma + f_D^2\tau^2 \tag{3.162}$$

and

$$\tan\alpha(d, \tau) = \frac{f_D\tau \cos\gamma}{d/\lambda + f_D\tau \sin\gamma} \tag{3.163}$$

Therefore, in general, the space-time covariance function $\rho(d, \tau)$ cannot be factored into time $\rho_T(\tau) = \rho(0, \tau)$ and space $\rho_S(d) = \rho(d, 0)$ parts: $\rho(d, \tau) \neq \rho_T(\tau)\rho_S(d)$.

The full covariance function $\rho(d_T, d_R, \tau)$ between any two links of a MIMO system can be obtained by combining (3.161) and the canonical expansion of the joint AoA/AoD PDF. In particular, it can be represented as a series

$$\rho(d_T, d_R, \tau) = \sum_{k=0}^{\infty} \rho_k \rho_{Tk}(d_T, \tau) \rho_{Rk}(d_R, \tau) \tag{3.164}$$

Here, ρ_k is a canonical correlation between k-th pair of canonical variables in expansion 3.63 and partial covariances $\rho_{Tk}(d_T, \tau)$ and $\rho_{Rk}(d_R, \tau)$ are given by

$$\rho_k(d, \tau) = \int_{-\pi}^{\pi} \exp\left[j2\pi x(d, \tau)\sin(\theta + \alpha(d, \tau))\right] R_k(\theta)p(\theta) \tag{3.165}$$

Here $R_k(\theta) = P_k(\theta_T)$ and $R_k(\theta) = Q_k(\theta_R)$. Other parameters are defined in a similar manner through Equations (3.162)–(3.163) with parameters corresponding to either transmit or receive sides. In general, as could be expected, the covariance function depends on the distances d_T and d_R between two links on each side, the velocity of each side, and the correlation between sides.

3.11.2   Approximations

3.11.2.1   Time only covariance function

If $d_T = d_R = 0$ we recover the time covariance function of a single SISO link. In this case

$$x(d, \tau) = f_D \tau \tag{3.166}$$

and

$$\alpha(d, \tau) = \frac{\pi}{2} - \gamma \tag{3.167}$$

thus leading to the following representation of $\rho_k(0, \tau)$ in Equation (3.165):

$$\rho_k(\tau) = \int_{-\pi}^{\pi} \exp\left[j2\pi f_D \tau \cos(\theta - \gamma)\right] R_k(\theta)p(\theta)d\theta \tag{3.168}$$

If only one side of the link is in motion, the equation for SISO covariance function significantly reduces since if $f_D = 0$ on one of the sides, the corresponding term $\rho_k(\tau) = 0$ vanishes for $k > 1$, while $\rho_1(\tau) = 1$ due to orthogonality of canonical variables. Thus, the covariance function depends only on geometry (distribution of the angles and the antenna orientation) and speed of the moving side:

$$\rho(\tau) = \int_{-\pi}^{\pi} \exp\left[j2\pi f_D \tau \cos(\theta - \gamma)\right] p(\theta)d\theta \tag{3.169}$$

If both sides are moving but there is no correlation between the sides, the time covariance function of a SISO link factorizes in the product of two terms, each corresponding to only one side moving

$$\rho(\tau) = \rho_T(\tau)\rho_R(\tau) \tag{3.170}$$

since $\rho_k = 0$ for $k > 1$. This is widely known in the literature (Akki 1994; Akki and Haber 1986).

The expression (3.169) can be explicitly calculated if the PDF obeys the von Mises distribution

$$p(\theta) = \frac{\exp\left[\kappa \cos(\theta - \mu)\right]}{2\pi I_0(\kappa)} \tag{3.171}$$

In this case the covariance function evaluates to

$$\rho(\tau) = \frac{1}{I_0(\kappa)} I_0\left(\sqrt{\kappa^2 + 4\pi^2 f_D^2\tau^2 - j4\pi f_D\tau \cos(\mu - \gamma)}\right) \tag{3.172}$$

which not only allows for compact analytical expression but also allows us to study the transition of the covariance function from the case of very narrowbeam scattering ($\kappa > 3$) to uniform scattering $\kappa = 0$.

By expanding $\ln \rho(\tau)$ into a series as a function of τ one obtains

$$\ln \rho(\tau) = j\kappa_1 2\pi f_D\tau - \frac{\kappa_2}{2} 4\pi^2 f_D^2\tau^2 + \mathcal{O}(\tau^3) \tag{3.173}$$

where

$$\kappa_1 = \frac{I_1(\kappa)}{I_0(\kappa)} \cos(\mu - \gamma) \tag{3.174}$$

and

$$\kappa_2 = \frac{I_2(\kappa)I_0(\kappa) - I_1^2(\kappa)}{I_0^2(\kappa)} \cos^2(\mu - \gamma) + \frac{I_1(\kappa)}{\kappa I_0(\kappa)} \tag{3.175}$$

Therefore, the covariance function could be approximated as

$$\rho(\tau) \approx \exp(j\omega\tau)\mathrm{sinc}(a\tau) \tag{3.176}$$

where

$$\omega = 2\pi f_D \frac{I_1(\kappa)}{I_0(\kappa)} \cos(\mu - \gamma) \tag{3.177}$$

$$a = \sqrt{12} f_D \sqrt{\frac{I_2(\kappa)I_0(\kappa) - I_1^2(\kappa)}{I_0^2(\kappa)} \cos^2(\mu - \gamma) + \frac{I_1(\kappa)}{\kappa I_0(\kappa)}} \tag{3.178}$$

The approximation is worst in the case of uniform distribution $\kappa = 0$: $\rho(\tau) = J_0(2\pi f_D\tau)$, $\omega = 0$, and $a = \sqrt{6} f_D$. However, even in this case the approximation is quite accurate for $f_D\tau < 0.1$.

For small values of κ one can easily derive the following power series expansion for ω and a:

$$\omega \approx \kappa \pi f_D \cos(\mu - \gamma), \ a \approx \sqrt{6} f_D \qquad (3.179)$$

For large κ the term $I_1(\kappa)/I_0(\kappa) \rightarrow 1$ and, therefore, $\omega \rightarrow 2\pi f_D \cos(\mu - \gamma)$. More accurately, for $\kappa > 1$, one can write

$$\frac{I_1(\kappa)}{I_0(\kappa)} \sim 1 - \frac{1}{2\kappa}, \ \omega \approx 2\pi f_D \left(1 - \frac{1}{2\kappa}\right) \cos(\mu - \gamma) \qquad (3.180)$$

A similar expression can be derived for coefficient a. Keeping a single term in asymptotic expansion of (3.178) one can easily obtain that

$$a \approx \frac{\sqrt{12}}{\kappa} f_D \sin(\mu - \gamma) \rightarrow 0 \qquad (3.181)$$

As expected, the covariance function approaches a cosine with frequency $f_D \cos(\mu - \gamma)$ which corresponds to the correlation of a monochromatic wave with uniform random phase (Papoulis 1991), which can be associated with the line of sight or specular components.

3.11.2.2 Small lag τ approximation

In practical systems with a high data rate the value of the fading rate $f_D\tau \ll 1$ is relatively small (at most 0.1 for many applications) even over the duration of a block. Separation of antenna elements is at least half of the wavelength and therefore $2\pi d/\lambda \geq \pi$. Therefore, it is safe to assume that $f_D\tau \ll 2\pi d/\lambda$. Having this in mind, some simplifications of Equations (3.162)–(3.163) can be obtained. In particular

$$x(d, \tau) \approx 2\pi \frac{d}{\lambda} + f_D\tau \sin \gamma \qquad (3.182)$$

$$\alpha(d, \tau) \approx \frac{f_D\tau \cos \gamma}{2\pi d/\lambda} \qquad (3.183)$$

3.12 Examples: synthetic data and uniform linear array

In this section the proposed algorithms are applied to synthetic data. We choose a mixture of two von Mises PDFs as a model for AoA and AoD distributions.

$$p(\theta) = \alpha \frac{\exp[\kappa_1 \cos(\theta - \mu_1)]}{I_0(\kappa_1)} + (1 - \alpha)\frac{\exp[\kappa_2 \cos(\theta - \mu_2)]}{I_0(\kappa_2)} \qquad (3.184)$$

where $0 \leq \alpha \leq 1$ is the mixing parameter. The corresponding spatial covariance function is given by

$$
\rho(d) = \frac{\alpha}{I_0(\kappa_1)} I_0 \left(\sqrt{\kappa_1^2 - 4\pi^2 \frac{d^2}{\lambda^2} - j4\pi\kappa_1 \frac{d}{\lambda} \cos\mu_1} \right)
$$
$$
+ \frac{1-\alpha}{I_0(\kappa_1)} I_0 \left(\sqrt{\kappa_2^2 - 4\pi^2 \frac{d^2}{\lambda^2} - j4\pi\kappa_2 \frac{d}{\lambda} \cos\mu_2} \right) \tag{3.185}
$$

As a measure of difference between the two correlation matrices we chose the following quantity

$$
\varepsilon(\mathbf{R}_1, \mathbf{R}_2) = 1 - \frac{\text{trace } \mathbf{R}_1 \mathbf{R}_2}{\|\mathbf{R}_1\|_F \|\mathbf{R}_2\|_F} \tag{3.186}
$$

as suggested in (Herdin et al. 2005). Such a measure is appropriate since it reflects the eigenvector structure of both matrices, as shown in the reference (Herdin et al. 2005).

As an example of the application of the method suggested in Section 3.5.2 we consider the uniform linear array (ULA) of 8 antennas separated by distance

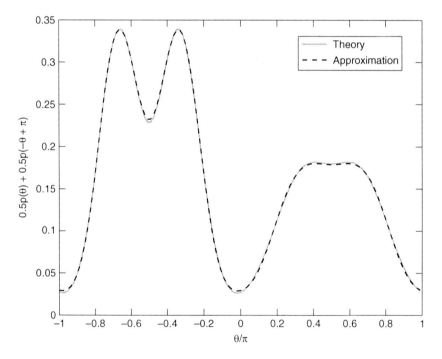

Figure 3.8   Estimation of $p(\theta)$ from the covariance matrix. Mixture of two von Mises distributions as given by Equation (3.184) with the following parameters $\alpha = 0.4$, $\kappa_1 = 4$, $\kappa_2 = 6$, $\mu_1 = \pi/3$ and $\mu_2 = -p/3$ The generating polynomial $h(z)$ is of order $N = 9$.

$d = 0.5\lambda$. Since there is ambiguity in the estimation of AoA/AoD for ULA (angles θ and $\pi - \theta$ are indistinguishable by ULA) we cannot expect that the resulting estimated PDF $\hat{p}(\theta)$ will match the proper PDF $p(\theta)$. However we do expect that

$$\hat{p}(\theta) = \frac{1}{2}\left[p(\theta) + p(\pi - \theta)\right] \tag{3.187}$$

the results of the numerical simulations are shown in Figure 3.8 and indicate a good agreement with Equation (3.187).

3.13 Approximation of a matrix by a Toeplitz matrix

It is clear from the measurement results that the covariance matrices of the MIMO channel could be non-Toeplitz. Without considering reasons for this fact we would like to approximate a non-Toeplitz positive definite $\mathbf{R}$ matrix of size $N \times N$ with a Toeplitz positive definite matrix $\mathbf{R}_T$. There are, of course, many ways to achieve such an approximation. One is simply to average elements of the corresponding main sub-diagonals as suggested (Wallace et al. 2003). Such an approximation is the best approximation in terms of the Frobenius norm, that is it is a solution of the unconstrained optimization problem

$$\mathbf{R}_T = \arg\min_{\mathbf{R}_T} ||\mathbf{R} - \mathbf{R}_T||_F \tag{3.188}$$

However, such an approximation does not guarantee preservation of the degree of diversity, which corresponds to a given $\mathbf{R}$. We suggest that a constrained optimization is used. In particular, we would like to preserve the Frobenius norm of $\mathbf{R}$ to ensure that the Nossek diversity measure $\Psi(\mathbf{R})$ remains the same. In other words we want to solve the following optimization problem

$$\mathbf{R}_T = \arg\min_{||\mathbf{R}_T||_F = ||\mathbf{R}||_F} ||\mathbf{R} - \mathbf{R}_T||_F \tag{3.189}$$

This can be solved using the method of the Lagrange multipliers (Korn and Korn 1967). It is easy to see that one can minimize a square of the Frobenius norm instead of the norm itself. In addition, minimization can be done sub-diagonal-wise. Therefore, we focus on optimization of a single entry r_k of the covariance matrix $\mathbf{R}_T$ corresponding to its k-th sub-diagonal.

Let $a_i = r_{i,k+i}$, $i = 1, \ldots N - k$ be elements of the k-th sub-diagonal and $b_i = r_{k+i,i}$, $i = 1, \ldots N - k$ be elements of the k-th sub-diagonal of the matrix $\mathbf{R}$. Also we define a new vector $\mathbf{c} = \mathbf{c}_R + j\mathbf{c}_I$ of length $2N - 2k$ with elements

$$\begin{aligned} c_i &= a_i, \quad i = 1, \ldots, N - k \\ c_i &= b_i^*, \quad i = N - k + 1, \ldots, 2N - 2k \end{aligned} \tag{3.190}$$

Our goal is to choose element $r_t = x + jy$ such that the mean square error

$$\varepsilon = \sum_{i=1}^{N-k} |a_i - r_t|^2 + \sum_{i=1}^{N-k} |b_i - r_t^*|^2 = \sum_{i=1}^{2N-2k} |c_i - r_t|^2 \tag{3.191}$$

while

$$(2N - 2k)\, |r_t|^2 = \sum_{i=1}^{N-k} |a_i|^2 + \sum_{i=1}^{N-k} |b_i|^2 = \sum_{i=1}^{2N-2k} |c_i|^2 \qquad (3.192)$$

The equivalent Lagrange multiplier problem thus becomes the unconstrained minimization of the following function of three variables

$$f(x, y, \lambda) = \varepsilon + \lambda \left(x^2 + y^2 - \sum_{i=1}^{2N-2k} |c_i|^2 \right)$$

$$= \sum_{i=1}^{2N-2k} (c_{Ri} - x)^2 + \sum_{i=1}^{2N-2k} (c_{Ii} - y)^2 + \lambda \left(x^2 + y^2 - \sum_{i=1}^{2N-2k} |c_i|^2 \right) \qquad (3.193)$$

Taking derivatives of $f(x, y, \lambda)$ with respect to x and y and equating them to zero we can find conditions for the local optimum. This results in the following expressions for x and y

$$x = \frac{1}{2N - 2k + \lambda} \sum_{i=1}^{2N-2k} c_{Ri} \qquad (3.194)$$

$$y = \frac{1}{2N - 2k + \lambda} \sum_{i=1}^{2N-2k} c_{Ii} \qquad (3.195)$$

or, recalling that $r_t = x + jy$ and $c_i = c_{Ri} + jc_{Ii}$

$$r_t = \frac{1}{2N - 2k + \lambda} \sum_{i=1}^{2N-2k} c_i \qquad (3.196)$$

The parameter λ has to be chosen in such a way that the constraint (3.192) is satisfied.

3.14 Asymptotic expansions of diversity measure

The following integral and its asymptotic have been derived

$$F(a, b; L) = \int_0^L \mathrm{sinc}(ax)\mathrm{sinc}(bx) \left(1 - \frac{x}{L} \right) dx$$

$$= \frac{1}{2\pi^2 Lab} \left[\ln \frac{b - a}{b + a} + \pi L(a + b)\, \mathrm{Si}\,\pi L(a + b) \right.$$

$$- \pi L(b - a)\, \mathrm{Si}\,\pi L(b - a) + \mathrm{Ci}\,\pi L(a + b)$$

$$\left. -\mathrm{Ci}\,\pi L(b - a) - 2\sin\pi La \sin\pi Lb \right] \qquad (3.197)$$

$$F(a, a; L) = \int_0^L \text{sinc}^2(ax) \left(1 - \frac{x}{L}\right) dx = \frac{1}{2\pi^2 a^2 L}$$
$$\times \left[\sin^2(\pi L a) + \text{Ci}(2\pi L) + 2\pi L a \text{Si}(\pi L a) - \ln(2\pi a L) - \gamma\right]$$

$$(3.198)$$

3.15 Distributed scattering model

A model which allows us to separately control diversity and rank properties was suggested in (Gesbert et al. 2002). The suggested model is based on the assumption that both the receive and the transmit antenna arrays are surrounded by a cloud of scatterers and the propoagation between the antennas can be separated into the following basic steps, as shown in Figure 3.9. It is assumed that there is an equal number S of scatterers on each side. The model discussed below is well suited in at least three of the following scenarios: a) a keyhole-like scenario when two rings of scattterers are separated by a large distance R; b) amplify and forward relays; c) propagation via a diffracting street corner (Salo et al. 2006).

1. Illumination of local scatterers by the transmit antenna array. This could be described by a $S \times N_T$ random matrix $\mathbf{H}_1 = \mathbf{G}_T \mathbf{R}_T^{1/2}$ where the $S \times N_T$ matrix $\mathbf{G}_T$ contains i.i.d. zero-mean complex Gaussian random variables and $\mathbf{R}_T$ is transmit side correlation matrix. The diameter of the scattering cluster is D_T and is assumed to be much larger than the antenna dimension.

2. Free space propagation between clusters. This could be described by a $S \times S$ matrix $\mathbf{\Phi}$ with deterministic elements since any magnitude and phase random variation could be absorbed into the correlation and phase variation on both the receive and transmit sides. Distance between the scattering clusters (transmit-receive range) is $L \gg \max D_R, D_T$,

3. Finally, the signal transfer from the receive scattering cluster to the received antenna array could be described by a $N_R \times S$ matrix term $\mathbf{H}_R = \mathbf{R}_R^{1/2} \mathbf{G}_R$ where $\mathbf{G}_R$ is matrix of i.i.d. zero-mean complex Gaussian elements. The diameter of the scattering cluster is D_R and is assumed to be much larger than the antenna dimension.

Combining items 1-3 one can arrive at the following distributed scattering model, suggested in (Gesbert et al. 2002)

$$\mathbf{H} = \frac{1}{S} \mathbf{R}_R^{1/2} \mathbf{H}_1 \mathbf{\Phi} \mathbf{H}_T \mathbf{R}_T^{1/2} \tag{3.199}$$

One can derive a vector equivalent of the Equation (3.199) as

$$\mathbf{h} = \text{vec } \mathbf{H} = \left(\mathbf{R}_T^{1/2} \otimes \mathbf{R}_R^{1/2}\right) \left(\mathbf{H}_T^T \otimes \mathbf{H}_R\right) \text{vec } \mathbf{\Phi} \tag{3.200}$$

The full correlation matrix $\mathbf{R}_h$ corresponding to this model is then given by

$$\mathbf{R}_h = \mathcal{E}\{\mathbf{h}\mathbf{h}^H\} = \left(\mathbf{R}_T^{1/2} \otimes \mathbf{R}_R^{1/2}\right) \mathbf{R}_S \left(\mathbf{R}_T^{1/2} \otimes \mathbf{R}_R^{1/2}\right) \tag{3.201}$$

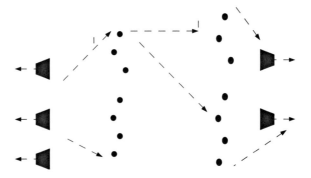

Figure 3.9   Propagation scenario, corresponding to distributed scattering model.

where

$$\mathbf{R}_S = \mathcal{E}\left\{\mathbf{H}_T^T \otimes \mathbf{H}_R \, \boldsymbol{\Phi}\boldsymbol{\Phi}^H \, \mathbf{H}_R^* \otimes \mathbf{H}_R^H\right\} \tag{3.202}$$

The following conclusions can be drawn for the model given by (3.199) (Gesbert et al. 2002):

- The model is symmetric in structure as could be expected from the propagation scenario.

- The spatial correlation on either side of the link is controlled by the spatial distribution of local scatterers, and is therefore fully controlled by one side correlation matricies $\mathbf{R}_T$ and $\mathbf{R}_R$.

- It is indicated by Equation (3.201) that even if spacing between antennas is significant such that $\mathbf{R}_T = \mathbf{I}_{N_T}$ and $\mathbf{R}_R = \mathbf{I}_{N_R}$ there is still correlation between both sides of the link, governed by the matrix $\mathbf{R}_S$. Thus the model exhibits a non-Kronecker structure.

- It can be seen that the matrix $\mathbf{H}$ could be rank defficient, that is rank $\mathbf{R}_S <$ min N_T, N_R. This is the case when propagation between the link sides does not support a sufficient amount of orthogonal modes. This may appear if the condition

$$\frac{4D_R D_T}{R\lambda} > \frac{(N_T - 1)(N_R - 1)}{N_R}$$

  is violated.

4

Modeling of wideband multiple channels

Most modern communication systems operate in a relatively wide band of frequencies, which allows us to identify various multiple path components (MPC) (Almers et al. 2007; Bello 1963) in the channel impulse response. A comprehensive review of modeling of wideband channels in the SISO case can be found in (Patzold 2002). In this chapter we focus on the particularities of MIMO models. A comprehensive review of recent developments can be found in (Almers et al. 2007; Asplund et al. 2006; Molisch et al. 2006; SCM Editors 2006) and other publications.

Significant progress in numerical simulation of electromagnetic waves in complex environments, coupled with an ever increasing body of measurements, gives a solid foundation to a class of geometric-based stochastic scattering models (GSCM), which are of main interest in this text.

Different versions of the GSCM differ mainly in the proposed scatterer distributions for their numerous parameters. The simplest GSCM can be derived by imposing that the scatterers are spatially uniformly distributed over a certain propagation region. Contributions from distant scatterers carry much less power due to free space propagation losses and can be excluded from consideration. In addition, scattering from objects also contributes significantly to loss of signal power, therefore it is often possible to consider so-called single-bounce geometrical models. Scattering centers can be placed in regular geometric shapes, such as rings or ellipses, giving rise to a large number of models and corresponding statistics of delay, power, and angular profiles (Paulraj et al. 2003). An alternative approach suggests placing the scatterers randomly around the MS (Almers et al. 2007).

Non-geometrical stochastic models describe paths from Tx to Rx by statistical parameters only, without reference to the geometry of a physical environment.

Wireless Multi-Antenna Channels: Modeling and Simulation, First Edition.
Serguei L. Primak and Valeri Kontorovich.
© 2012 John Wiley & Sons, Ltd. Published 2012 by John Wiley & Sons, Ltd.

There are two classes of stochastic non-geometrical models reported in the literature. The first one uses clusters of MPCs and is generally called the extended Saleh–Valenzuela model since it generalizes the temporal cluster model developed in [69]. The second one (known as the Zwick model) treats MPCs individually.

4.1   Standard models of channels

In order to evaluate the performance of different MIMO systems and algorithms a number of reference MIMO channel models are standardized. Each such model provides the designer with reproducible channel conditions, which can be how established reproducible channel conditions are defined. With physical models this means one can specify a channel model, reference environment, and parameter values for these environments. With analytical models, parameter sets representative of the target scenarios need to be prescribed. Five examples of such reference models are proposed within 3GPP (SCM Editors 2006), IST-WINNER [31], COST 259 (Asplund et al. 2006; Molisch et al. 2006), COST 273, IEEE 802.16a,e, and IEEE 802.11n.

4.1.1   COST 259/273

Several COST[1] initiatives were dedicated to wireless communications. COST 259 flexible personalized wireless communications and COST 273 towards mobile broadband multimedia networks (2001–2005) were dedicated to the development of channel models that include directional characteristics[2] of radio propagation and are thus suitable for the simulation study of adaptive antennas and MIMO systems.

The COST 259 directional channel model (DCM) (Molisch et al. 2006) provides a model for the delay and angle distributions on both sides of the link for 13 generic radio environments. It accounts for the relationships between BS-MS-distance, delay dispersion, angular spread, and other parameters.

Each radio environment is described by external parameters (e.g., BS position, radio frequency, average BS and MS height) and their distributions, which are called global parameters. The determination of the global parameters is through a mix of geometric and stochastic approaches. Each radio environment contains a number of propagation environments, which are defined as an area over which the local parameters are randomly generated realizations of the global parameters and describe the instantaneous channel behavior. The model generates impulse responses in the time domain.

While this model is general, there are two major restrictions of the COST 259 DCM: it assumes stationary scatterers and requires a lot of scatterers to properly

---

[1] European cooperation in the field of scientific and technical research.
[2] That is, describe angular dependence of scattering on each side of the link.

represent exponential power delay profiles. This makes it impossible to use such a model for indoor scenarios where some scatterers (people) are mobile. It also adds computational complexity to the simulator.

The COST 273 channel model generalizes COST 259 in a number of ways. First of all it allows significantly different radio environments for each side of the link as far as the antenna location is concerned (above or below rooftop level). It also introduces new scenarios, such as peer-to-peer, and the scattering cluster is modeled as two separate angular spreads.

4.1.2   3GPP SCM

The spatial channel model (SCM) (SCM Editors 2006) was developed by 3GPP/3GPP2 to be a common reference for evaluating different MIMO concepts in outdoor environments at a center frequency of 2 GHz and a system bandwidth of 5 MHz. The SCM consists of a calibration model and a system-simulation model. The former is an over-simplified channel model whose purpose is to check the correctness of simulation implementations while the latter is used for system evaluation. The simulation model is a physical model comprising three different environments (urban macrocell, suburban macrocell, and urban microcell). The model structure and simulation methodology are identical for all three environments, with the parameters such as angular spread, delay spread, and so on being environment dependent.

The simulation model combines both geometrical and stochastic components. Mobile stations are placed at random within a given cell, while the orientation of the antenna array and the direction of movement within the cell are chosen at random from a uniform distribution. The model specifies how to evaluate the average propagation loss depending on the evironment (Hata model for macrocells, and Walsh–Ikegami model for microcells). The model uses 6 taps with randomly chosen delays from a prescribed PDF.

Each tap shows an angular dispersion at the BS and the MS, implemented by representing each tap by a number of subpaths that all have the same delay, but different DOAs (and DODs). Angular dispersion is represented by a Gaussian distribution with the mean angle defined by the location of the scatterer with respect to the BS and MS, while variance is related to the angular spread of the scatterer. Each tap is represented by a Rician fading. The simulations are performed in short segments (called drops) with parameters randomly changing from drop to drop according to prescribed statistical distributions.

Antenna radiation patterns, antenna geometries, and orientations can be chosen arbitrarily thereby allowing for a wide range of antenna configurations. In addition the simulation model has several optional features: (i) a polarization model, (ii) far scatterer clusters, (iii) a LoS component for the microcellular case, and (iv) a modified distribution of the angular distribution at the MS, which emulates propagation in an urban street canyon.

4.1.3 WINNER channel models

The channel models developed in the IST-WINNER[3] project are related to both the COST 259 model and the 3GPP SCM model described above. The WINNER models are based on the GSCM principle, the drop concept, and the generic approach to model all scenarios with the same generic structure. System level simulations use generic multilink models. Parameters of those models are chosen are based on the expanded measurement campaign in European countries.

The bandwidth up to 100 MHz is covered by introducing an intracluster delay spread. Center frequencies of 5 GHz are included by defining corresponding pathloss functions. A MATLAB implementation of the SCME is available on the project's website http://www.ist-winner.org. A reduced version of this model was adopted for standardization of the 3GPP long term evolution (LTE).

4.2 MDPSS based wideband channel simulator

The sum of sinusoids (SoS) or sum of cisoids (SoC) simulators (Patzold 2002; SCM Editors 2006) are a popular way of building channel simulators both in the SISO and MIMO case. However, this approach is not a very good option when prediction is considered since it represents a signal as a sum of coherent components with a large prediction horizon (Papoulis 1991). In the communication we develop an approach which allows us to avoid this difficulty. The idea of a simulator combines the representation of the scattering environment advocated in (Almers et al. 2007; Asplund et al. 2006; Molisch 2004; Molisch et al. 2006; SCM Editors 2006) and the approach for a single cluster environment used in (Alcocer et al. 2005a; Fechtel 1993) with some important modifications (Alcocer et al. 2005a,b, 2006; Kontorovich 2006a; Norway, 2006b; Parra et al. 2002; Xiao et al. 2005; Yip and Ng 1997).

4.2.1 Geometry of the problem

Let us first consider a single cluster scattering environment, shown in Figure 7.1. It is assumed that both sides of the link are equipped with multi-element linear array antennas and both are mobile. The transmit array has N_T isotropic elements separated by distance d_T while the receive side has N_R antennas separated by distance d_R. Both antennas are assumed to be in the horizontal plane; however an extension on the general case is straightforward. The antennas are moving with velocities v_T and v_R respectively such that the angle between corresponding broadside vectors and the velocity vectors are α_T and α_R respectively. Furthermore, it is assumed that the impulse response is sampled at the rate F_{st} and the channel is sounded with the rate F_s impulse responses per second. The carrier frequency is f_0. Practical values will be given in Section 4.4.

The space between the antennas consists of a single scattering cluster whose center is seen at the the azimuth ϕ_{0T} and co-elevation θ_T from the receive side and

---

[3] See the following website http://www.ist-winner.org for the most recent update.

the azimuth ϕ_{0R} and co-elevation θ_R on the transmit side. The angular spread in the azimuthal plane is $\Delta\phi_T$ on the receive side and $\Delta\phi_R$ on the transmit side. No spread is assumed in the co-elevation dimension to simplify calculations and due to assumed low array sensitivity to the co-elevation spread. We also assume that $\theta_R = \theta_T = \pi/2$ to shorten equations. Corresponding corrections are rather trivial. The angular spread on both sides is assumed to be small compared to the angular resolution of the arrays due to a large distance between the antennas and the scatterer:

$$\Delta\phi_T \ll \frac{2\pi\lambda}{(N_T - 1)d_T}, \quad \Delta\phi_R \ll \frac{2\pi\lambda}{(N_R - 1)d_R}. \tag{4.1}$$

The cluster is also assumed to produce a certain delay spread $\Delta\tau$ variation of the impulse response due to its finite dimension. This spread is assumed to be relatively small, not exceeding a few sampling intervals $T_s = 1/F_{st}$.

4.2.2 Statistical description

It is well known that the angular spread (dispersion) in the impulse response leads to spatial selectivity (Fleury 2000) which can be described by the corresponding covariance function

$$\rho(d) = \int_{-\pi}^{\pi} \exp\left(j2\pi\frac{d}{\lambda}\phi\right) p(\phi)d\phi \tag{4.2}$$

where $p(\phi)$ is the distribution of the AoA or AoD. Since the angular size of the clusters are assumed to be much smaller than the antennas' angular resolution, one can further assume the following simplifications: a) the distribution of AoA/AoD are uniform and b) joint distribution $p_2(\phi_T, \phi_R)$ of AoA/AoD is just

$$p_2(\phi_T, \phi_R) = p_{\phi_T}(\phi_T)p_{\phi_R}(\phi_R) = \frac{1}{\Delta_{\phi_T}} \cdot \frac{1}{\Delta_{\phi_R}} \tag{4.3}$$

It was shown in (Salz and Winters 1994) that corresponding spatial covariance functions are modulated sinc functions

$$\rho(d) \approx \exp\left(j\frac{2\pi d}{\lambda}\sin\phi_0\right) \mathrm{sinc}\left(\Delta\phi\frac{d}{\lambda}\cos\phi_0\right) \tag{4.4}$$

The correlation function of the form 4.4 gives rise to a correlation matrix between antenna elements which can be decomposed in terms of frequency modulated discrete prolate spheroidal sequences (MDPSS) (Alcocer et al. 2005a; Sejdic et al. 2008; Slepian 1978):

$$\mathbf{R} \approx \mathbf{WU\Lambda U}^H \mathbf{W}^H = \sum_{k=0}^{D} \lambda_k \mathbf{u}_k \mathbf{u}_k^H \tag{4.5}$$

where $\mathbf{\Lambda} \approx \mathbf{I}_D$ is the diagonal matrix of size $D \times D$ (Slepian 1978), $\mathbf{U}$ is $N \times D$ matrix of the discrete prolate spheroidal sequences and $\mathbf{W} = \mathrm{diag}\{\exp(j2\pi d/\lambda \sin nd_A)\}$. Here d_A is the distance between the antenna elements, N number of antennas, $1 \leq n \leq N$, and $D \approx \lceil 2\Delta\phi \frac{d}{\lambda} \cos \phi_0 \rceil + 1$ is the effective number of degrees of freedom generated by the process with the given covariance matrix $\mathbf{R}$. For narrow-spread clusters the number of degrees of freedom is much less than the number of antennas $D \ll N$ (Slepian 1978). Thus, it could be inferred from Equation (4.5) that the desired channel impulse response $\mathbf{H}(\omega, \tau)$ could be represented as a double sum or tensor product

$$\mathbf{H}(\omega, t) = \sum_{n_t=1}^{D_T} \sum_{n_r=1}^{D_R} \sqrt{\lambda_{n_t} \lambda_{n_r}} \mathbf{u}_{n_r}^{(r)} \mathbf{u}_{n_t}^{(t)^H} h_{n_t, n_r}(\omega, t) \tag{4.6}$$

In the extreme case of a very narrow angular spread on both sides, $D_R = D_T = 1$ and $\mathbf{u}_1^{(r)}$ and $\mathbf{u}_1^{(t)}$ are well approximated by the Kaiser windows (Thomson 1982). The channel corresponding to a single scatterer is of course a rank 1 channel given by

$$\mathbf{H}(\omega, t) = \mathbf{u}_1^{(r)} \mathbf{u}_{n_t}^{(t)^H} h_{n_t, n_r}(\omega, t) \tag{4.7}$$

considering the shape of the functions $\mathbf{u}_1^{(r)}$ and $\mathbf{u}_1^{(t)}$ one can conclude that in this scenario angular spread is achieved by modulating the amplitude of the spatial response of the channel on both sides. It is also worth noting that the representation (4.6) is the Karhunen–Loeve series (van Trees 2001) in the spatial domain and therefore produces the smallest number of terms needed to represent the process selectivity in the spatial domain. It is also easy to see that such a modulation becomes important only when the number of antennas is significant.

Similar results can be obtained in frequency and Doppler domains. Indeed, if an excess delay τ is associated with the cluster under consideration and $\Delta\tau$ is the delay spread at the time of arrival of signals from this cluster; and in addition it is desired to provide a proper representation of the process in the bandwidth $[-W : W]$ using N_F equally spaced samples; assuming that the variation of power is relatively minor within the $\Delta\tau$ delay window, we once again recognize that variation of the channel in frequency domain can be described as a sum of modulated DPSS of length N_F and the time bandwidth product $W\Delta\tau$. The number of MDPSS needed for such a representation is approximately $D_F = 2W\Delta\tau + 1$ (Slepian 1978):

$$\mathbf{h}(\omega, t) = \sum_{n_f=1}^{D_f} h_{n_f}(t) \mathbf{u}_{n_f}^{(\omega)} \tag{4.8}$$

Finally, in the Doppler domain, the mean resulting Doppler spread can be calculated as

$$f_D = \frac{f_0}{c} [v_T \cos(\phi_{T0} - \alpha_T) + v_R \cos(\phi_{R0} - \alpha_R)] \tag{4.9}$$

The angular extent of the cluster from the sides causes the Doppler spectrum to widen by the amount

$$\Delta f_D = \frac{f_0}{c} [v_T \Delta\phi_T v_T | \sin(\phi_{T0} - \alpha_T) |$$
$$+ v_R \Delta\phi_R | \sin(\phi_{R0} - \alpha_R) |] \tag{4.10}$$

Once again, due to the small angular extent of the cluster it could be assumed that the widening of the Doppler spectrum is relatively narrow and no variation within the Doppler spectrum is of importance. Therefore, if it is desired to simulate the channel on the interval of time $[0 : T_{max}]$ then this could be accomplished by summing $D = 2\Delta f_D T_{max} + 1$ MDPS:

$$\mathbf{h}_d = \sum_{n_d=0}^{D} \xi_{n_d} \sqrt{\lambda_{n_d}} \mathbf{u}_{n_d}^{(d)} \tag{4.11}$$

where ξ_{n_d} are independent zero-mean complex Gaussian random variables of unit variance.

Finally, the derived representation could be summarized in tensor notions as follows. Let $\mathbf{u}_{n_t}^{(t)}$, $\mathbf{u}_{n_r}^{(r)}$, $\mathbf{u}_{n_f}^{(\omega)}$ and $\mathbf{u}_{n_d}^{(d)}$ are the DPSS corresponding to the transmit, receive, frequency, and Doppler time dimensions of the signal with the "domain-dual domain" products given by $\Delta\phi_T \frac{d}{\lambda}\cos\phi_{T0}$, $\Delta\phi_R \frac{d}{\lambda}\cos\phi_{R0}$, $W\Delta\tau$, and $T_{max}\Delta f_D$ respectively. Then a sample of a MIMO frequency selective channel with corresponding characteristics could be generated as

$$\mathcal{H}_4 = \mathcal{W}_4 \odot \sum_{n_t}^{D_T}\sum_{n_r}^{D_R}\sum_{n_f}^{D_F}\sum_{n_d}^{d} \sqrt{\lambda_{n_t}^{(t)}\lambda_{n_r}^{(r)}\lambda_{n_f}^{(\omega)}\lambda_{n_d}^{(T)}}\xi_{n_t,n_r,n_f,n_d} \, .$$

$$_1\mathbf{u}_{n_r}^{(r)} \times {}_2\mathbf{u}_{n_r}^{(r)} \times {}_3\mathbf{u}_{n_f}^{(\omega)} \times {}_4\mathbf{u}_{n_d}^{(d)} \tag{4.12}$$

where $\mathcal{W}_4$ is a tensor composed of modulating sinusoids

$$\mathcal{W}_4 = {}_1\mathbf{w}^{(r)}{}_2\mathbf{w}^{(t)}{}_1\mathbf{w}^{(\omega)}{}_1\mathbf{w}^{(d)} \tag{4.13}$$

$$\mathbf{w}^{(r)} = \left[1, \exp\left(j2\pi\frac{d_R}{\lambda}\right), \cdots, \exp\left(j2\pi\frac{d_R}{\lambda}(N_R-1)\right)\right]^T$$

$$\mathbf{w}^{(t)} = \left[1, \exp\left(j2\pi\frac{d_T}{\lambda}\right), \cdots, \exp\left(j2\pi\frac{d_T}{\lambda}(N_T-1)\right)\right]^T \tag{4.14}$$

$$\mathbf{w}^{(\omega)} = \left[1, \exp\left(j2\pi\Delta F\tau\right), \cdots, \exp\left(j2\pi\Delta F(N_F-1)\right)\right]^T$$

$$\mathbf{w}^{(d)} = \left[1, \exp\left(j2\pi\Delta f_D T_s\right), \cdots, \exp\left(j2\pi\Delta f_D(T_{max}-T_s)\right)\right]^T \tag{4.15}$$

and $\odot$ is the Hadamard (element-wise) product of two tensors (van Trees 2002).

4.2.3   Multi-cluster environment

The generalization of the model suggested in Section (4.2) is straightforward. The channel between the transmitter and the receiver is represented as a set of clusters, each described as in Section (4.2). The total impulse response is a superposition of the independently generated impulse response tensors from each cluster

$$\mathcal{H}_4 = \sum_{k=0}^{N_c-1} \sqrt{P_k}\mathcal{H}_4(k), \quad \sum_{k=1}^{N_c} P_k = P \tag{4.16}$$

where N_c is the total number of clusters, $\mathcal{H}_4(k)$ is a normalized response from the k-th cluster $||\mathcal{H}_4(k)||_F^2 = 1$, and $P_k \geq 0$ represents the relative power of the k-th cluster and P is the total power.

It is important to mention here that such a representation does not need to correspond to a physical cluster distribution. It rather reflects interplay between signals radiated, arriving from certain directions with a certain excess delay, ignoring the particular mechanism of propagation. Therefore it is possible, for example, to have two clusters with the same AoA and AoD but different excess delay. Alternatively, it is possible to have two clusters which correspond to the same AoD and excess delay but have very different AoA.

Equations (4.12) and (4.16) reveal the connection between the sum of cisoids approach (SCM Editors 2006) and the suggested algorithms: one can consider (4.12) as a modulated cisoid. Therefore, the simulator suggested above could be considered a sum of modulated cisoids simulator.

In addition to the space dispersive components, the channel impulse response may contain a number of highly coherent components, which can be modeled as pure complex exponents. Such components describe either direct LoS path or specularly reflected rays with a very small phase diffusion in time. Therefore Equation (4.16) should be modified to account for such components:

$$\mathcal{H}_4 = \sqrt{\frac{1}{1+K}} \sum_{k=0}^{N_c-1} \sqrt{P_{ck}}\mathcal{H}_4(k) + \sqrt{\frac{K}{1+K}} \sum_{k=0}^{N_s-1} \sqrt{P_{sk}}\mathcal{W}_4(k) \qquad (4.17)$$

Here N_s is the number of specular components including LoS and K is a generalized Rice factor describing the ratio between powers of specular P_{sk} and non-coherent/diffusive components P_{ck}

$$K = \frac{\sum_{k=0}^{N_s-1} P_{sk}}{\sum_{k=0}^{N_c-1} P_{ck}} \qquad (4.18)$$

While the distribution of the diffusive component is Gaussian by construction, the distribution of the specular component may not be Gaussian. A more detailed analysis is beyond the scope of this book and will be considered elsewhere. We also leave the question of identifying and distinguishing coherent and non-coherent components to a separate manuscript.

4.2.4   Simulation of dynamically changing environment

Modern communication systems operate in highly mobile environments. As a result, the propagation and scattering environment is often subject to changes. Therefore, it is important to model both processes described by local scattering (small time-scale variations) as well as dynamically changing scattering environments.

Some provisions of such modeling were introduced in (Asplund et al. 2006; Molisch et al. 2006). In general, the propagation environment is considered on two levels: radio environment (RE) (Asplund et al. 2006; Molisch et al. 2006) which describes the multitude of possible channel realizations via their statistical properties, and a particular propagation scenario (PS) which describes a single realization from an RE. Transitions between the two environments could be modeled using a switching function $\zeta(t)$ as discussed below. It is assumed that during a single radio communication session the system remains in a single RE (such as pico-, micro-, or macrocell) but particular propagation scenarios change. Therefore, it is important to be able to generate relatively short segments which correspond to the particular PS and smoothly combine than in a longer realization. This can be accomplished as follows. Let the first propagation scenario be dominant on $0 \leq ts_1 =\leq T_1 \gg \tau_{max}$ and the second propagation scenario be dominant on the time interval $T_2 \leq t \leq T_3$, $s_2 = T_3 - T_2 \gg \tau_{max}$, $0 < T_1 < T_2 < T_3$. However, on the interval $T_1 \leq t \leq T_2$ both propagation scenarios exist simultaneously in such a way that the first gradually disappears while the second gradually appears. Let $\mathbf{H}_1(\omega, t)$ and $\mathbf{H}_2(\omega, t)$ be samples of the MIMO channel time-varying transfer function defined on the intervals $[0 : T_2]$ and $[T_1 : T_3]$ corresponding to the first and the second propagation scenarios respectively. The sample $\mathbf{H}(\omega, t)$ of the channel defined on the interval is then given by

$$\mathbf{H}(\omega, t) = \begin{cases} \mathbf{H}_1(\omega, t) & 0 \leq t \leq T_1 \\ \zeta(t)\mathbf{H}_1(\omega, t) + (1 - \zeta(t))\mathbf{H}_2(\omega, t) & T_1 \leq t \leq T_2 \\ \mathbf{H}_2(\omega, t) & T_2 \leq t \leq T_3 \end{cases} \quad (4.19)$$

and

$$\zeta(t) = \frac{1}{\pi} \arctan \frac{2t - T_1 - T_2}{T_2 - T_1} \quad (4.20)$$

It is argued in (Molisch et al. 2006) that the particular form (4.20) of $\zeta(t)$ is advantageous since it approximates the Fresnel integrals. In addition we assume that the residency time s in each particular propagation environment is drawn from the exponential distribution with the mean $\bar{s}$ inversely proportional to the velocity of the vehicular and proportional to the average scale of the environment (Blaunstein et al. 2006). The transition time is also assumed to be drawn from an exponential distribution but with much shorter mean time.

4.3   Measurement based simulator

While some measurements are available to modelers, the sheer volume of data to be processed, the high variability of the measurement environment, and their non-stationarity poses a serious challenge to extracting parameters for simulations.

Overall there are two approaches which can be taken. For very wideband measurements one can attempt to extract cluster parameters (Vuokko et al. 2005) from the measurement using various techniques, such as ESPRIT (van Trees 2002), SAGE (Fleury et al. 1999), and so on. The results of such estimation can be directly used to describe a propagation scenario.

However, such accurate measurements are generally unavailable, especially during actual communication sessions. The measurements, collected through the channel estimation via pilot symbols or data assisted estimation, are usually noisy and are much coarser. In addition, online processing using SAGE algorithms can be extremely computationally costly. In such a situation one can only obtain very limited information about the actual scattering geometry. However, some insight into the corresponding angular, Doppler, and frequency spectra can be gained.

Since the number of antennas available in practical measurements is relatively small, the angular resolution suffers, especially on the stationary side of the link. In order to somewhat alleviate this problem we consider the following orthogonal transformation of the MIMO time-varying transfer function:

$$\mathbf{H}_T(\omega, t) = \left[\mathbf{h}_1, \mathbf{h}_2, \cdots, \mathbf{h}_{N_T} \right] = \mathbf{H}(\omega, t)\mathbf{U}_T \tag{4.21}$$

where $\mathbf{U}_T$ has been defined earlier. It is easy to see that the columns of $\mathbf{H}_T(\omega, t)$ are statistically independent and, therefore, generation of a MIMO channel could be separated into the generation of N_T independent SIMO channels with the following aggregation

$$\mathbf{H}(\omega, t) = \mathbf{H}_T(\omega, t)\mathbf{U}_T^H \tag{4.22}$$

Let us stress here that while identification of AoD without appealing to high resolution techniques such as MUSIC, ESPRIT, and SAGE is impossible, relatively accurate estimates of the transmit correlation matrix $\mathbf{R}_T$ can be obtain through time, frequency, and receive antenna averaging.

As the next step one can estimate the corresponding power delay profile $P_{n_t}(\tau)$ for each SIMO channel by averaging over all receive antennas and time samples. At this stage we assume that $P_{n_t}(\tau)$ can be well approximated[4] by a sum M_{n_t} of almost rectangular functions, each defined on interval $[\tau_{n_t,m} - \Delta\tau_{n_t,m}/2 : \tau_{n_t,m} + \Delta\tau_{n_t,m}/2]$, $m = 0, \cdots, M - 1$ as shown in Figure 4.1.

This could be considered a decomposition of a SIMO channel impulse response into a sum of virtual clusters, each corresponding to a unique interval of excess delays. Let emphasize here that these virtual clusters are not necessary associated with any physical scattering from compact clusters. This is a significant difference to methods such as SAGE where delay clustering is usually associated with an actual physical scattering center.

Let $\mathbf{h}_{\mathbf{n}_t,m}(\tau, t)$, $\tau \in [\tau_{n_t,m} - \Delta\tau_{n_t,m}/2 : \tau_{n_t,m} + \Delta\tau_{n_t,m}/2]$ be a portion of n_t-th virtual SIMO channel corresponding to m-th virtual delay cluster. The corresponding time varying frequency response could then be considered as a superposition of frequency flat virtual SIMO channels $\mathbf{h}_{n_t,m}^{(k)}(t)$

$$\mathbf{h}_m(\omega, t) = \exp(-j\omega\tau_{n_t,m}) \times \sum_{k=0}^{K_{n_t,m}-1} \mathbf{u}_k(\omega; N_F, W\Delta\tau_{n_t}^{(m)})\mathbf{h}_{n_t,m}^{(k)}(t) \tag{4.23}$$

---

[4] An alternative scenario, when such an approximation is rather impractical, will be treated in (Primak 2008b).

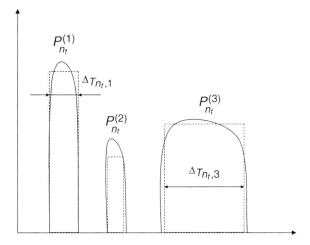

Figure 4.1   Approximation of PDP with rectangular functions.

where N_F is a number of frequency samples in the band W $\mathbf{u}_k(\omega; N_F, \Delta\tau_{n_t}^{(m)})$ is k-th DPSS of length N_F and time-bandwidth product $W\Delta\tau_{n_t}^{(m)}$. Virtual channels $\mathbf{h}_{n_t,m}^{(k)}(t)$ could be estimated from $\mathbf{h}_m(\omega, t)$ by the orthogonal projection.

It is worth noting that since AoA defines the contribution to both angular and Doppler spectrum, low resolution of the antenna array on the receiver side can be improved by effectively creating a larger virtual rectangular array of size $(N_R - 1)d_R/\lambda \times Lf_D/F_s$ in the wavelengths. This is the reason why we avoid decomposing the channel response on the receiver side as is traditionally done (Alcocer et al. 2005a; Costa and Haykin 2006). Instead, we use a two-dimensional multitaper approach to extract a Doppler spectrum (Percival and Walden 1993). This approach allows us to improve the Doppler (and as a result, angular) resolution compared to just SISO processing or processing based on the capon method (van Trees 2002). In addition, it allows us to resolve the ambiguity of the AoA associated with a linear array processing. As a result of such estimation, a joint time of arrival–Doppler spectrum plane can be obtained for each SIMO channel. This plane could be partitioned using rectangular tiles as in (Haghighi et al. 2010). Each tile can be considered a virtual cluster with parameters defined by the cluster geometry.

4.4   Examples

Fading channel simulators (Jeruchim et al. 2000) can be used for different purposes. The goal of the simulation often defines not only the suitability of a certain method but also dictates the choice of parameters. One of the possible goals of simulation is to isolate a particular parameter and study its effect on the system performance. Alternatively, various techniques are needed to avoid the problem of using the same model for both scenario simulation and analysis of the same scenario.

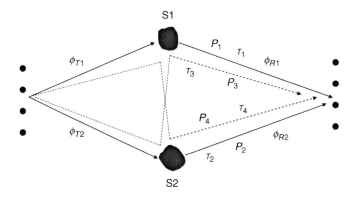

Figure 4.2   Geometry of a single cluster problem.

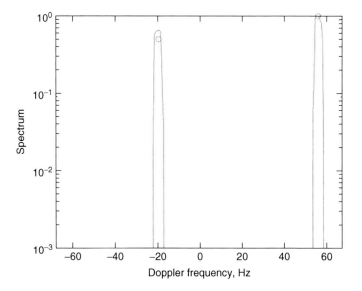

Figure 4.3   PSD of the two cluster channel response.

4.4.1   Two cluster model

The first example we consider here is a two-cluster model shown in Figure 4.2.

This geometry is the simplest non-trivial model for frequency selective fading. However, it allows us to study the effects of such parameters as angular spread, delay spread, correlation between sites on the channel parameters, and system performance. The results of the simulation are shown in Figures 4.3–4.4. In this example we chose $\phi_{T1} = 20°$, $\phi_{T2} = 20°$, $\phi_{R1} = 0°$, $\phi_{R2} = 110°$, $\tau_1 = 0.2\,\mu s$, $\tau_2 = 0.4\,\mu s$, $\Delta\tau_1 = 0.2\,\mu s$, $\Delta\tau_2 = 0.4\,\mu s$. As can be seen from the presented PSD

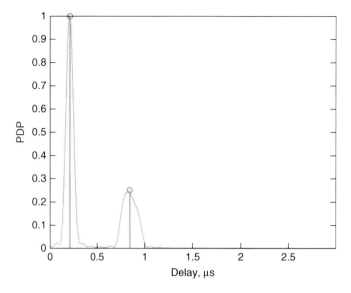

Figure 4.4   PDP of the two cluster channel response.

and PDP the proposed simulator provides a rather good agreement between the experimental and synthetic data.

4.4.2   Environment specified by joint AoA/AoD/ToA distribution

The most general geometrical model of a MIMO channel utilizes joint distribution $p(\phi_T, \phi_R, \tau)$, $0 \le \phi_T < 2\pi$, $0 \le \phi_R < 2\pi$, $\tau_{min} \le \tau \le \tau_{max}$, of AoA, AoD, and time of arrival (ToA). A few such models can be found in the literature (Algans et al. 2002; Andresen and Blaunstein 2003; Asplund et al. 2006; Blaunstein et al. 2006; Kaiser et al. 2006; Molisch et al. 2006). Theoretically, this distribution completely describes the statistical properties of the MIMO channel. Since the resolution of the antenna arrays on both sides is finite and a finite bandwidth of the channel is utilized, the continuous distribution $p(\phi_T, \phi_R, \tau)$ can be discredited to produce narrow "virtual" clusters centered at $[\phi_{Tk}, \phi_{Rk}, \tau_k]$ and with spread $\Delta\phi_{Tk}$, $\Delta\phi_{Rk}$, and $\Delta\tau_k$ respectively and the power weight

$$P_k = \frac{P}{4\pi^2(\tau_{max} - \tau_{min})}$$

$$\times \int_{\tau_k - \Delta\tau_k/2}^{\tau_k + \Delta\tau_k/2} d\tau \int_{\phi_{Tk} - \Delta\phi_{Tk}/2}^{\phi_{Tk} + \Delta\phi_{Tk}/2} d\phi_T \int_{\phi_{Rk} - \Delta\phi_{Rk}/2}^{\phi_{Rk} + \Delta\phi_{Rk}/2} p(\phi_T, \phi_R, \tau)d\phi_R$$

$$(4.24)$$

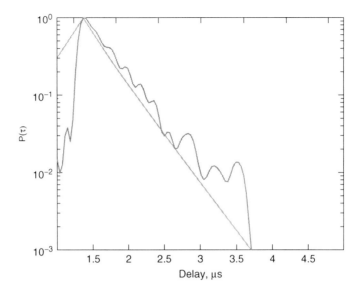

Figure 4.5   Simulated power delay profile for the example of Section 4.4.2.

We defer discussions about how to partition each domain, assuming for now that it could be done in such a way that the modeling of each "virtual" cluster obtained by such a partition is appropriate in the frame discussed in Section 4.2.

As an example let us consider the following scenario, described in (Blaunstein et al. 2006). In this case the effect of the two street canyon propagation results in two distinct angles of arrival $\phi_{R1} = 20°$ and $\phi_{R2} = 50°$, AoA spreads roughly of $\Delta_1 = \Delta_2 = 5°$, and exponential PDP corresponding to each AoA (see Figures 4.5 and 4.6 in (Blaunstein et al. 2006)). In addition, an almost uniform AoA on the interval $[60 : 80°]$ corresponds to early delays. Therefore, a simplified model of such an environment could be presented by

$$p(\phi_R, \tau) = \sqrt{P_1}\frac{1}{\Delta_1}\exp\left(-\frac{\tau - \tau_1}{\tau_{s1}}\right)u(\tau - \tau_1)$$

$$+\sqrt{P_2}\frac{1}{\Delta_2}\exp\left(-\frac{\tau - \tau_2}{\tau_{s2}}\right)u(\tau - \tau_2)$$

$$+\sqrt{P_3}\frac{1}{\Delta_3}\exp\left(-\frac{\tau - \tau_3}{\tau_{s3}}\right)u(\tau - \tau_3) \qquad (4.25)$$

where $u(t)$ is the unit step function, τ_{sk} and $k = 1, 2, 3$ describe the rate of decay of the PDP. By inspection of Figures 4.5–4.6 in (Blaunstein et al. 2006) we choose $\tau_1 = \tau_2 = 1.2$ ns, $\tau_3 = 1.1$ ns and $\tau_{s1} = \tau_{s2} = \tau_{s3} = 0.3$ ns. Similarly, by inspection of the same figures we assume $P_1 = P_2 = 0.4$ and $P_3 = 0.2$.

To model exponential PDP with unit power and average duration τ_s we represent it with a set of $N \geq 1$ rectangular PDP of equal energy $1/N$. The k-th virtual cluster then extends on the interval $[\tau_{k-1} : \tau_k]$ and has magnitude $P_k = 1/N\Delta\tau_k$

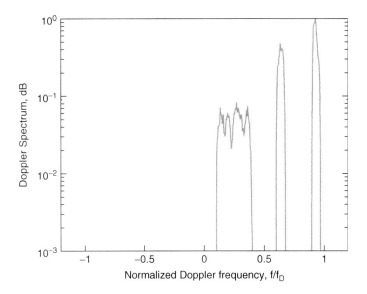

Figure 4.6 Simulated Doppler power spectral density for the example of Section 4.4.2.

where $\tau_0 = 0$

$$\tau_k = \tau_s \ln \frac{N - k}{N}, \quad k = 1, \dots, N - 1 \tag{4.26}$$

$$\tau_N = \tau_{N-1} + \frac{1}{N \tau_{N-1}}, \quad k = N \tag{4.27}$$

$$\Delta \tau_k = \tau_k - \tau_{k-1} \tag{4.28}$$

Results of numerical simulation are shown in Figures 4.5 and 4.6. It can be seen that a good agreement between the desired characteristics is obtained.

Similarly, the same technique could be applied to the 3GPP (SCM Editors 2002) and COST 259 (Asplund et al. 2006) specifications.

4.4.3 Measurement based simulator

4.4.3.1 High resolution data

As an example let us consider the results presented in (Li and Ho 2005). For dataset 84 described in the paper, five clusters have been consistently identified and their parameters are summarized in the Table 4.1. Since only AoA are specified, we assume uniform AoD distribution. The results of the simulation are in Figure 4.7.

Table 4.1   Approximated parameters of the scattering environment extracted from set 84 in (Li and Ho 2005).

Cluster	τ, ns	$\Delta\tau$, ns	ϕ_R	$\Delta\phi_R$
1	80	40	80°	20°
2	110	30	130°	10°
3	100	20	220°	15°
4	80	30	280°	15°
5	170	20	290°	10°

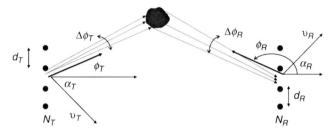

Figure 4.7 Simulated ToA/AoA distribution based on the values listed in Table above.

4.5   Appendix A: simulation parameters

The parameters of the simulator are summarized below in Tables 4.2–4.4.

Table 4.2   Parameters of the transmitter and receiver.

Parameter	Notation
Tr. Antenna Number	N_T
Rx. Antenna Number	N_R
Distance between Tr antennas	d_T
Distance between Rx antennas	d_R
Tr velocity	v_T
Rx velocity	v_R
Tr v angle with broadside	α_T
Rx v angle with broadside	α_R
Carrier frequency	f_0
Frequency band, one-sided	W
Simulation duration	T_s
Sampling Frequency Doppler	F_s
Sampling Frequency Delay	F_{st}

Table 4.3   Parameters of a single cluster.

Parameter	Notation
Mean Tr. angle	ϕ_{0T}
Mean Rx. angle	ϕ_{0R}
Tr. angle spread	$\Delta\phi_T$
Rx. angle spread	$\Delta\phi_R$
Mean excess delay	τ
Excess delay spread	$\Delta\tau$
Cluster power	P_k

Table 4.4   Parameters of a specular component.

Parameter	Notation
AoD	ϕ_{0T}
AoA	ϕ_{0R}
Power	P_k

5

Capacity of communication channels

5.1 Introduction

As in consideration of the SISO channel (Benedetto and Biglieri 1999) it is important to consider three different scenarios of the MIMO channel:

1. The channel matrix $\mathbf{H}$ is deterministic and constant.

2. The channel matrix $\mathbf{H}$ is a random matrix process chosen according to some probability distribution and time variation law: the value of the channel matrix changes significantly over the period of a communication session.

3. $\mathbf{H}$ is a random matrix drawn from a certain distribution but remains fixed during a communication session.

The first scenario is usually present only in wired links, however, methodology developed for such links is instrumental in considering both scenarios 2) and 3). The second scenario corresponds to a sufficiently long communication session, such that the channel varies numerous times over all possible states. Therefore, the quality of communication in such a case could be described via time-averaging of instanteneous characteristics obtained by considering the scenario 1). The consideration of the scenario 3) while still based on the results of 1) requires a different approach: in particular one may speak about the probability of achieving, (or failing to achieve) a certain level of quality of communications over the session lifetime. It also requires more information than just a simple average: higher moments or rate exponents are often desirable features (Teletar 1999), (Moustakas and

Wireless Multi-Antenna Channels: Modeling and Simulation, First Edition.
Serguei L. Primak and Valeri Kontorovich.
© 2012 John Wiley & Sons, Ltd. Published 2012 by John Wiley & Sons, Ltd.

Simon 2007). In this chapter, the scenario 2) constitutes the main focus of investigation, while the scenario 1 is considered for clarifying the corresponding mathematical apparatus.

5.2   Ergodic capacity of MIMO channel

5.2.1   Capacity of a constant (static) MIMO channel

The capacity of a proper MIMO channel was first considered in (Foschini and Gans 1998) and (Teletar 1999), although the former is most credited for the development. Let us consider frequency flat fading with N_T transmit and N_R receive antennas. In this case the relationship between the transmitted **s** and the received **r** symbols could be written as

$$\mathbf{r} = \mathbf{Hs} + \mathbf{n} \tag{5.1}$$

It is often assumed that the noise term **n** is comprised of independent complex zero-mean proper Gaussian random processes of the same variance:

$$\mathcal{E}\left\{\mathbf{nn}^H\right\} = \mathbf{I}, \ \mathcal{E}\left\{\mathbf{nn}^T\right\} = \mathbf{I} \tag{5.2}$$

while the average power of the transmitted signal is limited by a constant P

$$\mathcal{E}\left\{\mathbf{s}^H\mathbf{s}\right\} = \mathcal{E}\left\{\mathrm{tr}\,\mathbf{ss}^H\right\} = \mathrm{tr}\,\mathcal{E}\left\{\mathbf{ss}^H\right\} = \mathrm{tr}\,\mathbf{Q} \leq P \tag{5.3}$$

Here **Q** is the correlation matrix of the transmitted signal.

Similar to the original Shannon idea of the channel capacity (Cover and Thomas 2002), one can maximize mutual information $I(\mathbf{r}, \mathbf{s})$ between the received and the transmitted signals

$$C = \max_Q I(\mathbf{r}, \mathbf{s}) = \max_Q \left[\mathcal{H}(\mathbf{r}) - \mathcal{H}(\mathbf{r}|\mathbf{s})\right] = \max_Q \left[\mathcal{H}(\mathbf{r}) - \mathcal{H}(\mathbf{n})\right] \tag{5.4}$$

where $\mathcal{H}(\mathbf{r})$ is the entropy of the signal vector **r**.

It is shown in (Teletar 1999) that one can limit considerations to zero-mean signals $\mathcal{E}\mathbf{s}$ since the addition of a constant term does not add to the mutual information. In this case **Q** is the covariance matrix of the transmitted signal. Furthermore, the received signal **r** also has zero-mean and the covariance matrix $\mathbf{R_r} = \mathbf{HQH}^H$. Following the well known results on the maximum entropy property of Gaussian processes (Cover and Thomas 2002; Teletar 1999) it is shown that in order to maximize $I(\mathbf{r}, \mathbf{s})$ the process **s** must be a circularly symmetric complex Gaussian process. In this case

$$I(\mathbf{r}, \mathbf{s}) = \log\det\left(\mathbf{I} + R_r\right) = \log\det\left(\mathbf{I} + \mathbf{HQH}^H\right) = \log\det\left(\mathbf{I} + \mathbf{QH}^H\mathbf{H}\right) \tag{5.5}$$

The Wishart matrix $\mathbf{H}^H\mathbf{H}$ is Hermitian, therefore it could be diagonalized as $\mathbf{W} = \mathbf{U\Lambda U}$ where elements of the diagonal matrix $\mathbf{\Lambda}$ are non-negative. Since $\tilde{\mathbf{Q}}$ and $\mathbf{Q}$ are simultaneously non-negative definite and trace $\tilde{\mathbf{Q}} = \mathrm{trace}\,\mathbf{Q}$ maximization over

$\mathbf{Q}$ is equivalent to that over $\tilde{\mathbf{Q}}$. Therefore,

$$I(\mathbf{r}, \mathbf{s}) = \log \det (\mathbf{I} + R_r) = \log \det \left(\mathbf{I} + \mathbf{\Lambda}^{1/2} \mathbf{Q} \tilde{\mathbf{\Lambda}}^{1/2} \right)$$

$$\leq \log \prod_{k=1}^{M} \left(1 + \tilde{Q}_{kk} \lambda_k \right) \tag{5.6}$$

with equality only when $\tilde{\mathbf{Q}}$ is diagonal.[1] Here $M = \min\{N_R, N_T\}$. The required optimization subject to the power constraint $\operatorname{tr} \tilde{\mathbf{Q}} \leq P$ could be accomplished using Lagrangian multipliers. Indeed, introducing function

$$F(\lambda_1, \cdots, \lambda_M, \mu) = \log \prod_{k=1}^{M} \left(1 + \tilde{Q}_{kk} \lambda_k \right) - \mu \left(\sum_{k=1}^{M} \lambda_k - P \right) \tag{5.7}$$

and differentiating it with respect to λ_k one obtains a system of equations

$$\frac{\partial F}{\partial \lambda_k} - \mu = \frac{1}{1 + \tilde{Q}_k \lambda_k} - \mu = 0 \tag{5.8}$$

which has "water-filling"-type solutions (Cover and Thomas 2002)

$$\tilde{Q}_k = \left(\mu - \lambda_k^{-1} \right)^{+} \tag{5.9}$$

where

$$x^{+} = \begin{cases} x, & \text{if } x \geq 0 \\ 0 & \text{otherwise} \end{cases}$$

and μ is chosen to satisfy

$$\sum_{k}^{M} \tilde{Q}_k = \operatorname{tr} \mathbf{Q} = P.$$

Finally, the expression for the channel capacity becomes

$$C = \log \prod_{k=1}^{M} \left(1 + \tilde{Q}_{kk} \lambda_k \right) = \sum_{k=1}^{M} \log \left(1 + \left[\mu - \lambda_k^{-1} \right]^{+} \lambda_k \right) \tag{5.10}$$

It also worth noting that if no power reallocation at the transmitter is considered, the mutual information between the received and the transmitted signal is given simply by

$$C_{NCSIT} = \log \det \left(\mathbf{I}_{N_T} + \mathbf{H}^H \mathbf{H} \right) = \log \det \left(\mathbf{I}_{N_R} + \mathbf{H} \mathbf{H}^H \right)$$

$$= \sum_{k=1}^{M} \log (1 + \lambda_k) \tag{5.11}$$

---

[1] Here we have used a well known matrix inequality

$$\det (\mathbf{I} + \mathbf{A}) \leq \prod_{k} (1 + a_{kk})$$

which can be found in (Gantmacher 1959).

An alternative derivation of the capacity formula could be based on the unitary transformation of the transmit and received signals. Let the SVD decomposition of the channel matrix be $\mathbf{H} = \mathbf{U}\mathbf{\Lambda}^{1/2}\mathbf{V}^H$ where $\mathbf{U}$ and $\mathbf{\Lambda}$ coincide with the eigendecomposition of the Wishart matrix $\mathbf{W}$ defined earlier. Left mutiplying both parts of Equation (5.1) by $\mathbf{U}^H$ one can rewrite it as

$$\tilde{\mathbf{r}} = \mathbf{U}^H \mathbf{r} = \mathbf{\Lambda}^{1/2}\mathbf{V}^H \mathbf{s} + \mathbf{U}^H \mathbf{n} = \mathbf{\Lambda}^{1/2}\tilde{\mathbf{s}} + \tilde{\mathbf{n}} \tag{5.12}$$

The latter reveals that the MIMO channel could be interpreted as $M = \min N_T, N_R$ non-zero parallel (independent) channels with the channel gain $\lambda_k^{1/2}$ and unit variance noise.[2] As a result, the capacity of such a channel is a sum of the capacities of each individual channel, with the total power allocated according to the water-filling algorith (Cover and Thomas 2002).

More detailed investigation of the MIMO channel capacity formulae (5.10) or (5.11) shows that the way to achieve such capacity is to transmit independent information flows over virtual parallel channels with or without corresponding power loading. This is a new feature of MIMO channels, known as spatial multiplexing, and is not available in SISO, MISO, or SIMO systems. More detailed investigation is considered below.

5.2.2 Alternative normalization

Instead of normalization, used in the previous section, it is often convenient to normalize the channel matrix $\mathbf{H}$ instead of noise. In this case propagation losses are recalculated to the transmitted side and are absorbed by the total transmitted power $P = \mathrm{tr}\,\mathbf{R}_{ss}$. The variance of noise in each receive antenna is assumed to be σ_n^2 so that $\mathcal{E}\{\mathbf{n}\mathbf{n}^H\} = \sigma_n^2 \mathbf{I}_{N_R}$ and

$$\sum_{n_t=1}^{N_T} |h_{n_r n_t}|^2 = N_T, \quad n_r = 1, \cdots, N_R \tag{5.13}$$

and

$$\|\mathbf{H}\|_F^2 = \mathrm{tr}\,\mathbf{H}\mathbf{H}^H = \sum_{n_t=1}^{N_T} |\sum_{n_r=1}^{N_R} h_{n_r n_t}|^2 = N_T N_R \tag{5.14}$$

In this notation, the average signal-to-noise ratio (SNR) γ per receive antenna can be defined as

$$\gamma = \frac{P}{\sigma_n^2} \tag{5.15}$$

and the expression for the system capacity in the new notation becomes as follows

$$C = \log\det\left(\mathbf{I}_{N_R} + \frac{\gamma}{N_T}\mathbf{H}\mathbf{Q}\mathbf{H}^H\right) = \sum_{k=1}^{M} \log\left(1 + \frac{1}{\sigma_n^2}[\lambda_k \mu - \sigma_n^2]^+\right) \tag{5.16}$$

---

[2] The covariance matrix of noise is preserved under a unitary full rank transformation $\tilde{\mathbf{n}} = \mathbf{U}^H \mathbf{n}$: $\mathcal{E}\{\tilde{\mathbf{n}}\tilde{\mathbf{n}}^H\} = \mathbf{U}^H \mathcal{E}\{\mathbf{n}\mathbf{n}^H\}\mathbf{U} = \mathbf{I}_{N_R}$.

where

$$P_k = \left[\mu - \frac{\sigma_n^2}{\lambda_k} \right]^+, \quad \sum_{k=1}^{M} P_k = P \qquad (5.17)$$

In order to investigate the capacity of sparse channels, yet another normalization has been suggested in (Zhang et al. 2006).

5.2.3 Capacity of a static MIMO channel under different operation modes

It was mentioned earlier in Section 5.2.1 that the full capacity of a MIMO channel is realized by water-filling power loading of independent spatial transmission modes. However, it is useful to consider some less optimal forms of transmission to seek further insight into the performance of systems over MIMO channels (Vucetic and Yuan 2002). We will use the normalizations presented in the first part of the section 5.2.2.

5.2.3.1 SISO channel

In this case $N_T = N_R = 1$, $\mathbf{H} = h = 1$, and the capacity formula (5.16) reduces to the well known Shannon formula for the AWGN channel:

$$C_{SISO} = \log(1 + \gamma) \sim \begin{cases} \log \gamma & \text{if } \gamma \gg 1 \\ \gamma & \text{if } \gamma \ll 1 \end{cases} \qquad (5.18)$$

Thus, the capacity increases linearly for small SNR γ and slows down to a logarithmic growth for large γ. Increasing the capacity of SISO channels is thus very energy demanding.

5.2.3.2 SIMO channel

For the SIMO channel one has to set $N_T = 1$ and $\mathbf{H} = \mathbf{h}$ is a $N_R \times 1$ vector. Therefore, the matrix $\mathbf{W} = \mathbf{h}\mathbf{h}^H$ has rank 1, $\lambda_1 = \mathbf{h}^H\mathbf{h} = ||\mathbf{h}||_F^2 = N_R$, and $\lambda_k = 0$ for $k > 1$. Therefore, the capacity of such a link is given by

$$C_{SIMO} = \log\left(\mathbf{I}_{N_R} + \gamma \mathbf{h}\mathbf{h}^H\right) = \log(1 + \gamma ||\mathbf{h}||_F^2)$$
$$= \log(1 + N_R \gamma) = \log(1 + \gamma_{eff}) \qquad (5.19)$$

Equation (5.19) shows that an increase of capacity in SIMO channels is obtained via MRC combining of signals in all antennas. For very small values of SNR, $\gamma N_R \ll 1$ increase in number of antennas N_R linearly increases capacity, however, even for moderate SNR such an increase is only logarithmic: $C_{SIMO} \sim \log(\gamma N_T)$. Alternatively, it could be said that the increase in capacity is achieved through utilizing received diversity, provided by the SIMO channel compared to a SISO channel.

It is also important to realize how the capacity is achieved in the SIMO case: since the signal in all antennas is the same transmitted signal scaled by the channel coefficients, the combined signal is a linear combination

$$y = \mathbf{w}^H \mathbf{r} = \mathbf{w}^H \mathbf{h}s + \mathbf{w}^H \mathbf{n}, \quad \mathbf{w}^H \mathbf{h} = 1 \tag{5.20}$$

where

$$\mathbf{w} = \frac{1}{\sqrt{N_R}} \mathbf{h} \tag{5.21}$$

The latter could be considered either spatial matching filtering or beamforming (van Trees 2002).

5.2.3.3   MISO channels

In the case of multiple transmit antennas $N_R = 1$ and $\mathbf{H} = \mathbf{h}^T$ is $1 \times N_T$ vector-row. Without power control the transmit power is simply divided between N_T antennas only to be added later in the receive antenna with equal gain. Therefore, the net capacity of a MISO scheme is just

$$C_{MISO} = \log\left(1 + \frac{\gamma}{N_T} \mathbf{h}\mathbf{h}^H\right) = \log\left(1 + \frac{\gamma}{N_T} N_T\right)$$

$$= \log(1 + \gamma) = C_{SISO} \tag{5.22}$$

Thus, for a fixed MISO channel no capacity improvement could be achieved without power loading.[3] If the channel is known to the transmitter, the power loading could be performed. Since in this case $\mathbf{W} = \mathbf{H}\mathbf{H}^H = \lambda_1 \mathbf{u}\mathbf{u}^H$ has rank 1 all power must be allocated to a single virtual mode defined by the eigenvector $\mathbf{u} = \mathbf{h}/\sqrt{N_T}$. Since $\lambda_1 = \|\mathbf{h}\|_F^2 = N_T$, Equation (5.16) results in the following expression for the capacity of a MISO channel with power control

$$C_{MISO+P} = \log(1 + N_T \gamma) > C_{SISO} \tag{5.23}$$

Once again, the increase in capacity is achieved by (transmit) beamforming.

5.2.4   Ergodic capacity of a random channel

5.2.4.1   Constant transmit power

Let us consider a SISO scenario when the duration of a communication session T is substantially longer than the scale of the channel variation[4] τ_c: $T \ll \tau_c$. In this case, during the session the channel experiences all its possible states multiple times. In addition, if the channel itself is an ergodic process.[5] Then the average, or ergodic, capacity of the channel without power control could be defined as

$$\bar{C}_{SISO} = \int_0^\infty \log(1 + \gamma) p_\gamma(\gamma) d\gamma \tag{5.24}$$

---

[3] However, in the fading environment the transmit diversity allows for an increase of capacity compared to the SISO fading channel.

[4] The correlation interval or the coherence time could be a good measure of such a scale.

[5] Note that both $T \ll \tau_c$ and the ergodicity of the channel are required to define the concept of ergodic capacity.

where

$$\gamma = \frac{P\sigma_h^2}{\sigma_n^2} \tag{5.25}$$

is the instantaneous SNR and $p_\gamma(\gamma)$ is the PDF of the instantaneous SNR. By the Jensen inequality $\bar{C}$ does not exceed the capacity of the AWGN channel with the same average SNR $\bar{\gamma}$

$$\bar{C} = \int_0^\infty \log(1+\gamma)p_\gamma(\gamma)d\gamma \leq \log\left(1 + \int_0^\infty \gamma p_\gamma(\gamma)d\gamma\right)$$
$$= \log(1+\bar{\gamma}) \tag{5.26}$$

For example, for the Nakagami fading

$$\bar{C}(m, \Omega) = \int_0^\infty (1+\gamma)\frac{1}{\Gamma(m)}\left(\frac{m}{\bar{\gamma}}\right)^m \gamma^{m-1} \exp\left(-\frac{m\gamma}{\bar{\gamma}}\right) \tag{5.27}$$

5.2.4.2   Power control in a SISO channel

If information about the channel state is instantaneously available at both the receiver and the transmitter, this could be exploited to improve capacity C

$$C(\bar{\gamma}) = \int_0^\infty \log\left(1 + \frac{S(\gamma)}{P}\gamma\right)p_\gamma(\gamma)d\gamma \tag{5.28}$$

of the channel via some power control function $S(\gamma)$ such that

$$\int_0^\infty S(\gamma)p_\gamma(\gamma)d\gamma \leq P \tag{5.29}$$

There are at least three power allocation strategies that can be considered (Goldsmith and Varaiya 1997): a) water-filling power control; b) channel inversion; and c) partial channel inversion.

In the case of the water-filling algorithm it is easy to show that the optimal power allocation must satisfy the following mechanism:

$$\frac{S(\gamma)}{P} = \begin{cases} \frac{1}{\gamma_0} - \frac{1}{\gamma} & \text{if } \gamma \geq \gamma_0 \\ 0 & \text{otherwise} \end{cases} \tag{5.30}$$

where the cut-off SNR γ_0 satisfies the following condition

$$\int_0^\infty \left(\frac{1}{\gamma_0} - \frac{1}{\gamma}\right)p_\gamma(\gamma)d\gamma = 1 \tag{5.31}$$

This technique provides an optimal water-filling-type solution to the power allocation problem, however it requires timely information about the channel state. The effect of delayed and imperfect information is considered in Section 5.3. The corresponding capacity C_{opt} is given by

$$C_{opt} = \int_{\gamma_0}^\infty \log\frac{\gamma}{\gamma_0}p_\gamma(\gamma)d\gamma \tag{5.32}$$

The channel inversion technique (Goldsmith and Varaiya 1997) enforces constant SNR[6] by adjusting the power proportionally to the degradation of the power of the channel, that is

$$\frac{S(\gamma)}{P} = \frac{1}{\gamma \kappa} \qquad (5.33)$$

where κ can be found from the average power constraint (5.29)

$$\kappa = \int_0^\infty \frac{1}{\gamma} p_\gamma(\gamma) d\gamma = \mathcal{E}\{1/\gamma\} \qquad (5.34)$$

and the average capacity

$$C_{opt} = \log\left(1 + \frac{1}{\kappa}\right) \qquad (5.35)$$

While such a power allocation certainly simplifies the design of the receiver it is not practical due to the limited power range of any transmitter. Instead, the authors of (Goldsmith and Varaiya 1997) have suggested using a truncated channel inversion

$$\frac{S(\gamma)}{P} = \begin{cases} \frac{1}{\kappa\gamma} & \text{if } \gamma \geq \gamma_0 \\ 0 & \text{otherwise} \end{cases} \qquad (5.36)$$

where γ_0 is chosen to maximize the capacity. No analytical expression can be found (Goldsmith and Varaiya 1997).

Since in SIMO systems the optimal strategy is the coherent MRC combining, the question about the capacity of such channels could be resolved by using the results of the SISO channel with the PDF $p_\gamma(\gamma)$ representing the output of the MRC combiner.

5.2.5 Ergodic capacity of MIMO channels

Under the same assumptions regarding the duration of the communication session and the speed of fading one can define the ergodic capacity $\bar{C}$ of a MIMO channel as the expectation of the instantaneous capacity $C(t)$ over the fading distribution $p(\mathbf{H})$:

$$\bar{C} = \mathcal{E}\left\{\mathbf{I}_{N_R} + \frac{\bar{\gamma}}{N_T}\mathbf{H}\mathbf{Q}\mathbf{H}^H\right\} = \int_{R^{2N_R N_T}}\left\{\mathbf{I}_{N_R} + \frac{\bar{\gamma}}{N_T}\mathbf{H}\mathbf{Q}\mathbf{H}^H\right\} d\mathbf{H} \qquad (5.37)$$

where $\bar{\gamma}$ is the average SNR per receive antenna. (Zhong et al. 2009).

5.2.6 Asymptotic analysis of capacity and outage capacity

When the number of antennas is large on both sides of a link $M = \min(N_T, N_R) \gg 1$, useful asymptotic results could be obtained by utilizing so-called random matrix theory (RMT) (Chuah et al. 2002; Lozano and Tulino 2002; Miller 2000; Moustakas et al. 2003b; Müller 2004; Verdu 1998).

---

[6] Therefore constant BER for a fixed transmission rate, which is important in some real time applications.

Instead of considering the random matrix $\mathbf{H}$ as $N_T N_R$ i.i.d. random variables, one can treat it as N_R realizations of size N_T samples of a scalar random process h. In other words, H_{kl} could be treated as the l-th sample. As a result, any linear form $\mathbf{a}^H \mathbf{Hb}$ or vectors $\mathbf{a}^H \mathbf{H}$, $\mathbf{Hb}$ can be treated as respective averages of h with respect to realization and/or time.[7] Interestingly enough, as the size of the matrix $\mathbf{H}$ increases, distribution of its eigenvalues approaches some fixed distribution. While detailed investigation can be found in a number of publications such as (Mehta 1991; Tulino and Verdú 2004), we only summarize the most important conclusions, similar to ones presented in (Müller 2004). At this stage we also define the function Tr($\mathbf{H}$, called the normalized trace of a matrix $\mathbf{H}$, as

$$\text{Tr}(\mathbf{H}) = \lim_{N \to \infty} \frac{1}{N} \text{trace}(\mathbf{H}) \tag{5.38}$$

This identity could be readily used for the calculation of moments $M_{n,\lambda}$ of the eigenvalues of $\mathbf{H}$ since

$$M_{m,\lambda} = \frac{1}{N} \sum_{k=1}^{M} \lambda_k^m = \frac{1}{N} \text{trace } \mathbf{H^m} \to \text{Tr}(\mathbf{H}) \tag{5.39}$$

5.2.6.1 Quarter-circle law

Let $\mathbf{H} = \mathbf{U}\boldsymbol{\Lambda}_H \mathbf{U}^H$ be a square $N \times N$ matrix with zero mean i.i.d. entries of variance $1/N$ and let $\mathbf{Q} = \sqrt{\mathbf{HH}^H} = \mathbf{U}\boldsymbol{\Sigma}\mathbf{U}^H$. Here $\boldsymbol{\Lambda}_H = \text{diag}\{\lambda_k\}$ and $\boldsymbol{\Sigma} = \text{diag}\{\sigma_k = |\lambda_k|\}$, $k = 1, \cdots, N$. Furthermore, let $\mathcal{L}$ represent a set containing the eigenvalues of $\mathbf{Q}$ or, equivalently, the singular values of $\mathbf{H}$. The empirical distribution $P_{\mathbf{Q}}(x)$ of the eigenvalues of $\mathbf{Q}$ is defined as a relative number of eigenvalues not exceeding the threshold x, that is

$$P_{\mathbf{Q}}(x) = \frac{1}{N} \text{count}(\sigma_k < x, \ \sigma_k \in \mathcal{L}) \tag{5.40}$$

converges to a deterministic distribution with PDF $p_{\mathbf{Q}}(x)$ given by the so-called quarter-circle law

$$p_{\mathbf{Q}}(x) = \begin{cases} \frac{1}{\pi}\sqrt{4 - x^2} & x \in [0, 2] \\ 0 & \text{otherwise} \end{cases} \tag{5.41}$$

Since the eigenvalues of the matrix $\mathbf{Q}^2 = \mathbf{HH}^H$ are simply squares of the corresponding eigenvalues of the matrix $\mathbf{Q}$, their distribution could be found via direct transformation of the density $p_Q(x)$ (Papoulis 1991):

$$p_{Q^2}(x) = \frac{p_Q(\sqrt{x})}{2\sqrt{x}} = \begin{cases} \frac{1}{2\pi}\sqrt{\frac{4-x}{x}} & x \in [0, 4] \\ 0 & \text{otherwise} \end{cases} \tag{5.42}$$

---

[7] This is known as the *self-averaging* property of functions of large random matrices (Müller 2004).

5.2.6.2 Deformed quarter-circle law

Let $\mathbf{H} = \mathbf{U}\mathbf{\Lambda}\mathbf{V}^H$ be a $N \times K$ matrix of aspect ratio $\beta = K/N$. The elements of the matrix are zero-mean i.i.d. elements of variance $1/N$. In this case the empirical distribution of the eigenvalues of the matrix $\mathbf{P} = \sqrt{\mathbf{H}\mathbf{H}^H} = \mathbf{U}\sqrt{\mathbf{\Lambda}\mathbf{\Lambda}^H}\mathbf{U}^H$ converges to a deterministic PDF as both N and K approach infinity at equal rate (that is, the aspect ratio β remains constant). The PDF in question is given by the so called Marčenko–Pastur law (Tulino and Verdú 2004)

$$p_{\mathbf{P}}(x) = \begin{cases} \frac{\sqrt{4\beta - (x^2 - 1 - \beta)^2}}{\pi \sqrt{x}} & |1 - \sqrt{\beta}| \le x \le 1 + \sqrt{\beta} \\ [1 - \beta]^+ \, \delta(x) & \text{otherwise} \end{cases} \tag{5.43}$$

5.2.6.3 Expressions for moments of ergodic capacity in an asymptotic case

If a large MIMO system is considered, $N_T, N_R \to \infty$, its capacity is a sum of capacities of partial channels. Under very mild conditions this leads to a model that instantaneous capacity is distributed according to a Gaussian law with the mean $\bar{C}$ coinciding with the ergodic capacity and some variance σ_c^2. Using the distribution of the eigenvalues of the matrix $\mathbf{H}\mathbf{H}^H$ as specified by the deformed semi-circle law, the authors of (Teletar 1999) and (Tulino and Verdú 2004) have obtained the following expressions for per received antenna mean capacity $\bar{C}/N_R$ and variance σ_C^2 as

$$\frac{\bar{C}}{N_R} = \beta \ln \left(1 + \frac{\gamma}{\beta} - \frac{1}{4} F\left[\frac{\gamma}{\beta}, \beta \right] \right)$$

$$+ \ln \left(1 + \gamma - \frac{1}{4} F\left[\frac{\gamma}{\beta}, \beta \right] \right) - \frac{\beta}{\gamma} F\left[\frac{\gamma}{\beta}, \beta \right] \tag{5.44}$$

and

$$\sigma_C^2 = -\ln \left(1 - \beta \left[F\left(\frac{\gamma}{\beta}, \beta \right) \right]^2 \right) \tag{5.45}$$

where

$$F(x, y) = \left(\sqrt{x(1 + \sqrt{y})^2 + 1} - \sqrt{x(1 - \sqrt{y})^2 + 1} \right)^2 \tag{5.46}$$

and $\beta = N_T/N_R$ is the aspect ratio of the MIMO system.

Expressions for higher order moments (cumulants) can be found in the literature, see for example (Moustakas et al. 2003a). However, expressions become fairly complicated and hard to analyze. In addition, they are valid only in a limited range of small SNR.

It was noticed in (Tulino and Verdú 2004) that the asymptotic capacity exhibits the following properties: a) weak dependence of the capacity on a particular distribution of elements of the matrix $\mathbf{H}$ as long as the zero-mean i.i.d. assumption holds; b) A single time sample of MIMO systems contains all statistical information about the statistics of the channel (ergodicity of a sample); and c) convergence to the asymptotic limit is very fast, especially for the mean value (i.e, even 8×8 systems could be considered very large). Therefore, application of asymptotic expressions could be very useful in analyzing a system. For example, the Gaussian approximation of capacity has been used in (Loyka and Levin n.d.) to investigate finite SNR diversity-multiplexing trade-off.

5.3 Effects of MIMO models and their parameters on the predicted capacity of MIMO channels

The capacity of fading channels is a well-studied topic in communications theory under the assumption that perfect channel state information (CSI) is available at the receiver and possibly at the transmitter (Cover and Thomas 2002; Goldsmith and Varaiya 1997). For such ideal conditions, achievable data rates are bounded by the capacity and depend only on the first order statistics of the channel. In reality, however, the CSI has to be estimated from data sent by the transmitter and is therefore inherently inaccurate due to receiver noise, non-zero fading rates, and the limited amount of available measurements. Therefore, the achievable rates are limited in practice not only by noise, but also by channel estimation errors, which also depend on the fading rate.

The question of finding maximal achievable rates and optimal design of systems employing pilot symbol assisted estimation has been studied for quite a while with initial results obtained in (Cavers 1991). Most of the authors have focused on optimizing pilot placement based on the desired bit error rate for fixed rate transmission (Zhang and Ottersten 2003) or block adaptive modulation (Cai and Giannakis 2005) and assume that the same quality of estimates are available at all signal-to-noise ratio (SNR) values and all data symbol positions (Cai and Giannakis 2005; Øien et al. 2004). A new approach to solve this problem is proposed in (Abou-Faycal et al. 2005; Medard 2000) by finding a lower bound on the capacity of the channel with imperfect CSI. However, only the case of CSI at the receiver and a first-order Markov channel model was considered.

The main contributions of this section are as follows. Analytical expressions for the achievable data rates are derived for a number of pilot-based estimators based on prior knowledge of the channel covariance function. An optimal power control scheme for a channel with imperfect CSI is then proposed and its performance is evaluated. The optimal frame size is also studied for synthetic channels with varying angular spreads.

The rest of this chapter is organized as follows. Section 5.3.1 introduces the system model and analytical parameters related to the channel estimation quality that are used in the chapter. In Section 5.3.2, the optimal transmitter power allocation strategy for the case of imperfect CSI is developed and the corresponding optimal achievable rates are derived. Performance examples are presented in Section 5.3.3.

5.3.1   Channel estimation and effective SNR

5.3.1.1   System model

The frequency flat fading channel is modeled to be a complex zero-mean circularly symmetric Gaussian random process $h(t)$ with covariance function $R(\tau) = \sigma_h^2 \rho(\tau)$, $\rho(0) = 1$. The received signal $r(t)$ is given by

$$r(t) = h(t)s(t) + \xi(t), \quad t = nT_s, \ n \in \mathcal{N}^+. \tag{5.47}$$

Here, T_s is the bit duration and $s(nT_s)$ represents the signal transmitted during the n-th time interval. The additive white Gaussian noise (AWGN) $\xi(t)$ has the variance σ_n^2 and is independent of both $h(t)$ and $s(t)$. The transmitted signal has the variance $\sigma_s^2 = \gamma \sigma_n^2$, where γ denotes the data SNR. The time-varying channel is estimated at the receiver side using pilots periodically inserted into the transmitted data stream: one pilot per frame consisting of N symbols. The energy of the pilot is $E_p = \gamma_p T_s \sigma_n^2$, where the quantity γ_p represents the pilot SNR.

5.3.1.2   Estimation error

While it is common to distinguish between interpolation, estimation, and prediction based on the location of the symbols of interest (Papoulis 1991), in the following we use the term estimation for all three cases without loss of generality. The minimum mean-squared error (MMSE) estimator for the model assumed above is linear (Papoulis 1991)

$$\hat{h}(\tau) = \sum_{l=-L_1}^{L_2} \alpha_l(\tau) \left[h(lNT_s) \sqrt{\frac{E_p}{T_s}} + \xi(lNT_s) \right] \tag{5.48}$$

Here, $\tau = nT_s$, $n \in [1, N-1]$, L_1 and L_2 are the number of pilots in the past and future which can be used for estimation; $\alpha_l(\tau)$ are estimation filter coefficients. The notation $P(L_1, L_2)$ is used to denote an optimal estimation scheme which uses L_1 and L_2 past and future pilots, respectively. The weights $\alpha(\tau)$ for every prediction horizon τ can be found using the normal equation (see equation (14-65) on page 500 in (Papoulis 1991)). Using the principle of orthogonality of the estimation error and the observation (Papoulis 1991), we can obtain expressions for the normalized variance δ_τ of the estimation error $\varepsilon(\tau) = h(\tau) - \hat{h}(\tau)$ and the variance $\sigma_{\hat{h}}^2(\tau)$ of the estimate $\hat{h}(\tau)$

$$\delta_\tau = 1 - \sigma_{\hat{h}}^2(\tau)/\sigma_h^2 = \sigma_\varepsilon^2(\tau)/\sigma_h^2, \tag{5.49}$$

where $\sigma_\varepsilon^2(\tau)$ is the variance of the estimation error. The quantity δ_τ is called the CSI quality by some authors (Misra et al. 2006). Analytical expressions can be easily obtained for the following estimation schemes: $P(0, 0)$ (causal single pilot), $P(1, 0)$ (causal two pilots), and $P(0, 1)$ (non-causal two pilots). Moreover, we can show that for the $P(0, 0)$ scheme

$$\delta_\tau = 1 - \frac{\gamma_p}{1 + \gamma_p} |\rho(\tau)|^2, \tag{5.50}$$

while for the $P(0, 1)$ and $P(1, 0)$ schemes one has

$$\delta_\tau = 1 - \frac{\left(1 + \gamma_p^{-1}\right)\left(|\rho(\tau)|^2 + |\rho(s)|^2\right) - 2\Re\rho^*(T)\rho(\tau)\rho(s)}{\left(1 + \gamma_p^{-1}\right)^2 - |\rho(\tau)|^2}. \tag{5.51}$$

Here, $\Re(z)$ is the real part of z, $T = NT_s$, $s = T - \tau$ holds for the $P(0, 1)$ scheme, and $s = T + \tau$ holds for the $P(1, 0)$ scheme. It is important to note that δ_τ is a function of the pilot SNR γ_p, which was not taken into account in earlier publications (Yoo and Goldsmith 2006).

The best possible estimator takes into account all past and future samples, that is $L_1 = L_2 = \infty$. While impractical, this asymptotic case provides a lower bound on the estimation error that can be related to the properties of the fading signal. Due to overlap of the noise and signal power spectral densities, estimates that are based on even infinitely many samples in the past and future are subject to error. The normalized MMSE error δ is given by (van Trees 2001)

$$\delta = \frac{1}{\sigma_s^2} \frac{1}{2\pi} \int_{-\pi}^{\pi} \ln\left[\sigma_n^2 + S(\omega)\right] d\omega = \frac{\sigma_n^2}{\sigma_s^2} \frac{1}{2\pi} \int_{-\pi}^{\pi} \ln\left[1 + \frac{S(\omega)}{\sigma_n^2}\right] d\omega. \tag{5.52}$$

If a signal has power spectral density (PSD) $S(\omega)$ composed of K spectrally separable segments with uniform power density

$$S(\omega) = \sigma_s^2 \sum_{k=1}^{K} \frac{2\pi P_k}{\Delta\omega_k} \Pi(\omega_k, \Delta\omega_k) \tag{5.53}$$

then the corresponding error can be easily computed as

$$\delta = \frac{\sigma_n^2}{\sigma_s^2} \frac{1}{2\pi} \int_{-\pi}^{\pi} \ln\left[1 + \frac{\sigma_s^2}{\sigma_n^2} \sum_{k=1}^{K} \frac{2\pi P_k}{\Delta\omega_k} \Pi(\omega_k, \Delta\omega_k)\right] d\omega$$

$$= \sum_{k=1}^{K} P_k \frac{\Delta\omega_k}{2\pi \gamma P_k} \ln\left[1 + \gamma \frac{2\pi P_k}{\Delta\omega_k}\right] d\omega = \sum_{k=1}^{K} P_k \gamma_k^{-1} \ln(1 + \gamma_k). \tag{5.54}$$

Here $\Pi(\omega_k, \Delta\omega_k) = 1$ for $|\omega - \omega_k| \leq \Delta\omega_k/2$ and 0 otherwise, and $0 \leq P_k \leq 1$ is the fraction of the total signal power concentrated in the sub-band $[\omega_k - \Delta\omega_k/2; \omega_k + \Delta\omega_k/2]$.

If all the power is concentrated in a single sub-band of width $0 \leq \Delta\omega_1 \leq 2\pi$, $P_1 = 1$, then the MMSE error becomes

$$\delta_1(\Delta\omega_1, \gamma) = \frac{\Delta\omega_1}{2\pi\gamma} \ln\left(1 + \frac{2\pi\gamma}{\Delta\omega_1}\right). \tag{5.55}$$

Simple calculus shows that for a fixed SNR $\gamma > 0$, the error δ_1 in (5.55) is an increasing function of the bandwidth $\Delta\omega_1$ with minimum $\delta_{min} = \delta_1(0, \gamma) = 0$ and maximum $\delta_{max} = \delta_1(2\pi, \gamma) = \gamma^{-1} \ln(1 + \gamma)$. Thus, regardless of the SNR, if the process has zero bandwidth, it could be estimated without any error for any non-zero SNR. Otherwise, a residual estimation error is unavoidable and it increases with bandwidth due to the signal/noise overlap.

Since working with long frames is unrealistic (Abou-Faycal et al. 2005), it is safe to assume that the normalized covariance function is evaluated only for small lags and, therefore, can be approximated by a few terms of its Taylor series with respect to a small parameter $f_D T_S$:

$$\rho_n = \rho(nT_s) \approx 1 + \rho'(0)nf_D T_s + 0.5\rho''(0)n^2(f_D T_s)^2. \qquad (5.56)$$

Here f_D is the maximum Doppler spread encountered in the channel. This can be easily calculated for analytical models of the covariance function or from measurements.

For realistic measurements, values of $\rho'(0) = ja_1$ and $\rho''(0) = -a_2^2/2$ can be estimated from the power spectral density $S(\omega)$ by

$$\rho'(0) = \frac{j}{2\pi} \int_{-2\pi f_D}^{2\pi f_D} \omega S(\omega)d\omega, \quad \rho''(0) = -\frac{1}{2\pi} \int_{-2\pi f_D}^{2\pi f_D} \omega^2 S(\omega)d\omega. \qquad (5.57)$$

Note that PSD estimates are often available at the receiver. Keeping in mind that $S(\omega)$ is a real function, one can obtain the following simple approximation for $|\rho(nT_S)|^2$

$$|\rho(nT_s)|^2 \approx 1 + (a_1^2 + a_2^2)n^2 f_D^2 T_s^2. \qquad (5.58)$$

5.3.1.3 Effective SNR

An equivalent channel model, which takes into account the channel estimation error, can be expressed as a known channel $\hat{h}(\tau)$ plus an aggregate noise $\eta(\tau)$ model in the following form

$$r(\tau) = \hat{h}(\tau)s(\tau) + \varepsilon(\tau)s(\tau) + \xi(\tau) = \hat{h}(\tau)s(\tau) + \eta(\tau). \qquad (5.59)$$

Here, $\eta(\tau)$ represents the effects of AWGN and self-interference $\varepsilon(\tau)s(\tau)$. In general, $\eta(\tau)$ is not Gaussian distributed and is correlated with $s(t)$ and $h(t)$. However, it is shown in (Hassibi and Hochwald 2003; Medard 2000), that if $\eta(t)$ is assumed to be an AWGN process with variance $\sigma_\eta^2 = \sigma_\varepsilon^2 \sigma_s^2 + \sigma_n^2$, it represents the worst case scenario in terms of capacity and probability of error. The resulting "effective" SNR, or signal to self-interference plus noise ratio (SSINR), $\gamma_{eff}(\tau)$, can then be expressed as

$$\gamma_{eff}(\tau) = \frac{\sigma_{\hat{h}}^2(\tau)\sigma_s^2}{\sigma_\varepsilon^2(\tau)\sigma_s^2 + \sigma_n^2} = \frac{(1-\delta_\tau)\gamma}{1+\delta_\tau\gamma} \leq \gamma, \quad \delta_\tau \leq 1. \qquad (5.60)$$

Here, we made use of (5.49) and the fact that $\gamma = \sigma_h^2 \sigma_s^2/\sigma_n^2$ is the channel SNR. It can be seen that the transformation (5.60) is a one-to-one monotonic function with inverse given by

$$\gamma = \frac{\gamma_{eff}(\tau)}{1 - \delta_\tau\left(1 + \gamma_{eff}(\tau)\right)}. \qquad (5.61)$$

As the SNR γ approaches infinity, the effective SNR approaches a finite limit $\gamma_{eff\,\infty}(\tau) = 1/\delta_\tau - 1$. This causes a well-known error floor phenomenon (Proakis

2001) for a given modulation scheme. It also limits achievable data rates in channels with pilot assisted channel estimation.

It can be seen from (5.60) that the effective SNR, and therefore the achievable data rates, vary with position of the symbol in the data block, that is, the block fading assumption is often invalid. The authors of (Abou-Faycal et al. 2005) treat such channels as an array of $N - 1$ partial channels, obtained by sampling the original channel at the rate $1/NT_s$ and shifting by one symbol with respect to each other with each partial channel having a fixed effective SNR. The optimal communication strategy over such channels involves optimizations over each partial channel. The overall achievable rate is a sum of rates over all partial channels.

5.3.2 Achievable rates in Rayleigh channels with partial CSI

5.3.2.1 No CSI at the transmitter

Assuming that the noise $\eta(\tau)$ in (5.59) is AWGN, we can obtain a lower bound $C(\tau)$ for the achievable data rates in each partial channel without power control (that is, no CSI at the transmitter)

$$
C(\tau) = \int_0^\infty \ln\left(\frac{1+\gamma}{1+\delta_\tau\gamma}\right) p(\gamma)d\gamma
$$

$$
= \exp\left(\frac{1}{\bar{\gamma}}\right) \text{Ei}\left(-\frac{1}{\bar{\gamma}}\right) - \exp\left(\frac{1}{\delta_\tau\bar{\gamma}}\right) \text{Ei}\left(-\frac{1}{\delta_\tau\bar{\gamma}}\right)
$$

$$
= C_\infty(\bar{\gamma}) - C_\infty(\delta_\tau\bar{\gamma}). \tag{5.62}
$$

Here $\text{Ei}(x)$ is the exponential integral function (Abramowitz and Stegun 1965), $p(\gamma) = 1/\bar{\gamma} \exp(-\gamma/\bar{\gamma})$ is the probability density function (PDF) of the instantaneous SNR with average $\bar{\gamma}$. Thus, the lower bound is the difference between the capacity of a perfectly known Rayleigh fading channel with average SNR $\bar{\gamma}$ and the capacity of a perfectly known Rayleigh fading channel with average SNR $\bar{\gamma}\delta_\tau(\bar{\gamma})$. The explicit dependence of δ_τ on $\bar{\gamma}$ emphasizes the fact that these two quantities cannot be chosen independently to produce artificially loose bounds.

5.3.2.2 Partial CSI at the transmitter

If the transmitter power is adaptive, the optimal allocation of power $\Psi(\gamma)$ can be found as a strategy which maximizes the following integral

$$
C_{opt} = \max \int_0^\infty \ln\left(1 + \frac{\Psi(\gamma)}{\hat{\gamma}}\gamma_{eff}\right) p(\gamma)d\gamma \tag{5.63}
$$

subject to the average power constraint

$$
\int_0^\infty \Psi(\gamma)p(\gamma)d\gamma = \bar{\gamma}. \tag{5.64}
$$

Changing variables according to (5.60), the optimization becomes

$$
C_{opt} = \max \int_0^\infty \ln\left(1 + \frac{F(\gamma_{eff})}{\hat{\gamma}}\gamma_{eff}\right) p_e(\gamma_{eff})d\gamma_{eff} \tag{5.65}
$$

subject to the average power constraint

$$\int_0^\infty F(\gamma_{eff})p_e(\gamma_{eff})d\gamma_{eff} = \hat{\gamma}, \quad F(\gamma_{eff}) = \Psi\left(\frac{(1-\delta_\tau)\gamma}{1+\delta_\tau\gamma}\right), \tag{5.66}$$

where $p_e(\gamma_{eff})$ is the PDF of the effective SNR. Here, the solution is given by the water-filling algorithm (Goldsmith and Varaiya 1997)

$$\frac{F(\gamma_{eff})}{\bar{\gamma}} = \begin{cases} \frac{1}{\gamma_0} - \frac{1}{\gamma_{eff}} & \text{if } \gamma_{eff} > \gamma_0 \\ 0 & \text{if } \gamma_{eff} < \gamma_0 \end{cases} \tag{5.67}$$

where γ_0 is the solution of

$$\int_{\gamma_0}^\infty \left(\frac{1}{\gamma_0} - \frac{1}{\gamma_{eff}}\right)p_e(\gamma_{eff})d\gamma_{eff} = 1. \tag{5.68}$$

Changing the variable again according to (5.60) and (5.68), we find that the optimal power allocation strategy is

$$\frac{\Psi(\gamma)}{\bar{\gamma}} = \begin{cases} \frac{1}{\gamma_0} - \frac{\delta_\tau}{1-\delta_\tau} - \frac{1}{(1-\delta)\gamma} & \text{if } \gamma > \frac{\gamma_0}{1-\delta_\tau(1+\gamma_0)} \\ 0 & \text{if } \gamma < \frac{\gamma_0}{1-\delta_\tau(1+\gamma_0)} \end{cases} \tag{5.69}$$

where γ_0 is a positive solution of the following equation

$$\bar{\gamma}\left[1-\delta_\tau(1+\gamma_0)\right]\exp\left(-\frac{\gamma_0}{\left[1-\delta_\tau(1+\gamma_0)\right]\bar{\gamma}}\right)$$

$$- \gamma_0\text{Ei}\left(\frac{\gamma_0}{1-\delta_\tau(1+\gamma_0)\bar{\gamma}}\right) = \bar{\gamma}\gamma_0(1-\delta_\tau). \tag{5.70}$$

The corresponding optimal achievable rates are then

$$C_{opt} = \text{Ei}\left(\frac{\gamma_0}{\left[1-\delta_\tau(1+\gamma_0)\right]\bar{\gamma}}\right)$$

$$- \exp\left(\frac{1}{\bar{\gamma}\delta_\tau}\right)\text{Ei}\left(\frac{1-\delta_\tau}{\left[1-\delta_\tau(1+\gamma_0)\right]\delta_\tau\bar{\gamma}}\right). \tag{5.71}$$

The results of the numerical evaluation of C_{opt} with fixed δ_τ and different average SNR are presented in Figure 5.1. Let us emphasize here that we consider individual partial channels. While the average SNR remains the same for each partial channel, the variance of the estimation error varies from one partial channel to another. The same value of δ_τ may correspond to a different partial channel. It can be seen that the level of estimation error significantly affects the achievable rates, especially for high average SNRs. The saturation of the capacity as a function of average SNRs is due to self-interference induced by the error in estimating the channel. Such errors may result from a combination of two factors: the finite SNR of the pilot signal and the fading rate. The quality of different estimation schemes are considered in Figure 5.2. As for the perfect CSI case, the power adaptation provides some gain at lower average SNRs and almost no gain for average SNRs higher than 10 dB.

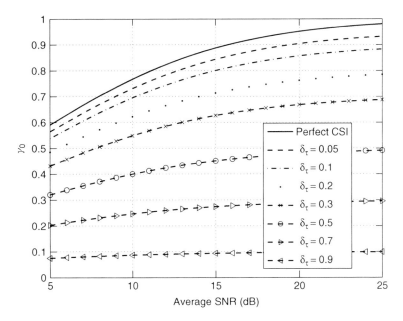

Figure 5.1   Optimal cut off SNR values for different values of δ_τ.

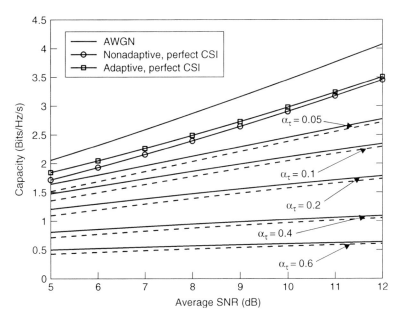

Figure 5.2   Improvement achieved through power adaptation. Solid lines: adaptive power allocation; dashed lines: constant power allocation.

5.3.2.3 Optimization of the frame length

The total achievable rates C_T for the channel can be obtained by summing the achievable rates of each partial channel, that is

$$C_T = N^{-1} \sum_{n=1}^{N-1} C(nT_s). \tag{5.72}$$

The factor $1/N$ is included to take into account the fact that each partial channel occupies only a fraction of the frame. Since for a frame of length 2 only half of the transmission time is used for data transmission, an increase in the frame length will result in a better utilization of the channel. However, longer frames lead to larger estimation errors, thus decreasing the achievable rates. The balance between these trends is achieved at some optimal frame length N_{opt}, which can be found numerically. A few examples are included in Section 5.3.3.

5.3.3 Examples

5.3.3.1 $P(0, 0)$ Estimation

In order to analytically evaluate the effects of pilot SNR and finite correlation intervals on the achievable rates, we consider the simplest case of $P(0, 0)$ estimation.

Each partial channel can be treated as a separate channel with perfect CSI and an effective SNR given by (5.60). Since perfect CSI is available, some sort of power adaptation can be exercised over each partial channel. At the same time, since different partial channels experience different average equivalent noise conditions, one may consider the optimal distribution of power among the partial channels.

Without power adaptation, the achievable rates are given by Equation (5.62), where, according to (5.50),

$$\delta_\tau = 1 - |\rho(\tau)|^2 \frac{\gamma_p}{1 + \gamma_p}. \tag{5.73}$$

The case of perfect CSI is recovered by setting $\gamma_p = \infty$ and $\rho = 1$, yielding the well-known expression for the capacity of a Rayleigh channel with perfect CSI (Goldsmith and Varaiya 1997) at the receiver:

$$C_\infty = \exp\left(\frac{1}{\bar{\gamma}}\right) \mathrm{Ei}\left(-\frac{1}{\bar{\gamma}}\right). \tag{5.74}$$

The effect of the channel correlation alone can be traced by setting $\gamma_p = \infty$ and $b = 1 - |\rho(\tau)|^2$, thus resulting in the following expression

$$C_\rho = C_\infty - \exp\left(\frac{1}{(1 - |\rho(\tau)|^2)\, \bar{\gamma}}\right) \mathrm{Ei}\left(-\frac{1}{(1 - |\rho(\tau)|^2)\, \bar{\gamma}}\right). \tag{5.75}$$

As expected, $C_\rho \to 0$ as $\rho \to 0$. The effect of imperfect estimation in very slow fading can be obtained by setting $\rho = 1$, resulting in

$$C_{\gamma_p} = C_\infty - \exp\left(\frac{1 + \gamma_p}{\bar{\gamma}}\right) \mathrm{Ei}\left(-\frac{1 + \gamma_p}{\bar{\gamma}}\right). \tag{5.76}$$

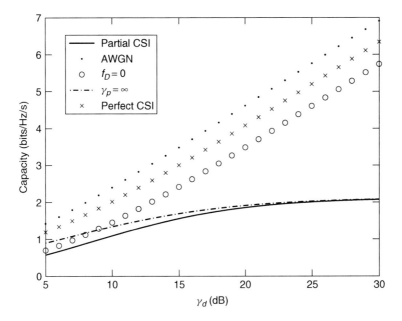

Figure 5.3   Channel capacities in different scenarios. $f_D T_s = 0.04$.

The negative terms in (5.75) and (5.76) represent a decrease in achievable rates due to imperfect correlation and due to the finite power of the pilot energy, respectively. The results are shown in Figure 5.3. This figure allows us to track the impact of two factors on the achievable rates. As expected, a constant, perfectly known channel provides the maximum capacity. A fading, but perfectly known channel results in only slightly smaller rates for all SNRs. Furthermore, if the channel is unknown and being estimated but does not vary in time ($f_D = 0$), additional degradation of the achievable rates can be observed. This degradation is almost the same for all levels of SNR, approximately 1 bit/sec/Hz. However, even if a fading rate $f_D T_s = 0.04$ is taken into account, it can be seen from Figure 5.3 that in a high SNR region the achievable rate saturates at a level significantly lower than that predicted by the zero rate fading model. This can be explained by the fact that at a high SNR level the estimation error is negligible compared to the error due to time selectivity ($f_D T_s > 0$). Thus, a further increase in the SNR does not improve the net estimation error and the performance saturates. In contrast, for low SNRs, the estimation error is dominated by noise and the performance is only marginally lower than that of zero rate fading.

Figure 5.4 demonstrates the dependence of the optimal frame length on the average SNR and fading rate. A more pronounced maximum corresponds to a faster fading scenario. It may appear strange that for the same value of $f_D T_s$ a lower SNR will result in longer optimal frames. However, for low SNRs, the effective SNR is dominated by the noise and not by self-interference: the optimum frame length is then defined by a balance between the loss in accuracy of the estimate

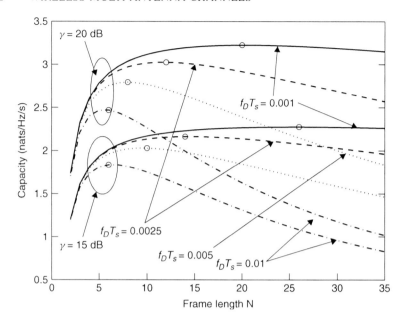

Figure 5.4   Optimal frame size for two values of average SNR, $\gamma_p = \gamma$ as function of fading rate and the frame length.

and the utilization of the power, which increases with the frame duration. In contrast, for a high average SNR, the effective SNR is dominated by self-interference, which increases with the frame size: the balance between power utilization and the increased interference is achieved for shorter frames. Similar conclusions can be reached for other estimation schemes.

5.3.3.2   Effect of non-uniform scattering

In this section we assume that the scattering environment forming the channel is such that scattering appears uniformly from only a single cluster, seen at the azimuth angle ϕ_0 ($\phi_0 = 0$ corresponds to the direction of the mobile movement) and angular spread Δ. The squared magnitude of the covariance function of the fading signal is then (Loyka 2005)

$$|\rho(\tau)|^2 = \text{sinc}^2\left(\Delta f_D \tau \cos \phi_0\right) \approx 1 - 0.33\,\Delta^2 \pi^2 f_D^2 \tau^2 \cos \phi_0^2. \qquad (5.77)$$

Using (5.77), one can obtain insight into how a cluster's angular spread and orientation affects the achievable rates. A smaller spread Δ results in a slower decay of $|\rho(\tau)|^2$ and in a slow increase in estimation error. As a result, the quality of CSI, δ_τ, is better for a narrower cluster and longer frames could be used: a wider angular spread corresponds to a wider fading bandwidth making the channel less predictable (Papoulis 1991). Similarly, orientation of the cluster plays a role in changing δ_τ since the increase in ϕ_0 from 0 to 90^0 leads to a slower decay of $|\rho(\tau)|$ and a slower growth of δ_τ. Optimal frame length designs for different angular spreads Δ and $P(0, 0)$ and $P(1, 0)$ estimators are shown in Figure 5.4. It can be seen that while the use of an additional pilot increases the overall quality of estimation and achievable rates, the optimal frame length is less sensitive to the estimation scheme compared to the rate of decay of the fading covariance function. A situation with multiple clusters and measurements could be similarly analyzed.

Finally, we summarize the finding of this section as follows. The achievable data rates of systems with pilot based channel estimates have been analyzed. Expressions for the maximum achievable data rates were derived in terms of the fading channel estimation quality. It has been shown how the classical water-filling algorithm should be modified for application in scenarios with imperfect CSI at the transmitter. It has been found that the gain achieved through power adaptation is more pronounced at low SNRs and for good estimates, but overall it is not significant. The impact of the estimation scheme on achievable data rates has been studied for physical channel models with varying angular spreads. It has been observed that for causal estimates, the optimal frame sizes are relatively short and the optimum is well pronounced. In the case of non-causal estimation, the optimum block size is at least twice as long and the corresponding maxima are relatively flat.

It is shown that a concept of effective SNR can be constructively used in order to evaluate achievable rates in fading channels with imperfect CSI at the receiver. It is shown that the achievable rates can be interpreted as the difference between the capacity of the channel which is known perfectly and the entropy of noise caused by self-interference. A modified water-filling algorithm for power adaptation is developed if the partial CSI is available at the transmitter side. It is shown that the cut-off instantaneous SNR in power control schemes depends on the quality of estimation and is lower than one calculated for the perfect CSI. In general, optimal communication over a channel with imperfect CSI can be achieved by treating it as a set of parallel channels with perfect CSI and fixed SNR, which in turn varies from partial channel to partial channel. Near capacity achieving coding could be applied to each partial channel, thus approaching achievable rates for the whole channel.

5.4   Time evolution of capacity

While investigation of ergodic, outage, or instantaneous capacity is now well understood in various scenarios, the time evolution of capacity is a much less studied subject. One of the first studies on the subject is presented in (Giorgetti et al. 2003). The problem is difficult to handle in its generality due to the complicated nature of function $\log(1 + x)$. Therefore it is important to focus on some limiting cases which allow analytical treatment. Once again, low SNR and high SNR regions seem to offer such a possibility.

5.4.1   Time evolution of capacity in SISO channels

Application of adaptive transmission schemes (ATS) plays a vital role in the area of wireless communication systems. Recently proposed ATS (Bai and Shami 2008;

Liu 2005) exploit knowledge of the underlying time-varying parameters of the wireless channel to adapt physical and/or medium access parameters of the channel to improve performance.

In the case of MIMO channels, knowledge of the temporal behavior of the eigenvalues of the channel matrix (and corresponding time variational properties of the capacity) are parameters of interest in the design of channel ATS. There have been a number of attempts to characterize quantities of interest such as the level-crossing rate (LCR), average fade duration (AFD) (Giorgetti et al. 2003; Kuo and Smith 2005; Smith et al. 2005) and time correlation of the channel capacity (S.Vaihunthan et al. 2005). Most of the publications assume rich scattering and the Jakes spectrum (Paulraj et al. 2003) of individual SISO channels. In addition, the covariance function of the capacity is often assumed just to follow that of the SISO channel (Smith et al. 2005; S.Vaihunthan et al. 2005). However, measurements (Bultitude et al. 2000; Burr 2003), conducted in urban environments revealed that the scattering process is often dominated by a small number of scatterers. Such channels, known as sparse (Burr 2003) are well investigated in terms of average capacity. A geometric-based MIMO model of sparse channels was presented in (Primak and Sejdic 2008). This paper considers first and second order statistics, including LCR and AFD, for these sparse MIMO channels. The results presented below are useful in understanding of effect of scattering environment on cross-layer design of MIMO aware adaptive systems.

5.4.2   SISO channel capacity evolution

Let us consider a stationary SISO Gaussian channel $h(t) = h_I(t) + j h_Q(t)$ with known covariance function $\rho_h(\tau) = \mathcal{E}\{h(t+\tau)h^*(t)\}$ and improper covariance function $\varrho_h(\tau) = \mathcal{E}\{h(t+\tau)h(t)\}$, $\rho_h(0) = 1$, $\varrho_h(0) = 0$. Our goal is to evaluate the covariance $\rho_C(\tau)$.[8]

For a small value of the average SNR $\bar{\gamma}$, the instantaneous capacity can be approximated as $C(t) \approx \gamma |h(t)|^2$, therefore, the corresponding covariance function is given by

$$
\begin{aligned}
\rho_C(\tau) &= \mathcal{E}\{C(t+\tau)C(t)\} - \mathcal{E}\{C(t)\}^2 \approx \mathcal{E}\left\{|h(t+\tau)|^2 |h(t)|^2\right\} - \mathcal{E}^2\left\{|h(t)|^2\right\} \\
&= \mathcal{E}\left\{h_I^2(t+\tau)h_I^2(t)\right\} + \mathcal{E}\left\{h_Q^2(t+\tau)h_Q^2(t)\right\} + \mathcal{E}\left\{h_Q^2(t+\tau)^2 h_I^2(t)\right\} \\
&\quad + \mathcal{E}\left\{h_I^2(t+\tau)^2 h_Q^2(t)\right\} - \mathcal{E}^2\left\{|h(t)|^2\right\} = 2\rho_{II}^2(\tau)\sigma_I^4 + 2\rho_{QQ}^2(\tau)\sigma_Q^4 \\
&\quad + 2\left[\rho_{QI}^2(\tau) + \rho_{IQ}^2(\tau)\right]\sigma_I^2\sigma_Q^2
\end{aligned}
\tag{5.78}
$$

Here we used the standard integral

$$
\int_{-\infty}^{\infty}\int_{-\infty}^{\infty} \frac{x^2 y^2}{2\pi\sigma_I\sigma_Q\sqrt{1-r^2}} \exp\left[-\frac{1}{2(1-r^2)}\left(\frac{x^2}{\sigma_I^2} - 2r\frac{x}{\sigma_I}\frac{y}{\sigma_Q} + \frac{y^2}{\sigma_Q^2}\right)\right]
$$

$$
= (1 + 2r^2)\sigma_I^2\sigma_Q^2
\tag{5.79}
$$

---

[8] Since $C(t)$ is a real process no improper correlation is required.

Furthermore

$$\rho_{II}(\tau) = \frac{1}{2}\Re\left[\rho(\tau) + \varrho(\tau)\right] \tag{5.80}$$

$$\rho_{QQ}(\tau) = \frac{1}{2}\Re\left[\rho(\tau) - \varrho(\tau)\right] \tag{5.81}$$

$$\rho_{QI}(\tau) = \frac{1}{2}\Im\left[\rho(\tau) + \varrho(\tau)\right] \tag{5.82}$$

$$\rho_{IQ}(\tau) = \frac{1}{2}\Im\left[\rho(\tau) - \varrho(\tau)\right] \tag{5.83}$$

In the case of the channel $h(t)$ is a circularly symmetric Gaussian process $\sigma_I = \sigma_Q = \sigma$, one obtains $\varrho(\tau) = 0$, $\rho_{II}(\tau) = \rho_{QQ}(\tau) = \Re\,\rho(\tau)/2$, $\rho_{IQ}(\tau) = \rho_{QI}(\tau) = \Im\,\rho(\tau)/2$ and the expression (5.78) reduces to

$$\rho_C(\tau) = 4\sigma^4\left(\rho_{II}^2 + \rho_{QI}^2\right) = |\rho(\tau)|^2 \tag{5.84}$$

Thus, for small average SNR, the covariance function of the capacity is a weighted sum of squared covariance and cross-covariance functions of the in-phase and quadrature components. In the case of circular channel it is just a square of the magnitude of the complex channel covariance function, and, as such, is defined by covariance of the channel instantaneous SNR $\gamma \sim |h|^2$.

It is well known that the level-crossing rate $LCR(c)$ of a random process $C(t)$ is proportional to the value of the second derivative of the covariance function $-\rho_C''(0)$ evaluated at zero (Papoulis 1991). Using corresponding Taylor series for $\rho(\tau)$ and $\varrho(\tau)$

$$\rho(\tau) = 1 + ja\tau - \frac{b^2}{2}\tau^2 + \mathcal{O}(\tau^3) \tag{5.85}$$

$$\varrho(\tau) = jc\tau - \frac{d^2}{2}\tau^2 + \mathcal{O}(\tau^3) \tag{5.86}$$

one obtains, after a simple algebra,

$$-\rho_C''(0) = b^2 - \frac{1}{2}(a+c)^2 \tag{5.87}$$

5.4.2.1   Example: Jakes fading

In the case of Jakes spectrum $\rho(\tau) = J_0(2\pi f_D\tau)$, the covariance of the capacity becomes $\rho_C(\tau) = \sigma_h^4 J_0^2(2\pi f_D\tau)$, and, therefore

$$\sqrt{-\frac{\rho_C''(0)}{2\pi}} = \sqrt{2\pi}\,f_D\sigma_h^2 \tag{5.88}$$

This example shows a well understood trend that the LCR is proportional to the maximum Doppler spread f_D.

5.4.2.2 Example: Single narrow cluster

Let us now assume that all scattering between a transmitter and a receiver can be attributed to a single scattering center seen at the angle ϕ_R at the receiver with a narrow angular spread Δ_T. In this case the covariance function of the fading channel $h(t)$ can be approximated as (Loyka 2005)

$$\rho(\tau) = \sigma_h^2 \operatorname{sinc}\left(\frac{\Delta_T}{\pi} f_D \tau \sin \phi_T \tau\right) \exp\left(j 2\pi f_D \cos \phi_T \tau\right) \qquad (5.89)$$

$$\sqrt{-\frac{\rho_C''(0)}{2\pi}} = \frac{\Delta_T |\sin \phi_T|}{\sqrt{3}} f_D \sigma_h^2 \qquad (5.90)$$

In addition to the previous example, it can be seen that the LCR also depends on the position of the scatter with respect to the receive antenna movement and its angular spread Δ_T via product $\Delta_T \sin \phi_T$.

5.4.2.3 Example: Two narrow clusters

In order to get deeper insight into the role of the scattering environment of the LCR of the capacity let us consider the situation where there are two clusters seen by the received antennas at angles ϕ_{R1} and ϕ_{R2}, contributing relative power P_1 and $P_2 = 1 - P_2$ respectively and having angular spread Δ_{T1} and Δ_{T2} respectively. In this case the covariance function $\rho_h(\tau)$ can be represented as a sum of two modulated sinc functions:

$$\rho(\tau) = P_1 \sigma_h^2 \operatorname{sinc}\left(\frac{\Delta_{T1}}{\pi} f_D \tau \sin \phi_{T1} \tau\right) \exp\left(j 2\pi f_D \cos \phi_{T1} \tau\right)$$
$$+ P_2 \sigma_h^2 \operatorname{sinc}\left(\frac{\Delta_{T2}}{\pi} f_D \tau \sin \phi_{T2} \tau\right) \exp\left(j 2\pi f_D \cos \phi_{T2} \tau\right) \qquad (5.91)$$

After simple algebra

$$\sqrt{-\frac{\rho_C''(0)}{2\pi}} = \frac{\Delta_{T1} |\sin \phi_{T1}|}{\sqrt{3}} f_D \sigma_h^2 + \frac{\Delta_{T2} |\sin \phi_{T2}|}{\sqrt{3}} f_D \sigma_h^2 \qquad (5.92)$$

5.5 Sparse MIMO channel model

Let us consider a situation where the number of transmit and receive antennas, N_T and N_R respectively, is high and the scattering environment is dominated by a small number $L \leq \min\{N_T, N_R\}$ of angularly separable scatters (clusters) (Bultitude et al. 2000). In this case, the MIMO channel matrix $\mathbf{H}(t)$ can be represented as (Primak and Sejdic 2008)

$$\mathbf{H}(t) = \sum_{l=1}^{L} \sqrt{P_l}\, \mathbf{a}_l \mathbf{b}_l^H \xi_l(t) \exp(j 2\pi f_D \cos \alpha_l t) \qquad (5.93)$$

Here α_l is the angle of arrival of the signal from the l-th scatterer, f_D is the maximum Doppler frequency, P_l is the relative power of the l-th path, $\sum_{l=1}^{L} P_l = 1$, $\mathbf{a}_l$, and $\mathbf{b}_l$ are the receive and transmit steering vectors of the l-th path and $\xi_l(t)$ is a complex Gaussian process of zero-mean and a covariance function $\rho_l(\tau) = E\left\{\xi_l(t + \tau)\xi_l^*(t)\right\}$ given by (Primak and Sejdic 2008)

$$\rho_l(\tau) = \sigma^2 \text{sinc} \left(2\Delta\alpha_l \cos\alpha_l f_D\tau\right) \exp(j2\pi f_D \cos\alpha_l\tau) \tag{5.94}$$

where $\Delta\alpha_l \ll 1$ is a narrow angular spread of the l-th cluster. The steering vectors are mutually orthogonal due to cluster separation in the angular domain, that is $\mathbf{a}_L^H\mathbf{a}_k = N_R\delta_{lk}$, $\mathbf{b}_L^H\mathbf{b}_k = N_T\delta_{kl}$. As a result, the matrix $\mathbf{W}(t) = \mathbf{H}(t)\mathbf{H}^H(t)$ can be expanded as

$$\mathbf{W}(t) = N_T \sum_{l=1}^{L} \mathbf{a}_l\mathbf{a}_l^H P_l|\xi_l(t)|^2 = \mathbf{AP}(t)\mathbf{A}^H \tag{5.95}$$

where the matrix $\mathbf{A}$, $\mathbf{AA}^H = N_R\mathbf{I}_{N_R}$ is composed of the steering vectors $\mathbf{a}_l$ and $\mathbf{P}(t) = \text{diag}\left\{P_lN_T|\xi_l(t)|^2\right\}$ is a diagonal matrix. By inspection of Equation (5.95) and the fact that $L \leq \min\{N_T, N_R\}$ one can conclude that it represents eigenvalue-decomposition of the matrix $\mathbf{W}$.

Let $\mathbf{H} = \mathbf{U\Sigma V}^\dagger$ be the SVD decomposition of the channel matrix $\mathbf{H}$. Here both $\mathbf{U}$ and $\mathbf{V}$ are unitary. Furthermore since ξ_l is a complex zero-mean Gaussian process with well-known (Papoulis 1991) statistical properties, the unordered eigenvalue density function corresponding to a single scatter is given as

$$p_{\Lambda_l}(\lambda_l) = \frac{1}{N_R N_T P_l} \exp\left(-\frac{\lambda_l}{N_R N_T P_l}\right) \tag{5.96}$$

Further, the ordered eigenvalue distribution ($\lambda_L \geq \lambda_{L-1} \geq \cdots \geq \lambda_1 \geq 0$) in the case of equal power from all clusters ($P_l = 1/L, l = 1, 2, \ldots, L$) of the kth eigenvalue can also be found as (Paulraj et al. 2003)

$$p_{\lambda_k}(\lambda) = k\frac{L!}{k!(L-k)!} F_x^{k-1}(\lambda)[1 - F_x(\lambda)]^{L-k} p_x(\lambda) \tag{5.97}$$

where

$$F_x(x) = 1 - \exp\left(-\frac{xL}{N_R N_T}\right) \tag{5.98}$$

and

$$p_x(x) = \frac{L}{N_R N_T} \exp\left(-\frac{xL}{N_R N_T}\right) \tag{5.99}$$

or by applying the Bapat–Beg theorem (Bapat and Beg 1989) when power from each cluster is not equal.

5.6   Statistical properties of capacity

5.6.1   Some mathematical expressions

This section deals with evaluation of integrals of the following kind

$$I_m = \int_0^\infty \ln^m(1+\gamma) p_\gamma(\gamma) d\gamma \qquad (5.100)$$

where $p_\gamma(\gamma)$ is the PDF of the SNR γ. Exact evaluation of the integral (5.100) could be based on the following identity

$$\ln^m(1+\gamma) = \lim_{v \to 0} \frac{d^m}{dv^m} (1+\gamma)^v \qquad (5.101)$$

Therefore, evaluation of I_m can be reduced to evaluation of

$$J_m = \int_0^\infty (1+\gamma)^v p_\gamma(\gamma) d\gamma \qquad (5.102)$$

followed by differentiation and limiting

$$I_m = \lim_{v \to 0} \frac{d^m}{dv^m} J_m \qquad (5.103)$$

In particular, assuming an exponential distribution $p_\gamma(\gamma) = \bar{\gamma}_l^{-1} \exp(-\gamma/\bar{\gamma}_l)$ one obtains

$$J_m = \exp\left(\frac{1}{\bar{\gamma}_l}\right) \bar{\gamma}_l^{\,v} \Gamma\left(1+v, \frac{1}{\bar{\gamma}_l}\right) \qquad (5.104)$$

$$\mu_C = I_1 = \exp\left(\frac{1}{\bar{\gamma}_l}\right) E_1\left(\frac{1}{\bar{\gamma}_l}\right) \qquad (5.105)$$

where $E_1(\cdot)$ is the exponential integral (Abramowitz and Stegun 1965), and

$$\sigma_C^2 + \mu_C^2 = I_2 = 2 \exp\left(\frac{1}{\bar{\gamma}_l}\right) G_{23}^{30}\left(\begin{array}{c} 1,1 \\ 0,0,0 \end{array}\middle| \frac{1}{\bar{\gamma}_l}\right) \qquad (5.106)$$

where $G_{pq}^{mn}\left(\begin{array}{c} a_1,\ldots,a_p \\ b_1,\ldots,b_q \end{array}\middle| z\right)$ is the Meijer G-function (Abramowitz and Stegun 1965).

To summarize, the mean μ_C and the variance σ_C^2 of the capacity C can be analytically calculated as:

$$\mu_C = \sum_{l=1}^L E\{C_l\} = \sum_{l=1}^L \exp\left(\frac{1}{\bar{\gamma}_l}\right) E_1\left(\frac{1}{\bar{\gamma}_l}\right) \qquad (5.107)$$

$$\sigma_C^2 = \sum_{l=1}^L \left[2\exp\left(\frac{1}{\bar{\gamma}_l}\right) G_{23}^{30}\left(\begin{array}{c} 1,1 \\ 0,0,0 \end{array}\middle| \frac{1}{\bar{\gamma}_l}\right) - E^2\{C_l\} \right] \qquad (5.108)$$

where $\bar{\gamma}_l = \gamma_0 N_R P_l$. Asymptotic analysis of (5.107) and (5.108) will be presented elsewhere.

Through a trivial transformation of random variables (Papoulis 1991), one can also show that the exact distribution of the capacity contribution by any single scatter is given by

$$p_{c,l}(x_l) = \frac{1}{\bar{\gamma}_l} \exp\left(-\frac{\exp(x_l) - 1}{\bar{\gamma}_l} + x_l\right) \tag{5.109}$$

and the capacity distribution to the contribution of L independent clusters is then

$$p_c(x) = p_{c,1}(x_1) * p_{c,2}(x_2) * \cdots * p_{c,L}(x_L) \tag{5.110}$$

where $*$ denotes the convolution operation. One can also approximate (5.110) as a Gaussian (Hochwald et al. 2004) random variable described completely through the mean and variance given in (5.107) and (5.108).

5.7 Time-varying statistics

Time-varying dynamics of the MIMO channel, LCR and AFD of capacity, and corresponding eigenvalues are of high importance in the design of adaptive systems (Bai and Shami 2008; Liu 2005). In this section we derive closed form expressions for these and related quantities.

5.7.1 Unordered eigenvalues

It follows from (5.94) that for any unordered eigenvalue, the correlation function of the envelope of ξ_l is given by

$$\rho_{env,l}(\tau) = \sigma^2 \mathrm{sinc}\,(2\Delta\alpha_l \cos\alpha_l f_D \tau) \tag{5.111}$$

It is well known (Papoulis 1991) that the LCR of a single eigenvalue with a density function given in (5.96) is exactly

$$N_\Lambda(\lambda) = 2\sqrt{\frac{\lambda}{N_R N_T P_l}} \exp\left(\frac{-\lambda}{N_R N_T P_l}\right)\sqrt{\frac{\beta_l}{2\pi}} \tag{5.112}$$

where for simplicity we have made use of the following notation

$$\beta_l = -\left.\frac{\partial^2}{\partial\tau^2}\{\rho_{env,l}(\tau)\}\right|_{\tau=0} = \frac{\sigma^2}{3}(2\pi\,\Delta\alpha_l\cos\alpha_l f_D)^2 \tag{5.113}$$

The corresponding AFD is then

$$AFD_\Lambda(\lambda) = \frac{F_\Lambda(\lambda)}{N_\Lambda(\lambda)} \tag{5.114}$$

where $F_\Lambda(\lambda)$ is the CDF of the exponential distribution given in (5.96).

5.7.2   Single cluster capacity LCR and AFD

It is possible to derive an exact expression for LCR and AFD for a single cluster environment. It is well known (Papoulis 1991) that one can express the level-crossing rate of the Rayleigh process ξ_l with $\sigma^2 = 0.5$ as

$$N_R(r) = p_\xi(r)\sqrt{\frac{\beta_l}{2\pi}} \tag{5.115}$$

where β_l is given in (5.113). After simple algebraic manipulation, one can show that the LCR and AFD of capacity in a single cluster environment is given by

$$N_{C_l}(c) = \sqrt{\frac{2\beta_l(\exp(c) - 1)}{\pi\bar{\gamma_l}}} \exp\left(-\frac{\exp(c) - 1}{\bar{\gamma_l}}\right) \tag{5.116}$$

and

$$AFD_{C_l}(c) = \frac{F_{C_l}(c)}{N_{C_l}(c)} \tag{5.117}$$

where $F_{C_l}(c)$ is the cumulative density function of (5.109).

5.7.3   Approximation of multi-cluster capacity LCR and AFD

It was noted in Section 5.6 that the contribution of individual clusters to the capacity is independent. However, this is not so for computation of LCR and AFD, therefore requiring direct use of the general Rice formula (Papoulis 1991). This requires knowledge of the joint PDF of the capacity which is not usually known (Smith et al. 2005). Since a derivative of a stationary process is often a symmetric function uncorrelated with the original process, it is often possible to approximate LCR as in the case of a Gaussian

$$N_C(c) = p_c(c)\sqrt{\frac{\sigma_{\dot{c}}^2}{2\pi}} \tag{5.118}$$

where $\sigma_{\dot{c}}^2$ is the variance of the derivative process. This equation becomes more accurate as the original process approaches Gaussian

Under the assumption of validity of (5.118) the approximation for $\sigma_{\dot{c}}^2$ is given by the following lemma:

Lemma 5.7.1 *The capacity derivative process $\dot{C}$ is approximately a zero-mean Gaussian random variable with a variance of*

$$\sigma_{\dot{c}}^2 \approx -4\exp(-1)\pi\sigma_c^2 \sum_{l=1}^{L} P_l\beta_l$$

First, we note that due to the non-negative nature of $|\xi_l(t)|$, there exists a one-to-one mapping between $C_l(t)$ and $|\xi_l|$ such that at any instant of time, $C_l(t) = \log(1 + \gamma_0 N_R P_l |\xi_l(t)|^2)$. Furthermore, the non-negative property of $|\xi_l(t)|$ implies that the average number of times the $|\xi_l(t)|$ process crosses a threshold r is equal to the average number of times the $C_l(t)$ process crosses a threshold c where $c = \log(1 + \gamma_0 N_R P_l |r|^2)$.

Under this observation, we suggest that that the maximum value of the LCR occurs in the same manner as the mapping between the underlying Rayleigh and capacity processes. Under our Gaussian capacity approximation, where the capacity variance is σ_c^2, we obtain a maximum distribution value of $1/\sqrt{2\pi\sigma_c^2}$. Using this feature-matching method, the ratio of the Rayleigh process (with $\sigma^2 = 0.5$) maximum to the Gaussian capacity is:

$$\kappa = \sqrt{2} \exp(-\frac{1}{2})\sqrt{2\pi\sigma_c^2}$$

We further note that the weighted correlation double derivative evaluated at zero is given by

$$\rho''(0) = -\sum_{l=1}^{L} P_l \beta_l$$

where all quantities have been previously described.

The approximated variance of the Gaussian approximation of the capacity derivative is then given by

$$\sigma_{\dot{c}}^2 \approx \kappa^2 \rho''(0) = -4\exp(-1)\pi\sigma_c^2 \sum_{l=1}^{L} P_l \beta_l$$

Using the LCR in (5.118), the AFD is simply

$$AFD_C(c) = \frac{F_C(c)}{N_C(c)} \tag{5.119}$$

Here $F_C(c)$ is a Gaussian CDF with mean and variance given above.

Table 5.1   Simulation parameters.

Parameter	Value	Parameter	Value
Number of Samples	20000	γ_0	$0dB$
F_s	$1250Hz$	v_R	$60km/hr$
Antenna Spacing	$\lambda/2$	f_C	$2.4GHz$
$\Delta\alpha_l, \forall l$	$5°$	$P_l, \forall l$	$1/L$
α_1	$0°$	α_2	$30°$
α_3	$60°$		

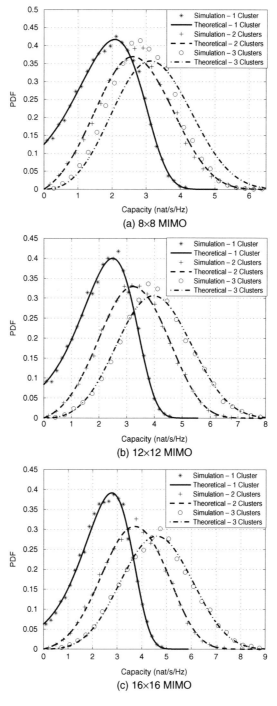

Figure 5.5   Capacity distribution.

5.7.4 Statistical simulation results

In order to evaluate the accuracy of our model and approximations, simulation analysis is performed. All default parameters are listed in Table 5.1. For brevity, only selected results are presented.

In Figure 5.5 we demonstrate the simulated and analytical distribution for various antenna configurations. We observe for the most part that the analytical distribution closely matches the simulated result. For the case of the 8×8 system with 3 clusters we observe that this is not exactly the case. In this case, the number of clusters relative to the number of antennas is increasing resulting in incomplete separability of the individual clusters. Further, for the 2 and 3 cluster case, we apply the KS-TEST with a significance level of 0.05 to verify the accuracy of the Gaussian approximation. The KS-TEST fails to reject this approximation under all scenarios, except for the case of 8×8 system with 3 clusters. The rationale behind this is as before.

In Figures 5.6(a)–5.6(c) and 5.6(d)–5.6(f) respectively we observe the LCR and AFD for the channel capacity. Trends of the LCR and AFD (exact and approximate) are as expected based on the accuracy of the expressions for the capacity distribution.

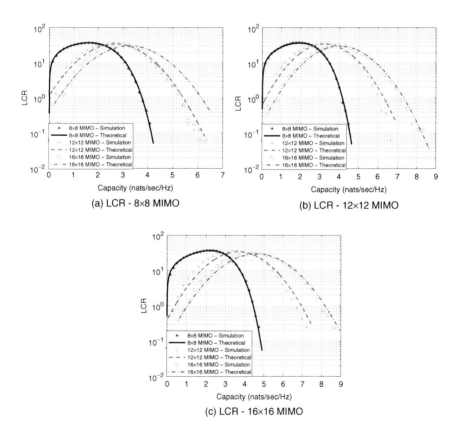

Figure 5.6   Time-varying statistics of capacity.

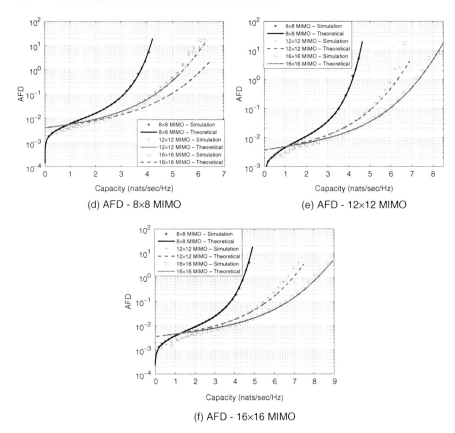

(d) AFD - 8×8 MIMO

(e) AFD - 12×12 MIMO

(f) AFD - 16×16 MIMO

Figure 5.6   (Continued)

6

Estimation and prediction of communication channels

6.1 General remarks on estimation of time-varying channels

Most of the current practical systems must rely on some sort of channel estimation (Proakis 2001). This includes estimation of the channel coefficients h_{nm} as in the case of coherent/partially coherent detection, estimation channel power $|h_{mn}|^2$, Doppler spread, velocity, level-crossing rates, or even finer structures of the channel, such as scattering clusters (Haghighi et al. 2010). The quality of estimation significantly effects the performance of systems (Hassibi and Hochwald 2003) and therefore it is important to evaluate both the estimation algorithms and their effect on system performance. This Chapter deals with issues related to this problem. It worth noticing that majority of classical algorithms rely on the Clark's model of the power spectral density (Jakes 1974). However, this is not necessarily the most common situation, especially in the MIMO settings. Therefore, special attention will be paid to the algorithms, designed for non-uniform AoA/AoD scattering environments.

6.2 Velocity estimation

6.2.1 Velocity estimation based on the covariance function approximation

One of the earliest algorithms, dealing with the velocity estimation in the non-uniform scattering environment was suggested in (Tepedelenlioğlu and

Wireless Multi-Antenna Channels: Modeling and Simulation, First Edition.
Serguei L. Primak and Valeri Kontorovich.
© 2012 John Wiley & Sons, Ltd. Published 2012 by John Wiley & Sons, Ltd.

Giannakis 2001). Here, the Authors have assumed that the AoA is distributed according to the von Mises–Tikhonov PDF

$$p(\theta) = \frac{1}{2\pi I_0(\kappa)} \exp\left[\kappa \cos(\theta - \alpha)\right], \quad \alpha \in (-\pi; \pi] \tag{6.1}$$

If the channel is further modeled as a Rician channel with the Rice factor K (Simon and Alouini 2000), the expression for the correlation function is given by

$$\frac{r_h(\tau)}{\sigma_h^2} = \frac{1}{K+1} \frac{J_0\left(\sqrt{-\kappa^2 + \omega_D^2 \tau^2 - 2j\kappa\omega_D\tau \cos\alpha}\right)}{I_0(\kappa)}$$
$$+ \frac{K}{K+1} \exp\left(-j\omega_D\tau \cos\tau\theta_0\right) \tag{6.2}$$

Here θ_0 is the direction of arrival of the LoS component and κ is the shape parameter of the distribution in Equation (6.1). The authors further capitalize on the fact mentioned earlier, that for small values of $\omega_D\tau$ all covariance functions could be approximated by a parabola. Therefore, if L point estimates $\hat{r}(lT_s)$, $l = 0, \cdots, L-1$ of the correlation functions are available, assuming $LT_s \ll 1$ (Tepedelenlioğlu and Giannakis 2001) one can find coefficients a_k of the approximating parabola $r_p(\tau) = a_0 + a_1\tau + a_2\tau^2$ by minimizing mean square error

$$\hat{\alpha}_k = \arg\min_{\alpha_k} \sum_{l=0}^{L-1} \left|\hat{r}_h(lT_s) - a_0 - a_1 l - a_2 l^2\right|^2 \tag{6.3}$$

Once coefficients α_0 and α_2 are estimated, one can immediately derive corresponding equations for the channel variance estimate $\hat{\sigma}_h^2 = \hat{\alpha}_0$ and the estimate of the second derivative of the covariance function $\hat{r}''(0) = 2\hat{a}_2/T_s^2$. Finally

$$\hat{\omega}_D = \sqrt{\frac{-2\hat{r}''(0)}{\hat{r}(0)}} \tag{6.4}$$

6.2.2   Estimation based on reflection coefficients

A potentially robust estimator based on the reflection coefficients (Kay 1993) has been suggested in (Baddour and Beaulieu 2005). Reflection coefficients[1] can be found from the estimates of the correlation function as

$$k_1 = -\frac{r(1)}{r(0)} \tag{6.5}$$

$$k_2 = -\frac{r(2) + k_1 r(1)}{r(0)\left(1 - |k_1|^2\right)} \tag{6.6}$$

$$k_3 = -\frac{r(3) + r(2)\left(k_1 + k_1 k_2\right) + r(1)k_2}{r(0)\left(1 - |k_1|^2\right)\left(1 - |k_2|^2\right)} \tag{6.7}$$

---

[1] Obtained also from Levinson–Durbin recursion (Kay 1993).

It is noted that reflection coefficients k_i have a useful asymptotic property if the Rician channel model is accepted:

$$\lim_{m \to 0} |k_m| = \cos(\pi f_D T_s) \tag{6.8}$$

where T_s is the sampling interval. Equation (6.8) gives rise to an estimator (Baddour and Beaulieu 2005)

$$\hat{v} = \frac{c}{\pi f_0} \arccos |k_m| \tag{6.9}$$

The authors of (Baddour and Beaulieu 2005) have suggested that the following estimator could be used

$$\hat{f}_D = \frac{\arccos |\hat{k}_2| + \arccos |\hat{k}_3|}{2\pi T_s} \tag{6.10}$$

In order to estimate the reflection coefficients the authors used the Burg algorithm (Kay 1993), which is more accurate for short data records.

Alternative algorithms could be found in the following references (Jingyu et al. 2005; Narasimhan and Cox 1999; Vatalaro and Forcella 1997). It is worth noting that if MIMO measurements are available, estimation of the velocity could be improved by averaging the estimation over individual links.

6.3 K-factor estimation

Estimation of the Rice K factor is an important part of the channel estimation since it has significant influence on numerous parameters of a system performance. In particular it affects link budget (Ref 1 in (Messier and Hartwell 2008)), adaptive receiver design (Ref 2 in (Messier and Hartwell 2008)), optimal loading of transmit diversity systems (Ref 3 in (Messier and Hartwell 2008)), DMT trade-off (Narasimhan 2006), and adaptive modulation and coding in MIMO (Ref 4 in (Messier and Hartwell 2008)). A great deal of effort has been spent on numerous techniques of estimation of the K factor (Ref 5-12 in (Messier and Hartwell 2008)).

6.3.1 Moment matching estimation

Since the Rice distribution

$$p(A) = \frac{2(K+1)A}{\Omega} \exp\left(-K - \frac{(K+1)A^2}{\Omega}\right) I_0\left(2\frac{K(K+1)}{\Omega}A\right) \tag{6.11}$$

has two parameters, K and Ω it is possible to use the so called moment matching technique to estimate these parameters (Abdi et al. 2001). This can be done either based on the first two moments or based on the second and the fourth moments. Based on a general expression for moments of the Rice distribution

$$m_k = \mathcal{E}\{A^k\} = \Omega^{k/2}\Gamma(k/2+1)\exp(-K) \cdot {}_1F_1(k/2+1; 1; K) \tag{6.12}$$

it is easy to see that

$$\mu = \frac{\mathcal{E}\{A\}}{\mathcal{E}\{A^2\}} = \frac{\pi}{2} \frac{\exp(-K/2)}{\sqrt{K+1}} \left[(K+1)I_0(K/2) + K I_1(K/2) \right] \tag{6.13}$$

and

$$\gamma = \frac{\text{var}\{A^2\}}{\mathcal{E}\{A^2\}^2} \frac{2K+1}{(K+1)^2} \tag{6.14}$$

It is interesting that both these quantities depend only of K; therefore K could be estimated independently on the power parameter Ω.

While it is expected that the estimator based on (6.13) has a smaller asymptotic variance as based on lower order moments,[2] evaluation of K from μ requires numerical solution of Equation (6.13) and could be computationally burdensome in practical systems. On the other hand, the estimator, built on Equation (6.14), while less accurate, allows for analytical inversion

$$K = \frac{\sqrt{1-\gamma}}{1 - \sqrt{1-\gamma}} \tag{6.15}$$

More detailed investigation of the same problem can be found in (Tepedelenlioğlu et al. 2003).

6.3.2   I/Q based methods

If both in-phase and the quadrature components $x(n) = x_I(n) + j x_Q(n)$ are available for measurements, the estimation of the K factor could be improved as shown in (Baddour and Willink 2007; Tepedelenlioğlu et al. 2003). The authors of (Tepedelenlioğlu et al. 2003) suggest looking at the following statistics of the complex received signal $x(n)$

$$X_1(N) = \frac{1}{N} \sum_{n=0}^{N-1} x(n) \exp\left(-j\hat{\omega}_0 n\right) \tag{6.16}$$

$$X_2(N) = \frac{1}{N} \sum_{n=0}^{N-1} |x(n)|^2 = \hat{m}_2 \tag{6.17}$$

where

$$\hat{\omega}_0 = \arg\max_{\omega} \left| \frac{1}{N} \sum_{n=0}^{N-1} x(n) \exp\left(-j\omega n\right) \right| \tag{6.18}$$

---

[2] This is investigated numerically in (Abdi et al. 2001). It is shown through Monte-Carlo simulations that the estimator based on (6.13) performs very close to the Cramer–Rao lower bound. At the same time degradation of quality of estimation based on (6.14) is often negligible.

The corresponding estimate of K is given by

$$\hat{K} = \frac{|X_1(N)|^3}{X_2(N) - |X_1(N)|^2} \tag{6.19}$$

Alternatively, The authors of (Baddour and Willink 2007) have suggested the following algorithm for maximum-likelihood estimation of the Rician factor K

$$\hat{K}_{ML} = \frac{\left(\sum_{n=0}^{N-1} h_{In}\right)^2 + \left(\sum_{n=0}^{N-1} h_{Qn}\right)^2}{N \sum_{n=0}^{N-1} \left(h_{In}^2 + h_{Qn}^2\right) - \left(\sum_{n=0}^{N-1} h_{In}\right)^2 - \left(\sum_{n=0}^{N-1} h_{Qn}\right)^2} \tag{6.20}$$

This estimator is biased; in order to rectify the problem a simple scaling procedure can be used (Baddour and Willink 2007)

$$\hat{K}_{MML} = \frac{1}{N}\left[(N-2)\hat{K}_{ML} - 1\right] \tag{6.21}$$

Other estimators are also available in the literature.

6.4 Estimation of four-parametric distributions

Let us assume first that N i.i.d. complex samples $h_n = h_{In} + jQ_n$, $0 \le n \le N - 1$ of a four-parametric distribution, considered in Section 2.3.1 are available for purpose of parameters estimation. In this case log-likelihood Λ of the sample $\mathbf{h} = [h_0, h_1, \cdots, h_{N-1}]$ can be calculated as

$$\Lambda = -N \ln 2\pi - N \ln \sigma_I - N \ln \sigma_Q - \sum_{n=0}^{N-1} \frac{(h_{In} - m_I)^2}{2\sigma_I^2} - \sum_{n=0}^{N-1} \frac{\left(h_{Qn} - m_Q\right)^2}{2\sigma_Q^2} \tag{6.22}$$

Direct maximum-likelihood estimation (MLE) of the parameters m_I, m_Q, σ_I and σ_Q can be accomplished by setting partial derivatives of Λ to zero (Kay 1993). In particular

$$\frac{\partial \Lambda}{\partial m_I} = \sum_{n=0}^{N-1} \frac{(h_{In} - m_I)}{\sigma_I^2} = 0, \text{ therefore } \hat{m}_I = \frac{1}{N}\sum_{n=0}^{N-1} h_{In} \tag{6.23}$$

$$\frac{\partial \Lambda}{\partial m_Q} = \sum_{n=0}^{N-1} \frac{(h_{Qn} - m_Q)}{\sigma_Q^2} = 0, \text{ therefore } \hat{m}_Q = \frac{1}{N}\sum_{n=0}^{N-1} h_{Qn} \tag{6.24}$$

$$\frac{\partial \Lambda}{\partial \sigma_I} = -\frac{N}{\sigma_I}\sum_{n=0}^{N-1} \frac{(h_{In} - m_I)^2}{\sigma_I^3} = 0, \text{ therefore } \hat{\sigma}_I^2 = \frac{1}{N}\sum_{n=0}^{N-1}(h_{In} - \hat{m}_I)^2 \tag{6.25}$$

$$\frac{\partial \Lambda}{\partial \sigma_Q} = -\frac{N}{\sigma_Q}\sum_{n=0}^{N-1} \frac{(h_{Qn} - m_Q)^2}{\sigma_Q^3} = 0, \text{ therefore } \hat{\sigma}_Q^2 = \frac{1}{N}\sum_{n=0}^{N-1}(h_{Qn} - \hat{m}_Q)^2 \tag{6.26}$$

Estimation of q^2

The MLE of the parameter q^2 of the four-parametric distribution can be evaluated as

$$\hat{K} = \hat{q}^2 = \frac{\hat{m}_I^2 + \hat{m}_Q^2}{\hat{\sigma}_I^2 + \hat{\sigma}_Q^2} = \frac{N^2 X}{N^2 Y}$$

$$= \frac{\left(\sum_{n=0}^{N-1} h_{In}\right)^2 + \left(\sum_{n=0}^{N-1} h_{Qn}\right)^2}{N \sum_{n=0}^{N-1} \left(h_{In}^2 + h_{Qn}^2\right) - \left(\sum_{n=0}^{N-1} h_{In}^2\right)^2 - \left(\sum_{n=0}^{N-1} h_{Qn}^2\right)^2} \tag{6.27}$$

Here $X = \hat{m}_I^2 + \hat{m}_Q^2$ and $Y = \hat{\sigma}_I^2 + \hat{\sigma}_Q^2$. In other words, X is a sum of squared sampled means, while Y is a sum of biased variance estimates. While statistics of estimates (6.23)–(6.26) are well known in the literature, the statistics of $\hat{q}$ are not. In the case of equal variance the problem has been studied in (Baddour and Willink 2007); however the case $\sigma_I \neq \sigma_Q$ has not been yet been considered properly.

Since $\hat{m}_I$ and $\hat{m}_Q$ are independent Gaussian variables with the mean m_I and m_Q and the variance σ_I^2/N and σ_Q^2/N respectively, the characteristic of their squared sum is given by

$$\Theta_X(\omega) = \left[\frac{1}{(1 - j2\omega N^{-1}\sigma_I^2)(1 - j2\omega N^{-1}\sigma_Q^2)}\right]^{N/2}$$

$$\times \exp\left(\frac{j\omega m_I^2}{1 - j2\omega N^{-1}\sigma_I^2}\right) \exp\left(\frac{j\omega m_Q^2}{1 - j2\omega N^{-1}\sigma_Q^2}\right) \tag{6.28}$$

At the same time, the cumulant generating functions $\Psi(\omega) = \ln \Theta(\omega)$ (Papoulis 1991; Primak et al. 2004) can also be easily expressed as

$$\Psi(\omega) = -\frac{N}{2}\left[\ln\left(1 - j2\omega N^{-1}\sigma_I^2\right) + \ln\left(1 - j2\omega N^{-1}\sigma_Q^2\right)\right]$$

$$+ \frac{j\omega m_I^2}{1 - j2\omega N^{-1}\sigma_I^2} + \frac{j\omega m_Q^2}{1 - j2\omega N^{-1}\sigma_Q^2} \tag{6.29}$$

Expanding the cumulant generating function into a power series $\Psi_X(\omega)$ with respect to ω one obtains the following expression for the first two cumulants (the mean and the variance) of X:

$$m_X = m_I^2 + m_Q^2 + \frac{\sigma_I^2 + \sigma_Q^2}{N} \tag{6.30}$$

$$\sigma_X^2 = \frac{\sigma_I^4 + \sigma_Q^4}{N^3} + 2\frac{m_I^2\sigma_I^2 + m_Q^2\sigma_Q^2}{N^2} \tag{6.31}$$

The next step is to study statistics of the denumerator Y. Estimates $\hat{\sigma}_I^2$ and $\hat{\sigma}_Q^2$ are central squared distributed with $N - 1$ degrees of freedom and have variances

σ_I^2/N and σ_Q^2/N respectively. Assuming that $N - 1 = 2m + 1$ for simplicity one can obtain the following probability density of Y (Simon 2002)

$$p_Y(z) = \frac{\sqrt{\pi} N}{2\sigma_I \sigma_Q \Gamma(m + 1/2)} \left(\frac{Nz}{2|\sigma_I^2 - \sigma_Q^2|} \right)^m \exp\left(-\frac{\sigma_I^2 + \sigma_Q^2}{4\sigma_I^2 \sigma_Q^2} \right) I_m \left(\frac{\sigma_Q^2 - \sigma_I^2}{4\sigma_I^2 \sigma_Q^2} z \right)$$

(6.32)

with corresponding moments of order k given by

$$m_k = N^{-k/2} \frac{2^{2(k+m)+1}(2m + k)!(\sigma_I \sigma_Q)^{2(m+k)+1}}{(2m)!(\sigma_I^2 + \sigma_Q^2)^{2m+k+1}}$$

$$\times {}_2F_1 \left(m + \frac{k + 1}{2}, m + 1 + \frac{k}{2}; m + 1; \left[\frac{\sigma_I^2 - \sigma_Q^2}{\sigma_I^2 + \sigma_Q^2} \right]^2 \right)$$

(6.33)

In particular, we will be interested in moments of negative order $k = -1$ and $k = -2$ which coincide with the mean and the second moment of the random variable $Z = Y^{-1}$. In the case $k = -1$ Equation (6.33) becomes

$$m_{-1} = \frac{N}{N - 2} \frac{1}{\sigma_I^2 + \sigma_Q^2} \left(\frac{2\sigma_I \sigma_Q}{\sigma_I^2 + \sigma_Q^2} \right)^{N-3}$$

$$\times {}_2F_1 \left(\frac{N}{2} - 1, \frac{N}{2} - \frac{1}{2}; \frac{N}{2}; \left[\frac{\sigma_I^2 - \sigma_Q^2}{\sigma_I^2 + \sigma_Q^2} \right]^2 \right)$$

(6.34)

As a result, the mean value of the estimate $\hat{q}^2$ can be now obtained as

$$\mathcal{E}\{\hat{q}^2\} = m_X \cdot m_{-1} = q^2 \frac{N}{N - 2} \left(\frac{2\sigma_I \sigma_Q}{\sigma_I^2 + \sigma_Q^2} \right)^{N-3}$$

$$\times {}_2F_1 \left(\frac{N}{2} - 1, \frac{N}{2} - \frac{1}{2}; \frac{N}{2}; \left[\frac{\sigma_I^2 - \sigma_Q^2}{\sigma_I^2 + \sigma_Q^2} \right]^2 \right)$$

$$+ \frac{1}{N - 2} \left(\frac{2\sigma_I \sigma_Q}{\sigma_I^2 + \sigma_Q^2} \right)^{N-3} {}_2F_1 \left(\frac{N}{2} - 1, \frac{N}{2} - \frac{1}{2}; \frac{N}{2}; \left[\frac{\sigma_I^2 - \sigma_Q^2}{\sigma_I^2 + \sigma_Q^2} \right]^2 \right)$$

(6.35)

As in (Baddour and Willink 2007) the estimator is biased. In the limit of large N it becomes unbiased. It can be seen that if $\sigma_I^2 = \sigma_Q^2 = \sigma^2$ the estimator of q^2 becomes the estimator of the Rice factor K, and

$$\mathcal{E}\{\hat{K}\} = \frac{N}{N - 2} K + \frac{1}{N - 2} = \frac{NK + 1}{N - 2}$$

(6.36)

as in (Baddour and Willink 2007).

Estimation of β

Similarly, the estimator of β can be based on the estimates of I and Q components. In this case

$$\hat{\beta} = \frac{\hat{\sigma}_Q^2}{\hat{\sigma}_I^2} = \frac{\sigma_Q^2}{\sigma_I^2} \frac{\hat{\sigma}_Q^2/\sigma_Q^2}{\hat{\sigma}_I^2/\sigma_I^2} = \beta\eta \qquad (6.37)$$

Since both $\hat{\sigma}_Q^2/\sigma_Q^2$ and $\hat{\sigma}_I^2/\sigma_I^2$ have standardized central ξ^2 distribution with $N-1$ degrees of freedom, their ratio η is described according to the Fisher–Snedekor $\mathcal{F}$-distribution (Kay 1993).

$$p_\eta(x) = \frac{1}{x} B^{-1} \left(\frac{N-1}{2}, \frac{N-1}{2} \right) \left(\frac{x}{1+x} \right)^{N/2} \left(\frac{1}{1+x} \right)^{N/2} \qquad (6.38)$$

The mean value of such an estimator is equal to (Kay 1993)

$$\mathcal{E}\{\hat{\beta}\} = \beta \frac{N}{N-2} \beta \qquad (6.39)$$

Therefore, the estimator (6.37) is biased with the bias

$$\Delta\beta = \mathcal{E}\{\hat{\beta}\} - \beta = -\frac{2}{N-2} \beta \qquad (6.40)$$

It is also asymptotically unbiased. The variance of the biased estimator can also be easily found from the variance of the $\mathcal{F}$ distribution

$$\sigma_{\hat{\beta}}^2 = \frac{2}{(N-2)(N-4)} \beta^2 \qquad (6.41)$$

Finally, the unbiased estimator of β can be suggested

$$\hat{\beta}_u = \frac{N-2}{N} \frac{\hat{\sigma}_Q^2}{\hat{\sigma}_I^2} \qquad (6.42)$$

6.5   Estimation of narrowband MIMO channels

6.5.1   Superimposed pilot estimation scheme

In this section we consider estimation of flat fading MIMO channels. Convention-ally, pilot symbols are transmitted to assist in recovering the channel matrix **H** from the received signal. There are three major ways of inserting such pilots (Coldrey and Bohlin 2007)

1. conventional (time-multiplexed) pilot (CP) scheme;

2. overlayed pilots (OP) scheme;

3. superimposed pilots (SIP) scheme.

6.5.1.1   General parameters of SIP model

Both conventional and overlayed schemes can be considered as particular cases of the superimposed scheme (Coldrey and Bohlin 2007). In the SIP mode the transmitted block of duration $T_b{}^3$ is split into two parts of duration T_t and T_d such that $T_b = T_t + T_d$. During the first part of duration T_t both pilots and data symbols are transmitted, with power σ_{dt}^2 and σ_p^2 respectively. During the second stage of duration T_d only data is transmitted with power σ_d^2. The total power transmitted over this block is $P_b = (\sigma_t^2 + \sigma_{dt}^2)T_t + \sigma_d^2 T_d$ and therefore the average power σ_s^2 is given by

$$\sigma_s^2 = \frac{(\sigma_t^2 + \sigma_{dt}^2)T_t + \sigma_d^2 T_d}{T_t + T_d} \tag{6.43}$$

If $\sigma_d^2 = 0$ one recovers the CP scheme considered in (Hassibi and Hochwald 2003) and (Biguesh and Gershman 2006). The OP scheme is obtained when $T_d = 0$. For now we concentrate on the block-fading model which presumes that over the interval T_b of a block the channel matrix $\mathbf{H}$ does not change. However, from block to block the values of the channel matrix are independent (so-called block fading model). During simultaneous training symbols and data transmission, that is $0 \le t \le T_t$, the received signal could be represented as

$$\mathbf{X}_t = \mathbf{H}\left(\sqrt{\frac{\sigma_t^2}{N_T}}\mathbf{S}_t + \sqrt{\frac{\sigma_{dt}^2}{N_T}}\mathbf{S}_{dt}\right) + \mathbf{N}_t \tag{6.44}$$

while during the data transmission stage

$$\mathbf{X}_d = \mathbf{H}\sqrt{\frac{\sigma_d^2}{N_T}}\mathbf{S}_d + \mathbf{N}_d \tag{6.45}$$

Here $\mathbf{X}_t, \mathbf{N}_t \in C^{N_R \times T_t}$, $\mathbf{X}_d, \mathbf{N}_d \in C^{N_R \times T_d}$, $\mathbf{S}_{dt} \in C^{N_T \times T_t}$ and $\mathbf{S}_d \in C^{N_T \times T_d}$.

Statistical properties of the training sequence could be defined by the training (deterministic) "correlation" matrix

$$\mathbf{R}_{tr} = \mathbf{S}_t^H \mathbf{S}_t \tag{6.46}$$

while correlation properties properties of the data are defined as

$$\mathbf{R}_{dt} = \mathcal{E}\{\mathbf{S}_{dt}^H \mathbf{S}_{dt}\}, \quad \mathbf{R}_d = \mathcal{E}\{\mathbf{S}_d^H \mathbf{S}_d\} \tag{6.47}$$

These matrices are normalized such that

$$\mathrm{Tr}\{\mathbf{R}_{tr}\} = N_T T_t \tag{6.48}$$

---

[3] The duration of block T_b is defined in (Coldrey and Bohlin 2007) in terms of coherence time. While such assignment is reasonable from theoretical point of view, practical implementation dictates different sizes of the block. In addition, variation of the channel within the coherence time interval often must be taken into account.

and

$$\text{Tr}\{\mathbf{R}_{dt}\} = N_T T_t, \ \text{Tr}\{\mathbf{R}_d\} = N_T T_d \tag{6.49}$$

6.5.1.2 Project based formulation of SIP scheme

An equivalent model of signaling using superimposed pilots was presented in (Coldrey and Bohlin 2007) based on the representation of signals of length T. In this case the received signal $\mathbf{X} = [\mathbf{X}T \ \mathbf{X}_d]$ is written as

$$\mathbf{X} = \mathbf{HS} + \mathbf{N} \tag{6.50}$$

where the matrix $\mathbf{S}$ represents total transmitted signal over the frame duration T. This signal is composed of two parts

$$\mathbf{S} = \sqrt{\frac{\sigma_t^2}{N_T}} \mathbf{S}_t + \sqrt{\frac{\sigma_{de}^2}{N_T}} \mathbf{P} \mathbf{S}_{data} \tag{6.51}$$

(Dong et al. 2004; Hassibi and Hochwald 2003; Tong et al. 2004).

6.5.2 LS estimation

In the case when no a priori information about statistical properties of the channel matrix $\mathbf{H}$ is available, one can resort to estimation based on the least squares inversion of the received signal. In this case, the transmitter sends $N \geq N_T$ pilot signals (vectors) $\mathbf{p}_1, \mathbf{p}_2, ..., \mathbf{p}_N$, or, equivalently a matrix $\mathbf{P} = [\mathbf{p}_1, \mathbf{p}_2, \cdots \mathbf{p}_N]$ of pilot (training) signals. The received signal matrix can then written as

$$\mathbf{X} = \mathbf{HP} + \mathbf{N} \tag{6.52}$$

where $\mathbf{N}$ is $N_R \times N$ matrix of i.i.d. sensor complex Gaussian noise with variance σ_n^2 per antenna link. The LS channel estimate could be obtained using the Moore–Penrose pseudoinverse $\mathbf{P}^\dagger = \mathbf{XP}^H \left(\mathbf{PP}^H\right)^{-1}$ (Golub and van Loan 1996)

$$\hat{\mathbf{H}}_{LS} = \mathbf{XP}^\dagger = \mathbf{XP}^H \left(\mathbf{PP}^H\right)^{-1} \tag{6.53}$$

In order to find the optimal choice of training matrix $\mathbf{P}$ one has to minimize the following functional

$$J_{LS}(\mathbf{P}) = \mathcal{E}\left\{||\mathbf{H} - \hat{\mathbf{H}}_{LS}||_F^2\right\} \tag{6.54}$$

subject to the transmit power constraint

$$||\mathbf{P}||_F^2 = N\mathcal{P} \tag{6.55}$$

where $\mathcal{P}$ is the total radiated power per symbol duration.

It follows from (6.53) that $\mathbf{H} - \hat{\mathbf{H}}_{LS} = \mathbf{NP}^\dagger$ and, therefore, Equation (6.54) can be recast as

$$J_{LS}(\mathbf{P}) = \mathcal{E}\left\{||\mathbf{H} - \hat{\mathbf{H}}_{LS}||_F^2\right\} = \mathcal{E}\left\{||\mathbf{NP}^\dagger||_F^2\right\}$$

$$= \sigma_n^2 N_R \mathrm{tr}\left(\mathbf{P}^{\dagger H}\mathbf{P}^\dagger\right) = \sigma_n^2 N_R \mathrm{tr}\left(\left[\mathbf{PP}^H\right]^{-1}\right) \qquad (6.56)$$

since $\mathcal{E}\{\mathbf{NN}^H\} = \sigma_n^2 \mathbf{I}_{N_R}$. Thus, in order to minimize the error of the LS estimator one has to find such a training matrix $\mathbf{P}$ which minimizes the trace of $\left(\mathbf{PP}^H\right)^{-1}$ subject to constraint (6.55)

$$\min_{\mathbf{P}} \mathrm{tr}\left(\mathbf{PP}^H\right)^{-1} \quad \text{subject to} \quad \mathrm{tr}\,\mathbf{PP}^H = N\mathcal{P} \qquad (6.57)$$

It is shown in (Biguesh and Gershman 2006) that the optimal solution to this optimization problem (6.57) must satisfy the following identity

$$\mathbf{PP}^H = \frac{N}{N_T}\mathcal{P}\mathbf{I}_{N_T} \qquad (6.58)$$

In other words, an arbitrary matrix with orthogonal rows of the same power (squared norm) $\mathcal{P}$ will produce the same optimal result. So, there is an infinite number of receivers and statistic-independent training matrices which are optimal.[4] Further selection may depend on the implementation issue. For example, if the peak power of the transmitter is limited, all elements of the training matrix have to be of the same amplitude. This can be accomplished by choosing the discrete Fourier transform (DFT) matrix (Biguesh and Gershman 2006)

$$\mathbf{P} = \sqrt{\frac{\mathcal{P}}{N_T}}\begin{bmatrix} 1 & 1 & \cdots & 1 \\ 1 & \mathcal{W}_N & \cdots & \mathcal{W}_N^{N-1} \\ 1 & \mathcal{W}_N^{N_T-1} & \cdots & \mathcal{W}_N^{(N_T-1)(N-1)} \end{bmatrix} \qquad (6.59)$$

where $\mathcal{W}_N = \exp\left(j\frac{2\pi}{N}\right)$.

Alternatively, entries of $\mathbf{P}$ could be confined to symbols of a certain modulation alphabet, for example to BPSK symbols ± 1. In this case, the optimal training

---

[4] It is important to note the following fact. If the power budget $\mathcal{P}_F = N\mathcal{P}$ per frame is fixed then the length of the training sequence does not improve the quality of estimation. Indeed, if $\mathbf{P}$ is a training sequence of length $N > N_T$. Consider SVD decomposition of $\mathbf{P} = \mathbf{U}_P [\mathbf{\Lambda}_P \mathbf{0}] \mathbf{V}^H$ where $\mathbf{U}_P$ and $\mathbf{V}_P$ are $N_T \times N_T$ and $N \times N$ unitary matrices, $\mathbf{\Lambda}_P$ is a $N_T \times N_T$ diagonal matrix with elements whose absolute values are $\sqrt{N\mathcal{P}/N_T}$ and $\mathbf{0}$ is $N_T \times N - N_T$ matrix with zero elements. It is clear that the matrix $\bar{\mathbf{P}} = \mathbf{U}_T^H \mathbf{PV} = [\mathbf{\Lambda}_P \ \mathbf{0}]$ is also an optimal training sequence since it satisfies Equation (6.58). The last $N - N_T$ elements of this new training sequence are zeros and could be dropped from transmission. Therefore, the same estimation quality can be achieved by using only $N_T \times N_T$ training sequence $\mathbf{\Lambda}_P$. However, such a training scheme uses power $\sqrt{N/N_T}\mathcal{P} \geq \mathcal{P}$ and may not always be advantageous. It is also easy to see that any unitary transformation $\mathbf{U}^\mathbf{P}$ or $\mathbf{PV}$ of an optimal training sequence $\mathbf{P}$ is still a training sequence.

matrix may not exist for all values of N_T and N (Coon and Sandell 2007; Golomb and Gong 2005).

For any training sequence, satisfying (6.58) one can easily obtain expressions for the error $\Delta\mathbf{H}$ of the estimation

$$\Delta\mathbf{H} = \frac{N_R}{N\mathcal{P}}\mathbf{XP}^H - \mathbf{H} = \frac{N_R}{N\mathcal{P}}\mathbf{NP}^H \tag{6.60}$$

and the MSE value

$$J_{LS,\min} = \sigma_n^2 N_R \mathrm{tr}\left(\left[\mathbf{PP}^H\right]^{-1}\right) = \frac{\sigma_n^2}{\mathcal{P}}N_T N_R \frac{N_T}{N} = N_T N_R \frac{\sigma_n^2}{\mathcal{P}\beta} \tag{6.61}$$

The term $N_T N_R$ in Equation (6.61) represents the complexity of the estimation problem with respect to SISO with the same transmit power in terms of the number of coefficients to be estimated. The shape factor $\beta = N/N_T$ represents gain in SNR through repeated use of training symbols. In contrast to SISO, this coefficient does not need to be an integer.

The next step is to investigate the statistical properties of the error matrix $\Delta\mathbf{H}$ defined by Equation (6.60). We will focus on the vectorized version $\Delta\mathbf{h} = \mathrm{vec}(\Delta\mathbf{H})$ (van Trees 2002). Using the well known vector identity (van Trees 2002) and notation $\mathbf{n} = \mathrm{vec}(\mathbf{N})$ one obtains

$$\Delta\mathbf{h} = \mathrm{vec}(\Delta\mathbf{H}) = \mathrm{vec}\left(\frac{N_T}{N\mathcal{P}}\mathbf{NP}^H\right) = \frac{N_T}{N\mathcal{P}}\mathbf{P}^* \otimes \mathbf{I}\,\mathbf{n} \tag{6.62}$$

Since $\mathbf{n}$ is a vector of i.i.d. zero-mean Gaussian complex processes with the variance σ_n^2, the vector $\Delta\mathbf{h}$ is also a zero-mean Gaussian vector. Its covariance matrix $\mathbf{R}_{\Delta\mathbf{h}}$ can be calculated as follows

$$\mathbf{R}_{\Delta\mathbf{h}} = \mathcal{E}\left\{\Delta\mathbf{h}\Delta\mathbf{h}^H\right\} = \frac{N_T^2}{N^2\mathcal{P}}\mathbf{P}^* \otimes \mathbf{I}_{N_R}\,\mathcal{E}\left\{\mathbf{nn}^H\right\}\mathbf{P}^T \otimes \mathbf{I}_{N_R}$$

$$= \frac{N_T^2}{N^2\mathcal{P}^2}\sigma_n^2\,\mathbf{P}^* \otimes \mathbf{I}_{N_R}\mathbf{P}^T \otimes \mathbf{I}_{N_R} = \frac{N_T^2}{N^2\mathcal{P}^2}\sigma_n^2\,\mathbf{PP}^{H*} \otimes \mathbf{I}_{N_R} = \frac{N_T}{N\mathcal{P}}\mathbf{I}_{N_R N_T} \tag{6.63}$$

Thus the elements of the error matrix are also i.i.d zero-mean complex Gaussian random variables with variance $N_T/N\mathcal{P}$.

Inspection of Equation (6.62) shows that the error of estimation is independent of the channel value $\mathbf{H}$ (or $\mathbf{h}$). However, as a result, the estimated value of the channel $\hat{\mathbf{h}} = \mathrm{vec}(\hat{\mathbf{H}})$ is correlated with the error $\Delta\mathbf{h}$. Indeed

$$\mathcal{E}\left\{\hat{\mathbf{h}}\Delta\mathbf{h}^H\right\} = \mathcal{E}\left\{\Delta\mathbf{h}\Delta\mathbf{h}^H\right\} = \mathbf{R}_{\Delta\mathbf{h}} = \frac{N_T}{N\mathcal{P}}\mathbf{I}_{N_R N_T} \tag{6.64}$$

6.5.3 Scaled least-square (SLS) estimation

The LS estimate (6.53) could be improved in terms of MSE by scaling:

$$\hat{\mathbf{H}}_{SLS} = k_s\hat{\mathbf{H}}_{LS} = k_s\mathbf{XP}^\dagger = k_s\mathbf{XP}^H\left(\mathbf{PP}^H\right)^{-1} \tag{6.65}$$

where the scaling constant k_s has to be chosen to minimize MSE

$$J_{SLS} = \mathcal{E}\left\{\|\mathbf{H} - k_s\hat{\mathbf{H}}_{LS}\|_F^2\right\} = \text{tr}\left[\mathcal{E}\left\{\left(\mathbf{H} - k_s\hat{\mathbf{H}}_{LS}\right)^H\left(\mathbf{H} - k_s\hat{\mathbf{H}}_{LS}\right)\right\}\right]$$

$$= (1 - k_s)^2\,\text{tr}\,\mathbf{R}_T + k_s^2\sigma_n^2 N_R\,\text{tr}\right]\left(\mathbf{PP}^H\right)^{-1}\right] = (1 - k_s)^2\,\text{tr}\,\mathbf{R}_T + k_s^2 J_{LS}(\mathbf{P}) \tag{6.66}$$

Here $\mathbf{R}_T = \mathcal{E}\left\{\mathbf{H}^H\mathbf{H}\right\}$ is the transmit correlation matrix[5] and $\hat{\mathbf{H}}_{LS}$ is given by Equation (6.53). Minimization of (6.66) over k_s produces the optimal value of k_{s0}:

$$k_{s0} = \frac{\text{tr}\,\mathbf{R}_T}{\text{tr}\,\mathbf{R}_T + J_{LS}(\mathbf{P})} \tag{6.67}$$

The corresponding value of J_{SLS} can now be easily evaluated by using k_{s0} in Equation (6.66):

$$J_{SLS}(\mathbf{P}) = \frac{J_{LS}(\mathbf{P})\,\text{tr}\,\mathbf{R}_T}{\text{tr}\,\mathbf{R}_T + J_{LS}(\mathbf{P})} \leq J_{LS}(\mathbf{P}) \tag{6.68}$$

and

$$\hat{\mathbf{H}}_{SLS} = \frac{\text{tr}\,\mathbf{R}_T}{\text{tr}\,\mathbf{R}_T + \sigma_n^2 N_R\,\text{tr}\left[\left(\mathbf{PP}^H\right)^{-1}\right]}\mathbf{XP}^{\dagger} \tag{6.69}$$

It follows from Equation (6.68) that the minimum MSE of the SLS estimator, given that $\text{tr}\,\mathbf{R}_T$ and σ_n^2 are known, is achieved by minimizing $J_{LS}(\mathbf{P})$. In other words, the optimal choice of the training matrix is the same as for LS estimation and is provided by Equation (6.58). The minimal MSE of the SLS estimator is then just

$$J_{SLS,min} = \frac{N_T N_R \sigma_n^2\,\text{tr}\,(\mathbf{R}_T)}{\mathcal{P}\beta\,\text{tr}(\mathbf{R}_T) + N_T N_R \sigma_n^2} \tag{6.70}$$

The homogeneous transmit side scenario Equation (6.70) could be further simplified to produce

$$J_{SLS,min} = \frac{N_T N_R \sigma_n^2}{\mathcal{P}\beta + \sigma_n^2} = N_T N_R\frac{1}{1 + \beta\mathcal{P}/\sigma_n^2} \tag{6.71}$$

If the trace of the transmit correlation matrix $\text{tr}\,\mathbf{R}_T$ is not known a priori it could be estimated, for example as an LS-based consistent sample estimate (Biguesh and Gershman 2006)

$$\text{tr}\hat{\mathbf{R}}_T = \text{tr}\,\hat{\mathbf{H}}_{LS}^H\hat{\mathbf{H}}_{LS} = \text{tr}\,\mathbf{P}^{\dagger H}\mathbf{X}^H\mathbf{XP}^{\dagger} \tag{6.72}$$

---

[5] It is important to note that only the trace of $\mathbf{R}_T$ enters the calculations below. Estimation of this parameter is much more reliable than estimation of the complete correlation matrix. If a MIMO system uses the same antennas and polarization on the transmit side it is easy to see that $\text{tr}\,\mathbf{R}_T = N_T N_R$. We refer to this scenario as *a homogeneous transmit side* scenario.

This value could be further used in the SLS estimator described above and is known as the LS–SLS estimator (Biguesh and Gershman 2006).

It can easily be seen through inspection of Equation (6.68) that SLS and LS perform similarly for high SNR and differ in the region of low SNR, that is if

$$\operatorname{tr} \mathbf{R}_T \ll J_{LS}(\mathbf{P}) = N_T N_R \frac{\sigma_n^2}{\beta \mathcal{P}} \tag{6.73}$$

For the homogeneous transmit side it simplifies to

$$\frac{\beta \mathcal{P}}{\sigma_n^2} \ll 1 \tag{6.74}$$

6.5.4 Minimum MSE

Both LS and SLS estimation schemes are suboptimal since they do not explore full knowledge of the channel statistics. Following (Biguesh and Gershman 2006) and (Kay 1993) our goal is to find the best linear estimator

$$\hat{\mathbf{H}}_{MMSE} = \mathbf{X} \mathbf{W}_0 \tag{6.75}$$

where matrix $\mathbf{W}_0$ minimizes the MSE ε of the prediction

$$
\begin{aligned}
\varepsilon &= \mathcal{E} \|\mathbf{H} - \mathbf{X}\mathbf{W}\|_F^2 = \operatorname{tr} \mathcal{E} \left\{ (\mathbf{H} - \mathbf{X}\mathbf{W})^H (\mathbf{H} - \mathbf{X}\mathbf{W}) \right\} \\
&= \operatorname{tr} \mathbf{R}_T - \operatorname{tr} \mathbf{R}_T \mathbf{P} \mathbf{W} - \operatorname{tr} \mathbf{W}^H \mathbf{P}^H \mathbf{R}_H + \operatorname{tr} \mathbf{W}^H \left(\mathbf{P}^H \mathbf{R} T \mathbf{P} + \sigma_n^2 N_R \mathbf{I}_L \right) \mathbf{W} \quad (6.76)
\end{aligned}
$$

Using rules of matrix calculus (equations (A378)–(A389) in (van Trees 2002)) the optimal matrix $\mathbf{W}_0$ can be found as the solution of $\partial \varepsilon / \partial \mathbf{W} = 0$ which is shown to be (Biguesh and Gershman 2006)

$$\mathbf{W}_0 = \left(\mathbf{P}^H \mathbf{R}_T \mathbf{P} + \sigma_n^2 N_R \mathbf{I}_L \right)^{-1} \mathbf{P}^H \mathbf{R}_T \tag{6.77}$$

Therefore, the MMSE estimator of the MIMO channel is given by

$$\hat{\mathbf{H}}_{MMSE} = \mathbf{X} \left(\mathbf{P}^H \mathbf{R}_T \mathbf{P} + \sigma_n^2 N_R \mathbf{I}_L \right)^{-1} \mathbf{P}^H \mathbf{R}_T \tag{6.78}$$

Performance of such an estimator is measured by the trace $J_{MMSE} = \operatorname{tr} \mathbf{R}_E$ of the covariance matrix

$$\mathbf{R}_E = \mathcal{E} \{ \mathbf{E} \mathbf{E}^H \} = \left(\mathbf{R}_T^{-1} + \frac{1}{\sigma_n^2 N_R} \mathbf{P} \mathbf{P}^H \right)^{-1} \tag{6.79}$$

of error matrix $\mathbf{E} = \mathbf{H} - \hat{\mathbf{H}}_{MMSE}$. Therefore

$$J_{MMSE} = \operatorname{tr} \left(\mathbf{R}_T^{-1} + \frac{1}{\sigma_n^2 N_R} \mathbf{P} \mathbf{P}^H \right)^{-1} \tag{6.80}$$

Let the covarince matrix on the receive side be represented by its eigendecomposition

$$\mathbf{R}_T = \mathbf{U}_T \mathbf{\Lambda}_T \mathbf{U}_T^H \tag{6.81}$$

where $\mathbf{U}$ is a unitary eigenvector matrix and $\mathbf{\Lambda}_T$ is a diagonal matrix with positive entries, reflecting the fact that the covariance matrix is a positive definite matrix. In this case, the expression for the MSE can be somewhat simplified to produce

$$J_{MMSE} = \operatorname{tr}\left(\mathbf{U}_T\mathbf{\Lambda}_T^{-1}\mathbf{U}_T^H + \frac{1}{\sigma_n^2 N_R}\mathbf{P}\mathbf{P}^H\right)^{-1}$$

$$= \operatorname{tr}\left(\mathbf{\Lambda}_T^{-1} + \frac{1}{\sigma_n^2 N_R}\mathbf{U}_T^H\mathbf{P}\mathbf{P}^H\mathbf{U}_T\right)^{-1} = \operatorname{tr}\left(\mathbf{\Lambda}_T^{-1} + \tilde{\mathbf{P}}\tilde{\mathbf{P}}^H\right)^{-1} \quad (6.82)$$

where

$$\tilde{\mathbf{P}} = \frac{1}{\sqrt{\sigma_n^2 N_R}}\mathbf{Q}^H\mathbf{P} \quad (6.83)$$

The new power constraint for training symbols $\tilde{\mathbf{P}}$ can be easily obtained from the constraint (6.55)

$$\operatorname{tr}\tilde{\mathbf{P}}\tilde{\mathbf{P}}^H = \frac{1}{\sigma_n^2 N_R}\operatorname{tr}\mathbf{Q}^H\mathbf{P}\mathbf{P}^H\mathbf{Q} = \frac{N\mathcal{P}}{\sigma_n^2 N_R} \quad (6.84)$$

In order to minimize J_{MMSE} given by (6.82) and the constraint given by (6.84) one can resort to the following property of the trace of the positive definite matrices (Biguesh and Gershman 2006; Kay 1993)

Lemma 6.5.1 *For a positive definite matrix* $\mathbf{A}$ *of size* $L \times L$, *the following inequality holds:*

$$\operatorname{tr}\mathbf{A}^{-1} \geq \sum_{l=1}^{L}\frac{1}{a_{l,l}} \quad (6.85)$$

where $a_{l,l}$ *is the l-th diagonal element of the matrix* $\mathbf{A}$. *The equality holds if the matrix* $\mathbf{A}$ *is diagonal.*

Based on this property, the optimal traning sequence $\tilde{\mathbf{P}}$ must be such that $\tilde{\mathbf{P}}\tilde{\mathbf{P}}^H$ is also a diagonal matrix with non-negative elements:

$$\tilde{\mathbf{P}}\tilde{\mathbf{P}}^H = \operatorname{diag}\left(|\tilde{p}_1|^2, |\tilde{p}_2|^2 \cdots, |\tilde{p}_{N_T}|^2\right) \quad (6.86)$$

Minimization of J_{MMSE} with respect to the training sequence $\tilde{\mathbf{P}}$ and the constraint (6.84) can now be accomplished using the Lagrange multiplier method, that is by minimizing the following functional

$$L(\tilde{\mathbf{P}}, \mu) = J_{MMSE}(\tilde{\mathbf{P}}) + \mu\left[\operatorname{tr}\tilde{\mathbf{P}}\tilde{\mathbf{P}}^H - \frac{N\mathcal{P}}{\sigma_n^2 N_R}\right]$$

$$= \operatorname{tr}\left(\mathbf{\Lambda}_T^{-1} + \tilde{\mathbf{P}}\tilde{\mathbf{P}}^H\right)^{-1} + \mu\left[\operatorname{tr}\tilde{\mathbf{P}}\tilde{\mathbf{P}}^H - \frac{N\mathcal{P}}{\sigma_n^2 N_R}\right]$$

$$= \sum_{n_t=1}^{N_T}\left(\lambda_{n_t}^{-1} + |\tilde{p}_{n_t}|^2\right)^{-1} + \mu\sum_{n_t=1}^{N_T}\left(|\tilde{p}_{n_t}|^2 - \frac{N\mathcal{P}}{\sigma_n^2 N_R}\right) \quad (6.87)$$

Taking a derivative of $L(\tilde{\mathbf{P}}, \mu)$ with respect to $|\tilde{p}_{n_t}|^2$, for $n_t = 1, \cdots, N_T$ and setting it to zero one can obtain a system of equations

$$\left(\lambda_{n_t}^{-1} + |\tilde{p}_{n_t}|^2\right)^{-2} - \mu = 0, \quad n_t = 1, \cdots, N_T \tag{6.88}$$

which has the following water-filling type solution

$$\tilde{p}_{n_t} = \begin{cases} \sqrt{\kappa - \lambda_{n_t}^{-1}} & \text{if } \kappa = \frac{1}{\sqrt{\mu}}, \ \lambda_{n_t} > \mu \\ 0 & \text{otherwise} \end{cases} \tag{6.89}$$

The value of the parameter μ has to be adjusted such that the power constraint (6.84) is satisfied. If $N = N_T$ then the optimal training matrix $\tilde{\mathbf{P}}$ can be written as

$$\tilde{\mathbf{P}} = \left(\left[\kappa \mathbf{I}_{N_T} - \mathbf{\Lambda}^{-1}\right]^+\right)^{1/2} \tag{6.90}$$

where x^+ is defined as

$$x^+ = \begin{cases} x & \text{if } x \geq 0 \\ 0 & \text{otherwise} \end{cases} \tag{6.91}$$

Finally, the actual training sequence $\mathbf{P}$ transmitted is given by

$$\mathbf{P} = \sqrt{\sigma_n^2 N_R}\tilde{\mathbf{P}} = \sqrt{\sigma_n^2 N_R} \left(\left[\kappa \mathbf{I}_{N_T} - \mathbf{\Lambda}^{-1}\right]^+\right)^{1/2} \tag{6.92}$$

In the case of $N \geq N_T$ the training sequence $\mathbf{P}$ can be obtained from one given by Equation (6.90) by augmenting it with $N_T \times N - N_T$ zero matriz and post-multiplying by an arbitrary unitary matrix $\mathbf{U}$, that is

$$\mathbf{P} = \sqrt{\sigma_n^2 N_R}\tilde{\mathbf{P}} = \sqrt{\sigma_n^2 N_R} \left[\left(\left[\kappa \mathbf{I}_{N_T} - \mathbf{\Lambda}^{-1}\right]^+\right)^{1/2}, \ \mathbf{0}_{N_T.N-N_T}\right]\mathbf{U} \tag{6.93}$$

6.5.5   Relaxed MMSE estimators

The considerations of the previous section imply that the value of the channel transmit correlation matrix $\mathbf{R}_T$ is known exactly. Since this is hardly ever the case in practical applications, either no a priori knowledge of the correlation matrix can be assumed or such information is extracted from some sort of measurement. One way to relax the requirement of knowledge of the complete covariance matrix is to assume $\mathbf{R}_T = \alpha \mathbf{I}_{N_T}$ Equation (6.78) where a positive parameter α must be chosen to optimize the MSE of the estimator

$$\hat{\mathbf{H}} = \alpha \mathbf{X} \left(\alpha \mathbf{P}^H \mathbf{P} + \sigma_n^2 N_R \mathbf{I}_{N_T}\right) \mathbf{P}^H = \frac{\alpha}{\sigma_n^2 N_R} \mathbf{X} \left(\mathbf{I}_{N_T} - \frac{\alpha N_T}{\alpha N \mathcal{P} + \sigma_n^2 N_R N_T} \mathbf{P}^H \mathbf{P}\right) \mathbf{P}^H \tag{6.94}$$

Here the Woodbury inversion formula (van Trees 2002) has been used. Using (6.94) and assuming orthogonal training (6.58), the channel MSE could be computed as (Biguesh and Gershman 2006)

$$J_{RMMSE}(\alpha) = \mathcal{E}||\mathbf{H} - \hat{\mathbf{H}}||_F^2 = \frac{z^2\sigma_n^2 N_R N_T}{z^2 N\mathcal{P} + \sigma_n^2 N_T}$$

$$+ \left(\frac{N_R z N^2 \mathcal{P}^2}{N_T}\right) \left(\frac{\alpha N_T}{\alpha N\mathcal{P} + \sigma_N^2 N_R N_T} - \frac{z^2 N_T}{z^2 N\mathcal{P} + \sigma_n^2 N_T}\right)^2 \tag{6.95}$$

where

$$z^2 = \frac{\text{tr}\,\mathbf{R}_T}{N_T}$$

The minimum of J_{RMMSE} is achieved when the second bracket is equal to zero, that is if

$$\alpha_{\text{opt}} = z^2 N_R = \frac{\text{tr}\,\mathbf{R}_T}{N_T} \tag{6.96}$$

Accidentally, (6.96) provides the value of α which minimizes the approximation of $\mathbf{R}_T$ by $\alpha\mathbf{I}_T$ with respect to the Frobenius norm.

Taking into account the value (6.96) in Equation (6.94) the new estimator can be written as

$$\hat{\mathbf{H}_{RMMSE}}\mathbf{X}\left[\mathbf{P}^H\mathbf{P} + \frac{\sigma_n^2 M_R N_T}{\text{tr}\,\mathbf{R}_T}\mathbf{I}_{N_T}\right]^{-1}\mathbf{P}^H \tag{6.97}$$

assuming that $\text{tr}\,\mathbf{R}_T$ is known or estimated. At this point the training matrix $\mathbf{P}$ is still an arbitrary one. However, it reduces to the SLS estimator (6.69) in the orthogonal training (6.58).

Finally, it follows from (6.95) that the estimation error is

$$J_{RMMSE}(\alpha_{\text{opt}}) = \frac{\text{tr}\,\mathbf{R}_T\sigma_n^2 N_R N_T^2}{\text{tr}\,\mathbf{R}_T N\mathcal{P} + \sigma_n^2 N_R N_T^2} = \frac{\text{tr}\,\mathbf{R}_T N\mathcal{P}}{\text{tr}\,\mathbf{R}_T N\mathcal{P} + \sigma_n^2 N_R N_T^2}J_{LS} \leq J_{LS} \tag{6.98}$$

In other words, the RMMSE estimator produces better results in term of MSE error than the LS estimator, especially for low SNR when $\text{tr}\,\mathbf{R}_T N\mathcal{P} \leq \sigma_n^2 N_R N_T^2$.

The RMMSE estimator still requires knowledge of $\text{tr}\,\mathbf{R}_T$. If it is estimated from the training sequence, as in the case of LS-SLS, one can obtain the following LS-RMMSE algorithm (Biguesh and Gershman 2006) which relies only on the received data and knowledge of the orthogonal training sequence

$$\hat{\mathbf{H}}_{LS-RMMSE} = \frac{N_T\text{tr}\,\mathbf{XX}^H}{N\mathcal{P}\left[\text{tr}\,\mathbf{XX}^H + \sigma_n^2 N_R N_T\right]}\mathbf{XP}^H \tag{6.99}$$

6.6   Using frames for channel state estimation

Accurate and sparse representation of a moderately fast fading channel using bases functions is achievable when both channel and bases bands align. If a mismatch exists, usually a larger number of bases functions is needed to achieve the same accuracy. In this section, we propose a novel approach for channel estimation based on frames, which preserves sparsity and improves estimation accuracy. Members of the frame are formed by modulating and varying the bandwidth of discrete prolate spheroidal sequences (DPSS) in order to reflect various scattering scenarios. To achieve the sparsity of the proposed representation, a matching pursuit approach is employed. The estimation accuracy of the scheme is evaluated and compared with the accuracy of a Slepian basis expansion estimator based on DPSS for a variety of mobile channel parameters. The results clearly indicate that for the same number of atoms, a significantly higher estimation accuracy is achievable with the proposed scheme when compared to the DPSS estimator.

Estimation and interpolation of a moderately fast fading Rayleigh/Rice channel is an important problem in modern communications. If the channel characteristics are known, that is the channel autocorrelation function, then an approach based on the Wiener filter provides the optimum solution (van Trees 2001). However, such an ideal case is rare in real-life applications, and we require a more universal approach. In general, basis expansions are used in such situations and several different basis functions including Fourier bases and discrete prolate spheroidal sequences (DPSS) have been adopted for such problems (Percival and Walden 1993; Zemen and Mecklenbrauker 2005). Previous studies have found that accurate and sparse representations are usually obtained when both the bases and the channel under investigation occupy the same band (Zemen and Mecklenbrauker 2005). However, when the bandwidth of the basis function is mismatched and larger than that of the signal, a larger number of bases is required to approximate the channel with the same accuracy. To resolve this particular problem, it was suggested to use a bank of bases with different bandwidths (Zemen and Mecklenbrauker 2005). However, such a representation again ignores the fact that in some cases the band occupied by the channel is not necessarily centered around DC, but rather at some frequency different from zero. Hence, a larger number of bases is again needed for accurate and sparse representation.

A need clearly exists for some type of overcomplete, redundant bases which accounts for a variety of scenarios. Therefore, in this section we propose an overcomplete set of bases called modulated discrete prolate spheroidal sequences (MDPSS). Such a set of bases is also known as a frame (Kovacevic and Chebira 2007; Mallat 1999). The bases within the frame are obtained by modulation and variation of the bandwidth of DPSSs in such a way as to reflect various scattering scenarios. During construction of multiple bases it is assumed that at least an upper bound of the maximum Doppler frequency is known. Furthermore, in order to obtain a sparse representation of the channel using the MDPSS frame, a matching pursuit approach is employed (Mallat 1999). The proposed scheme is tested using the channel model presented in Section 6.6.1 for various scattering

scenarios. The results demonstrate that the MDPSS frame provides superior estimation accuracy compared to a Slepian basis expansion DPSS approach (Zemen
and Mecklenbrauker 2005).

6.6.1 Properties of the spectrum of a mobile channel

The covariance function, $\rho(\tau)$, of a SISO frequency flat mobile communication
channel and its associated power spectral density $S(\omega)$ are related to the distribution
of angle of arrival (AoA) $p(\theta)$ as (Jakes 1974)

$$\rho(\tau) = P \int_{-\pi}^{\pi} \exp\left(j2\pi f_D \tau \cos\theta\right) p(\theta) d\theta, \tag{6.100}$$

$$S(f) = \frac{P}{\sqrt{1 - (f/f_D)^2}} p\left(\arccos\frac{f}{f_D}\right)$$

$$+ \frac{P}{\sqrt{1 - (f/f_D)^2}} p\left(\pi - \arccos\frac{f}{f_D}\right). \tag{6.101}$$

Here, P is the total power of the received signal, $f_D = f_0 v/c$ is the maximum
Doppler shift of the carrier frequency f_0 corresponding to the velocity v of the
mobile; c is the speed of light. Jakes spectrum,

$$S(f) = \frac{P f_D}{\pi \sqrt{1 - (f/f_D)^2}}, \tag{6.102}$$

which is a widely used model by communication researchers, corresponds to
the uniform distribution $p(\theta) = 1/2\pi$ of the AoA (Jakes 1974). This spectrum
closely resembles a uniform spectrum for most of the frequencies, and therefore its
Karhunen–Loeve basis (van Trees 2001) is close to the one defined by sinc type
covariance function. Thus, it seems natural that representation in terms of DPSS
suggested in (Zemen and Mecklenbrauker 2005) produces very good results.

   However, real-world measurements reveal that the AoA can deviate significantly from that of a uniform distribution on $[-\pi : \pi]$. Often, the received signal
is represented well as a sum of signals arriving from a narrow band of angles,
corresponding to individual clusters (Salz and Winters 1994). In this case, the PSD
of the received signal could be well approximated by a piece-constant distribution
of power in frequency domain. Conversely, the covariance function could be well
represented by a sum of sinc type functions in the time domain (Salz and Winters
1994). In addition, in some radio environments such as dense urban environments
(Molisch et al. 2006), the received signal could be dominated by multiple specular
components, therefore creating a mixed spectrum of the received signal. In these
conditions, the use of simple DPSS may not be optimal, since proper expansions
will require a significant number of higher order DPSS with small eigenvalues
which, in turn will result in an ill-posed problem. Therefore, modifications to
Zemen's approach in (Zemen and Mecklenbrauker 2005) are required.

6.6.2   Frames based on DPSS

Throughout this section and the rest of the chapter, it is assumed that only N discrete samples of the channel are available and that they were obtained with a sampling period T. Hence, the discrete frequency, ν, represents a continuous frequency, f, normalized with sampling period, $\nu = fT$.

6.6.3   Discrete prolate spheroidal sequences

Given N, the kth DPSS, $v_k(n, N, W)$, for $k = 0, 1, ..., N - 1$ is defined as the real solution to the system of equations (Slepian 1978):

$$\sum_{m=0}^{N-1} \frac{\sin[2\pi W(n - m)]}{\pi(n - m)} v_k(m, N, W) = \lambda_k(N, W) v_k(n, N, W) \qquad (6.103)$$

with $\lambda_k(N, W)$ being the ordered non-zero eigenvalues of (6.103)

$$\lambda_0(N, W) > \lambda_1(N, W), ..., \lambda_{N-1}(N, W) > 0. \qquad (6.104)$$

The first $2NW$ eigenvalues are very close to 1 while the rest rapidly decay to zero (Slepian 1978). Interestingly enough, it has been observed that these quantities are also the eigenvalues of an $N \times N$ matrix $C(m, n)$ (Slepian 1978), where the elements of such a matrix are

$$C(m, n) = \frac{\sin[2\pi W(n - m)]}{\pi(n - m)} \quad m, n = 0, 1, ..., N - 1, \qquad (6.105)$$

and the vector obtained by time-limiting the DPSS, $v_k(n, N, W)$, is an eigenvector of $C(m, n)$. The DPSS are doubly orthogonal, that is, they are orthogonal on the infinite set $\{-\infty, ..., \infty\}$ and orthonormal on the finite set $\{0, 1, ..., N - 1\}$, that is,

$$\sum_{-\infty}^{\infty} v_i(n, N, W) v_j(n, N, W) = \lambda_i \delta_{ij} \qquad (6.106)$$

$$\sum_{n=0}^{N-1} v_i(n, N, W) v_j(n, N, W) = \delta_{ij}, \qquad (6.107)$$

where $i, j = 0, 1, ..., N - 1$.

6.6.3.1   Modulated discrete prolate spheroidal sequences

If the DPSS are used for channel estimation, then usually accurate and sparse representations are obtained when both the DPSS and the channel under investigation occupy the same frequency band (Zemen and Mecklenbrauker 2005). However, problems arise when the channel is centered around some frequency $|\nu_o| > 0$ and the occupied bandwidth is smaller than $2W$, as shown in Figure 6.1.

In such situations, a larger number of DPSS is required to approximate the channel with the same accuracy despite the fact that such narrowband channel is more predictable than a wider band channel (Proakis 2001). In order to find a

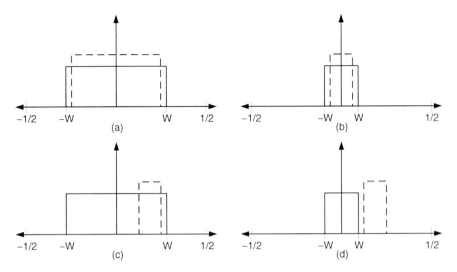

Figure 6.1   Comparison of the bandwidth for a DPSS (solid line) and a channel (dashed line): (a) both have a wide bandwidth; (b) both have narrow bandwidth; (c) a DPSS has a wide bandwidth, while the channel's bandwidth is narrow and centered around $v_0 > 0$; (d) both have narrow bandwidth, but centered at different frequencies.

better basis we consider so-called modulated discrete prolate spheroidal sequences (MDPSS), defined as

$$M_k(N, W, \omega_m; n) = \exp(j\omega_m n)v_k(N, W; n), \qquad (6.108)$$

where $\omega_m = 2\pi v_m$ is the modulating frequency. It is easy to see that MDPSS are also doubly orthogonal, obey the same Equation (6.103) and are bandlimited to the frequency band $[-W + v : W + v]$.

The next question that needs to be answered is how to properly choose the modulation frequency v. In the simplest case when the spectrum $S(v)$ of the channel is confined to a known band $[v_1; v_2]$, that is

$$S(v) = \begin{cases} \gg 0 \;\; \forall v \in [v_1, v_2] \text{ and } |v_1| < |v_2| \\ \approx 0 \qquad\qquad \text{elsewhere} \end{cases}, \qquad (6.109)$$

the modulating frequency, v_m, and the bandwidth of the DPSSs are naturally defined by

$$v_m = \frac{v_1 + v_2}{2} \qquad (6.110)$$

$$W = \left| \frac{v_2 - v_1}{2} \right|, \qquad (6.111)$$

as long as both satisfy:

$$|v_m| + W < \frac{1}{2}. \tag{6.112}$$

In practical applications the exact frequency band is known only with a certain degree of accuracy. In addition, especially in mobile applications, the channel is evolving in time. Therefore, only some relatively wide frequency band defined by the velocity of the mobile and the carrier frequency is expected to be known. In such situations, a one-band-fits-all approach may not produce a sparse and accurate approximation of the channel. To resolve this problem, it was previously suggested to use a band of bases with different widths to account for different speeds of the mobile (Zemen and Mecklenbrauker 2005). However, such a representation once again ignores the fact that the actual channel bandwidth $2W$ could be much less than $2v_D$ dictated by the maximum normalized Doppler frequency $v_D = f_D T$.

To improve the estimator robustness, we suggest the use of multiple bases, better known as frames (Kovacevic and Chebira 2007), precomputed in such a way as to reflect various scattering scenarios. In order to construct such multiple bases, we assume that a certain estimate (or rather its upper bound) of the maximum Doppler frequency v_D is available. The first few bases in the frame are obtained using traditional DPSS with bandwidth $2v_D$. Additional bases can be constructed by partitioning the band $[-v_D; v_D]$ into K sub-bands with the boundaries of each sub-band given by $[v_k; v_{k+1}]$, where $0 \leq k \leq K - 1$, $v_{k+1} > v_k$, and $v_0 = -v_D$, $v_{K-1} = v_D$. Hence, each set of MDPSS has a bandwidth equal to $v_{k+1} - v_k$ and a modulation frequency equal to $v_m = 0.5(v_k + v_{k+1})$. Obviously, a set of such functions again forms a basis of functions limited to the bandwidth $[-v_D; v_D]$. It is a convention in the signal processing community to call each basis function *an atom*. While a particular partition is arbitrary for every level $K \geq 1$, we can chose to partition the bandwidth in equal blocks to reduce the amount of stored precomputed DPSS, or to partition according to the angular resolution of the receive antenna, and so on, as shown in Figure 6.2.

Representation in the overcomplete basis can be made sparse due to the richness of such a basis. Since the expansion into simple bases is not unique, a fast,

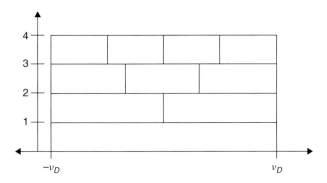

Figure 6.2   Sample partition of the bandwidth for $K = 4$.

convenient and unique projection algorithm cannot be used. Fortunately, efficient algorithms, known generically as pursuits (Mallat 1999) can be used and they are briefly described in the next section.

6.6.3.2 Matching pursuit with MDPSS frames

From the few approaches that can be applied for expansion in overcomplete bases, we choose the so-called matching pursuit (Mallat 1999). The main feature of the algorithm is that when stopped after a few steps, it yields an approximation using only a few atoms (Mallat 1999). The matching pursuit was originally introduced in the signal processing community as an algorithm that decomposes any signal into a linear expansion of waveforms that are selected from a redundant dictionary of functions (Mallat 1999). It is a general, greedy, sparse function approximation scheme based on minimizing the squared error, which iteratively adds new functions (i.e., basis functions) to the linear expansion. In comparison to a basis pursuit, it significantly reduces the computational complexity, since the basis pursuit minimizes a global cost function over all bases present in the dictionary (Mallat 1999). If the dictionary is orthogonal, the method works perfectly. Also, to achieve compact representation of the signal, it is necessary that the atoms are representative of the signal behavior and that the appropriate atoms from the dictionary are chosen.

The algorithm for the matching pursuit starts with an initial approximation for the signal, $\widehat{x}$, and the residual, R:

$$\widehat{x}^{(0)} = 0 \tag{6.113}$$

$$R^{(0)} = x \tag{6.114}$$

and it builds up a sequence of sparse approximation step-wise by trying to reduce the norm of the residue, $R = \widehat{x} - x$. At stage k, it identifies the dictionary atom that best correlates with the residual and then adds to the current approximation a scalar multiple of that atom, such that

$$\widehat{x}^{(k)} = \widehat{x}^{(k-1)} + \alpha_k \phi_k \tag{6.115}$$

$$R^{(k)} = x - \widehat{x}^{(k)}, \tag{6.116}$$

where $\alpha_k = \langle R^{(k-1)}, \phi_k \rangle / \|\phi_k\|^2$. The process continues until the norm of the residual $R^{(k)}$ does not exceed the required margin of error $\varepsilon > 0$: $\|R^{(k)}\| \leq \varepsilon$ (Mallat 1999). In our approach, a stopping rule mandates that the number of bases, χ_B, needed for signal approximation should satisfy $\chi_B \leq \lceil 2Nv_D \rceil + 1$. Hence, a matching pursuit approximates the signal using χ_B bases as

$$x = \sum_{n=1}^{\chi_B} \langle x, \phi_n \rangle \phi_n + R^{(\chi_B)}, \tag{6.117}$$

where ϕ_n are χ_B bases from the dictionary with the strongest contributions.

6.6.4   Numerical simulation

In this section, the performance of the MDPSS estimator is compared with the Slepian basis expansion DPPS approach (Zemen and Mecklenbrauker 2005) for a certain radio environment. The channel model used in the simulations is presented in Section 6.6.1 and it is simulated using the AR approach suggested in (Baddour and Beaulieu 2005). The parameters of the simulated system are the same as in (Zemen and Mecklenbrauker 2005): the carrier frequency is 2 GHz, the symbol rate used is 48600 $1/s$, the speed of the user is 102.5 km/h, 10 pilots per data block are used, and the data block length is $M = 256$. The number of DPSSs used in estimation is given by $\lceil 2Mv_D \rceil + 1$. The same number of bases is used for MDPSS, while $K = 15$ su-bands is used in generation of MDPSS.

As an introductory example, consider the estimation accuracy for the WSSUS channel with a uniform power angle profile (PAS) with central AoA $\phi_0 = 5$ degrees and spread $\Delta = 20$ degrees. We used 1000 channel realizations and Figure 6.3 depicts the results for the considered channel model. The mean-square errors (MSE) for both MDPSS and DPSS estimators have the highest values at the edges of the data block. However, the MSE for MDPSS estimator is several orders of magnitude lower than the value for the Slepian basis expansion estimator based on DPSS.

Next, let's examine the estimation accuracy for the WSSUS channels with uniform PAS, central AoAs $\phi_1 = 45$ and $\phi_1 = 75$, and spread $0 < \Delta \le 2\pi/3$. Furthermore, it is assumed that the channel is noisy. Figures 6.4 and 6.5 depict the results for $SNR = 10$ dB and $SNR = 20$ dB, respectively.

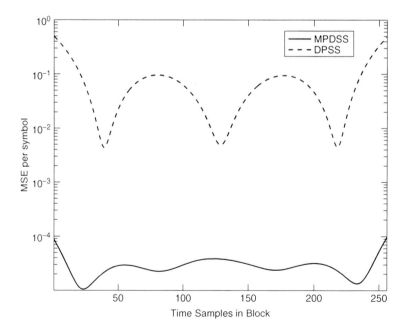

Figure 6.3   Mean-square error per symbol for MDPSS (solid) and DPSS (dashed) mobile channel estimators for the noise-free case.

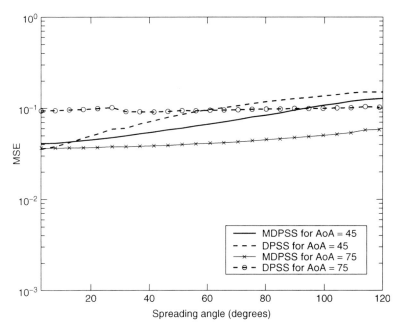

Figure 6.4   Dependence of the MSE on the angular spread Δ and the mean angle of arrival for SNR = 10 dB.

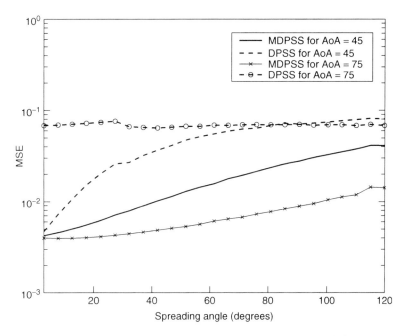

Figure 6.5   Dependence of the MSE on the angular spread Δ and the mean angle of arrival for SNR = 20 dB.

The results clearly indicate that the MDPSS frames are a more accurate estimation tool for the assumed channel model. For the considered angles of arrival and spreading angles, the MDPSS estimator consistently provided lower MSE in comparison to the Slepian basis expansion estimator based on DPSS. The advantage of the MDPSS stems from the fact that these bases are able to describe different scattering scenarios.

7

Effects of prediction and estimation errors on performance of communication systems

Estimation of a wireless channel is an important step in utilizing the capacity of the system which uses such a channel. Our main focus here is to evaluate the impact of multiple antennas on each side and geometry of the scattering environment on the quality of such an estimation. As will be seen there are opposite trends affecting such performance and it is difficult to account for all of them at once. Here we assume that an infinite number of the samples is available from the past of the MIMO channel, therefore, a simplified Kolmogorov–Szegö entropy function could be used to characterize the lower bound on the mean-square error (MSE) of one step prediction. Such an approach allows us to focus on the role of multiple antennas and sparsity of the scattering environment. Initial investigation of this problem has been partially conducted in (Svantesson and Swindlehurst 2003). However, this investigation focuses only on a ray model. Here we relate predictability with the angular width of scatterers and SNR in the system. We also investigate the effects of non-Kronecker structure on predictability.

Wireless Multi-Antenna Channels: Modeling and Simulation, First Edition.
Serguei L. Primak and Valeri Kontorovich.
© 2012 John Wiley & Sons, Ltd. Published 2012 by John Wiley & Sons, Ltd.

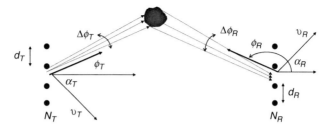

Figure 7.1   Geometry of a single cluster problem.

Basic model

General ULA formulation

At this stage we consider a flat fading scenario with uniform linear arrays on both sides as shown in Figure 7.1. There are N_T transmit antennas N_R receive antennas separated by distance d_T and d_R respectively. The receiver is moving with velocity v_R forming angle α with the broadside direction of the receive ULA while the transmitter remains static $v_T = 0$. The maximum Doppler spread is then $f_D = v_R/cf_0$ where c is the speed of light and f_0 is the carrier frequency.

We assume a 2-D scattering scenario for simplicity. This allows us to describe statistical properties of $N_R \times N_T$ channel impulse response $\mathbf{H}(t)$ in terms of joint AoA/AoD angular power spectrum $P_0 p(\phi_T, \phi_R)$. Here P_0 is the total power of the link and $p(\phi_T, \phi_R)$ is the joint AoA/AoD PDF. In particular, the cross-covariance function between two links, separated by distance x on the transmit side and distance y on the receive side can be written as (Jakes 1974)

$$R(x, y; \tau) = \frac{P_0}{N_T} \int_{-\pi}^{\pi} \int_{-\pi}^{\pi} p(\phi_T, \phi_R) \exp\left(j2\pi \frac{x}{\lambda} \sin \phi_T\right)$$
$$\times \exp\left[j2\pi \frac{y}{\lambda} \sin \phi_R + j2\pi f_D \cos(\phi_R - \alpha)\right] d\phi_T d\phi_R \quad (7.1)$$

Taking the Fourier transform with respect to τ one can obtain the following expression for the cross-power density $S(x, y; \omega)$

$$S(x, y; \omega) = \int_{-\infty}^{\infty} R(x, y; \tau) \exp(j\omega\tau) d\tau$$
$$= \frac{P_0}{N_T} \int_{-\pi}^{\pi} p(\phi_T, \alpha + \arccos \frac{\omega}{\omega_D}) \exp\left(j2\pi \frac{x}{\lambda} \sin \phi_T\right)$$
$$\times \exp\left[j2\pi \frac{y}{\lambda} \sin\left(\alpha + \arccos \frac{\omega}{\omega_D}\right)\right] \frac{d\phi_T}{\sqrt{\omega_D^2 - \omega^2}}$$
$$+ P_0 \int_{-\pi}^{\pi} p\left(\phi_T, \alpha - \arccos \frac{\omega}{\omega_D}\right) \exp\left(j2\pi \frac{x}{\lambda} \sin \phi_T\right)$$
$$\times \exp\left[j2\pi \frac{y}{\lambda} \sin\left(\alpha - \arccos \frac{\omega}{\omega_D}\right)\right] \frac{d\phi_T}{\sqrt{\omega_D^2 - \omega^2}} \quad (7.2)$$

Without loss of generality we can assume that $\alpha = 0$ to simplify calculations. In this case Equation (7.2) simplifies to the following expression

$$S(x, y; \omega) = \frac{P_0}{N_T} \int_{-\pi}^{\pi} p\left(\phi_T, \arccos \frac{\omega}{\omega_D}\right) \exp\left(j2\pi \frac{x}{\lambda} \sin \phi_T\right)$$

$$\times \exp\left[j2\pi \frac{y}{\lambda} \sqrt{1 - \frac{\omega^2}{\omega_D^2}}\right] \frac{d\phi_T}{\sqrt{\omega_D^2 - \omega^2}}$$

$$+ P_0 \int_{-\pi}^{\pi} p\left(\phi_T, -\arccos \frac{\omega}{\omega_D}\right) \exp\left(j2\pi \frac{x}{\lambda} \sin \phi_T\right)$$

$$\times \exp\left[-j2\pi \frac{y}{\lambda} \sqrt{1 - \frac{\omega^2}{\omega_D^2}}\right] \frac{d\phi_T}{\sqrt{\omega_D^2 - \omega^2}} \tag{7.3}$$

It is easy to see that for $x = y = 0$ this equation reduces to one presented in (Jakes 1974) for a SISO channel.

Some simple models

Further simplification of Equations (7.2) and (7.3) can be achieved only in a particular form of the joint distribution $p(\phi_T, \phi_R)$.

Kronecker model

One of the most used assumptions is to postulate that AoA and AoD are statistically independent,[1] that is $p(\phi_T, \phi_R) = p_T(\phi_T)p_R(\phi_R)$. In this case Equation (7.3) is reduced to

$$S_K(x, y; \omega) = \frac{P_0}{N_T} R_T(x) \left\{ \exp\left[j2\pi \frac{y}{\lambda} \sqrt{1 - \frac{\omega^2}{\omega_D^2}}\right] \frac{p_R\left(\arccos \frac{\omega}{\omega_D}\right)}{\sqrt{\omega_D^2 - \omega^2}} \right.$$

$$\left. + \exp\left[-j2\pi \frac{y}{\lambda} \sqrt{1 - \frac{\omega^2}{\omega_D^2}}\right] \frac{p_R\left(-\arccos \frac{\omega}{\omega_D}\right)}{\sqrt{\omega_D^2 - \omega^2}} \right\} \tag{7.4}$$

where $R_T(x)$ is the transmit correlation function

$$R_T(x) = \int_{-\pi}^{\pi} p_T(\phi_T) \exp\left(j2\pi \frac{x}{\lambda} \sin \phi_T\right) d\phi_T \tag{7.5}$$

Discrete ray model

Another important test model is known as the sum of discrete rays model. Such a model is widely used in so-called sum of cisoids simulators (Zheng and Tse 2002),

---

[1] This leads to the well known Kronecker model of MIMO channels (Pedersen et al. 2000).

(Patzold 2002), (Svantesson and Swindlehurst 2003). In this case

$$p(\phi_T, \phi_R) = \sum_{l=1}^{L} P_l \delta (\phi_T - \phi_{Tl}) \delta (\phi_R - \phi_{Rl}) \tag{7.6}$$

readily resulting in the following expression for the cross-spectral density

$$S_L(x, y; \omega) = \sum_{l=1}^{L} P_l \exp \left(j2\pi \frac{x}{\lambda} \sin \phi_{Tl} \right)$$

$$\times \exp \left(j2\pi \frac{x}{\lambda} \sin \phi_{Rl} \right) \delta (\omega - \omega_D \cos \phi_{Rl}) \tag{7.7}$$

Single narrow cluster

Although a single cluster model described below may not be a good model for a mobile wireless channel, it constitutes a nice building block for more complicated and representative distributions of AoA and AoD. For a case of a single narrow cluster, as shown in Figure 7.1, one can choose bivariate uniform distribution[2]

$$p_2(\phi_T, \phi_R) = \frac{1}{\Delta_T \Delta_R} \tag{7.8}$$

Here $\Delta_T \ll 2\pi$ and $\Delta_R \ll 2\pi$ describe the angular spread on the transmit and receive side around ϕ_{T0} and ϕ_{R0}.

Since we assume that the angular spreads are narrow, it is possible to assume that only one term in expression (7.3) could be kept, that is

$$S_1(x, y; \omega) = P_0 \int_{-\pi}^{\pi} p \left(\phi_T, \arccos \frac{\omega}{\omega_D} \right) \exp \left(j2\pi \frac{x}{\lambda} \sin \phi_T \right)$$

$$\times \exp \left[j2\pi \frac{y}{\lambda} \sqrt{1 - \frac{\omega^2}{\omega_D^2}} \right] \frac{d\phi_T}{\sqrt{\omega_D^2 - \omega^2}} \tag{7.9}$$

Using the same approximation as in (Zhao and Loyka 2004a), that is only keeping terms of the first order in Δ_T and Δ_R one obtains

$$S_1(x, y; \omega) \approx P_0 \exp \left(j2\pi \frac{x}{\lambda} \sin \phi_{T0} \right) \text{sinc} \left(\frac{\Delta_T x \cos \phi_{T0}}{\lambda} \right)$$

$$\times \exp \left(j2\pi \frac{y}{\lambda} \sin \phi_{R0} \right) \frac{d\phi_T}{\cos \phi_{R0}} \tag{7.10}$$

7.1   Kolmogorov–Szegö-Krein formula

The question of minimum prediction error given infinite observations is a well studied problem (Papoulis 1991; Poor 1994), especially in the one-dimensional

---

[2] Sub-index T indicates the transmitter side, the sub-index R indicates the receiver side.

case. Let a desired signal have power spectral density $S(\omega)$ and be embedded in WGN with variance σ_n^2. SNR $\gamma = \sigma_s^2/\sigma_n^2$ can be easily calculated in terms of $S(\omega)$ since

$$\sigma_s^2 = \frac{1}{2\pi} \int_{-\pi}^{\pi} S(\omega)d\omega, \quad S(\omega) = \sigma_s^2 S_0(\omega) \tag{7.11}$$

Due to the overlap of the noise and signal power spectral densities, the estimation of the signal even based on infinitely many samples in the past and the future is subject to an error. The normalized MMSE δ error is given by Kolmogorov–Szegö-Krein (KSK) formula (Poor 1994), (Papoulis 1991; van Trees 2001)

$$\delta = \frac{\sigma_n^2}{\sigma_s^2} \frac{1}{2\pi} \int_{-\pi}^{\pi} \ln\left[1 + \frac{S(\omega)}{\sigma_n^2}\right] d\omega = \frac{1}{\gamma} \frac{1}{2\pi} \int_{-\pi}^{\pi} \ln\left[1 + \gamma S_0(\omega)\right] d\omega \tag{7.12}$$

As expected, the normalized error approaches unity $\delta \to 1$ as $\gamma \to 0$ and $\delta \to 0$ as $\gamma \to \infty$ in consistence with the sampling theorem (Papoulis 1991) if $S(\omega)$ is bandlimited.

The multidimensional version of the KSK formula is less known. Let us consider a zero-mean random signal vector $\mathbf{x}(t)$ of size $M \times 1$ and described by its cross-power spectral density matrix $\mathbf{S}_{xx}(\omega)$ defined as the Fourier transform of its full covariance matrix

$$\mathbf{R}_{xx}(\tau) = \mathcal{E}\left\{\mathbf{x}(t+\tau)\mathbf{x}^*(t)\right\}, \quad \mathbf{S}_{xx}(\omega) = \mathcal{F}[\mathbf{R}_{xx}(\tau)] \tag{7.13}$$

Then, if this signal is embedded in a noise with a non-singular power spectral matrix $\mathbf{N}(\omega)$, the normalized MMSE is given by a multidimensional version of KSK formula

$$\delta = \frac{\operatorname{tr}\mathbf{N}(\omega)}{\operatorname{tr}\mathbf{R}_{xx}(0)} \frac{1}{2\pi} \int_{-\pi}^{\pi} \ln\det\left[\mathbf{I}_N + \mathbf{N}^{-1}(\omega)\mathbf{S}_{xx}(\omega)\right] d\omega \tag{7.14}$$

Assuming that the noise is a spatially and temporary white Gaussian noise with variance σ_n^2 per dimension, Equation (7.14) could be simplified to

$$\delta = \frac{M\sigma_n^2}{\operatorname{tr}\mathbf{R}_{xx}(0)} \frac{1}{2\pi} \int_{-\pi}^{\pi} \ln\det\left[\mathbf{I}_M + \frac{\operatorname{tr}\mathbf{R}_{xx}(0)}{\sigma_n^2} S_0(\omega)\right] d\omega \tag{7.15}$$

since in this case $\mathbf{N}(\omega) = \sigma_n^2 \mathbf{I}_N$. Here $\mathbf{I}_M$ is the $M \times M$ identity matrix and the normalized spectral matrix $\mathbf{S}_0(\omega)$ is defined as

$$\mathbf{S}_{xx}(\omega) = \operatorname{tr}\mathbf{R}_{xx}(0)\,\mathbf{S}_0(\omega) \tag{7.16}$$

Equation (7.15) is the key equation for the analysis of ultimate predictability of the MIMO channel. In this case we consider the vectorized version of the channel response matrix $\mathbf{H}(t)$ (van Trees 2002)

$$\mathbf{h}(t) = \operatorname{vec}\mathbf{H}(t) \tag{7.17}$$

which is obtained by stacking the columns of $\mathbf{H}(t)$. In this case $M = N_T N_R$ and the elements of the spectral matrix $\mathbf{S}_{hh}(\omega)$ are defined by means of Equation (7.2).

7.2   Prediction error for different antennas and scattering characteristics

7.2.1   SISO channel

In the case of the SISO channel Equation (7.12) could be directly applied. If signal $s(t) = \mathcal{F}^{-1}S(\omega)$ is composed of K spectrally separable segments of with uniform power density[3] as shown in Figure 7.2, the expression for the power spectral density is a sum of pulse function:

$$S_0(\omega) = \sigma_s^2 \sum_{k=1}^{K} \frac{2\pi P_k}{\Delta\omega_k} \Pi(\omega_k, \Delta\omega_k) \tag{7.18}$$

where

$$\Pi(\omega_k, \Delta\omega_k) = \begin{cases} 1 & \text{if } |\omega - \omega_k| \le \Delta\omega_k/2 \\ 0 & \text{otherwise} \end{cases} \tag{7.19}$$

and $0 \le P_k \le 1$ is the fraction of the total signal power concentrated in subband $[\omega_k - \Delta\omega_k/2; \omega_k + \Delta\omega_k/2]$, the corresponding error can be easily computed as follows.

$$\begin{aligned}
\delta &= \frac{\sigma_n^2}{\sigma_s^2} \frac{1}{2\pi} \int_{-\pi}^{\pi} \ln\left[1 + \frac{\sigma_s^2}{\sigma_n^2} \sum_{k=1}^{K} \frac{2\pi P_k}{\Delta\omega_k} \Pi(\omega_k, \Delta\omega_k)\right] d\omega \\
&= \frac{1}{2\pi\gamma} \sum_{k=1}^{K} \int_{-\pi}^{\pi} \ln\left[1 + \gamma \frac{2\pi P_k}{\Delta\omega_k} \Pi(\omega_k, \Delta\omega_k)\right] d\omega \\
&= \frac{1}{2\pi\gamma} \sum_{k=1}^{K} \int_{\omega_k - \Delta\omega_k/2}^{\omega_k + \Delta\omega_k/2} \ln\left[1 + \gamma \frac{2\pi P_k}{\Delta\omega_k}\right] d\omega \\
&= \sum_{k=1}^{K} P_k \frac{\Delta\omega_k}{2\pi\gamma P_k} \ln\left[1 + \gamma \frac{2\pi P_k}{\Delta\omega_k}\right] d\omega = \sum_{k=1}^{K} P_k \gamma_k^{-1} \ln(1 + \gamma_k) \tag{7.20}
\end{aligned}$$

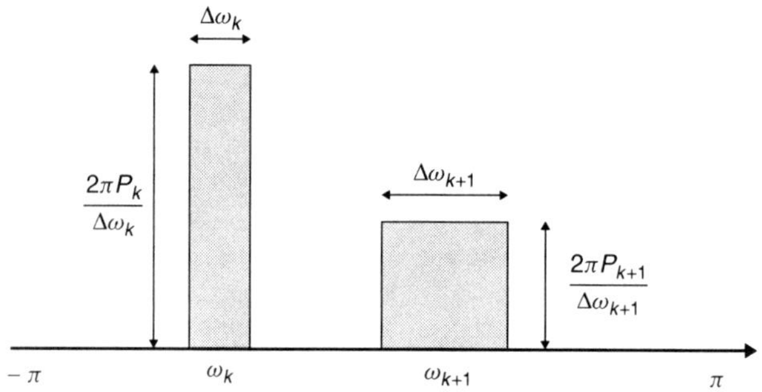

Figure 7.2   Power spectral density of a signal described by Equation (7.18).

---

[3] That is, the scattering is dominated by K narrow clusters.

where

$$\gamma_k = \gamma \frac{2\pi P_k}{\Delta \omega_k} \qquad (7.21)$$

is the SNR which can be achieved for the k-th spectral segment of the signal by using an ideal band-pass filter.[4] The net error is a weighted sum of errors associated with each individual sub-band. The weights are proportional to the fraction P_k of power of the signal in the k-th sub-band.

If all the power is concentrated in a single sub-band of width $0 \leq \Delta\omega_1 \leq 2\pi$, $P_1 = 1$ and the expression for MMSE becomes

$$\delta_1(\Delta\omega_1, \gamma) = \frac{\Delta\omega_1}{2\pi\gamma} \ln\left(1 + \frac{2\pi\gamma}{\Delta\omega_1}\right) \qquad (7.22)$$

Simple calculus shows that for a fixed SNR $\gamma > 0$, the error δ_1 in (7.22) is an increasing function of the bandwidth $\Delta\omega_1$ with the minimum $\delta_{min} = \delta(0, \gamma) = 0$ and the maximum $\delta_{max} = \delta(2\pi, \gamma) = \gamma^{-1} \ln(1 + \gamma)$. Thus, regardless of value of SNR, if the process has zero bandwidth (that is it is a single cissoid), it could be estimated without any error in any non-zero SNR if an infinite number of samples in the past are available. As bandwidth becomes non-zero, some residual estimation error is unavoidable and increases with bandwidth. The shape of function $\gamma^{-1} \ln(1 + \gamma)$ is shown in Figure 7.4. Evaluation of normalized MMSE error according to Equation (7.220) is shown in Figure 7.3.

For a fixed non-zero bandwidth $\Delta\omega_1 \leq 1$ the estimation error decreases from 1 at $\gamma = 0$ to zero as SNR increases, even for the full bandwidth $\Delta\omega_1 = 2\pi$. This should not come as a surprise since if $\gamma = \infty$ noise free samples are available at least the Nyquist rate and, therefore, perfect interpolation, is available in a classical sense.

Another interesting observation that can be made by analyzing Equation (7.20) is the fact that while the residual error δ depends on the bandwidth and the relative power of each sub-band it does not depend on the location of that sub-band as long as bands do not overlap. This is of course another consequence of the fact that we access infinitely many samples, and, therefore, can perform perfect band separation. When only a finite number of samples is available, such separation of sub-bands is not ideal and some inter-band interference causes further degradation of the quality of estimation. Since better suppression of inter-band interference could be observed when bands are separated by a bigger distance, we would speculate at this point that the prediction error increases as bands move closer together.

Assuming for simplicity that only one sub-band exists $\Delta\omega$ in the signal, we can partition that band into smaller non-overlapping sub-bands Δv_k each with relative power $P_k = \Delta v_k / \Delta\omega$. Direct substitution in (7.20) shows that such partition does not change the value of the irreducible error.[5]

---

[4] Since the infinite past is considered, there are enough taps to design a fixed number of ideal low-pass filters.

[5] Therefore, no more gain could be achieved by further reducing the width of the sub-bands by increasing the number of antennas or number of available samples.

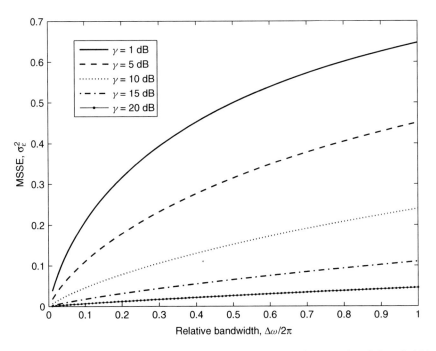

Figure 7.3   Normalized MMSE error as a function of normalized bandwidth $\Delta\omega_1/2\pi$ *and SNR as given by Equation (7.220).*

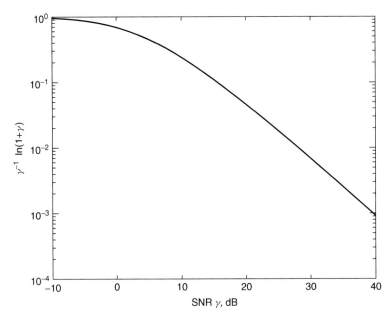

Figure 7.4   Function $\gamma^{-1}\ln(1+\gamma)$.

7.2.2   SIMO channel

Let us turn our attention to a SIMO configuration $N_T = 1$, $N_R > 1$. In this case one has to set $x = 0$ in Equation (7.3). At the same time the element $S_{kl}(\omega)$ of the cross-power spectral matrix $\mathbf{S_{hh}}(\omega)$ is the cross-spectrum between channels corresponding to the the k-th and l-th receive antennas, that is $y = (k - l)d_T$ in Equation (7.3). Furthermore, $x = 0$ dependence of the distribution of the AoD is removed from the equation. Thus

$$S_{kl}(0, y; \omega) = P_0 \exp\left[j2\pi \frac{(k - l)d_R}{\lambda} \sqrt{1 - \frac{\omega^2}{\omega_D^2}} \right] \frac{P_R\left(\arccos \frac{\omega}{\omega_D} \right)}{\sqrt{\omega_D^2 - \omega^2}}$$

$$+ P_0 \exp\left[-j2\pi \frac{(k - l)d_R}{\lambda} \sqrt{1 - \frac{\omega^2}{\omega_D^2}} \right] \frac{P_R\left(-\arccos \frac{\omega}{\omega_D} \right)}{\sqrt{\omega_D^2 - \omega^2}} \quad (7.23)$$

As we can see, in general, the spatial characteristics of the SIMO channel do not separate from its spectral/temporal characteristics and general analysis is complicated.

7.2.2.1   Single cluster

In order to gain some insight let as assume that the same model is used as in the Section 7.2.1. In particular let us consider a single cluster model first. In this case, Equation (7.23) is significantly simplified to produce

$$S_{kl}(0, y; \omega) = \exp\left[j2\pi \frac{(k - l)d_R}{\lambda} \sin \phi_{R0} \right] \frac{P_0}{\Delta_R \sin \phi_{R0}} \quad (7.24)$$

Thus, the cross-spectral matrix could be written in the following form

$$\mathbf{S_{hh}}(\omega) = \frac{P_0 N_R}{\Delta_R \sin \phi_{R0}} \mathbf{aa}^H, \quad |\omega - \omega_D \cos \phi_{R0}| \leq \frac{\Delta_R}{2} \sin \phi_{R0} \quad (7.25)$$

where

$$\mathbf{a} = \frac{1}{\sqrt{N_R}} \left[1, z, \cdots, z^{N_T - 1} \right]^H \quad (7.26)$$

$$z = \exp\left[j2\pi \frac{d_R}{\lambda} \sin \phi_{R0} \right] \quad (7.27)$$

Integration of Equation (7.25) over ω produces the following covariance matrix $\mathbf{R_{hh}}(0)$:

$$\mathbf{R_{hh}}(0) = P_0 N_R \mathbf{aa}^H, \quad \text{tr } \mathbf{R_{hh}}(0) = P_0 \quad (7.28)$$

Finally, we are able to use the KSK equation to obtain the variance of irreducible error of estimation

$$\delta = \frac{\sigma_n^2}{P_0} \frac{\Delta_T \sin \phi_{R0}}{2\pi} \ln\left[1 + \frac{P_0 N_R 2\pi}{\sigma_n^2 \Delta_R \sin \phi_{R0}} \right] \quad (7.29)$$

If we consider the MSE per receive antenna, one can conclude that $\delta = N_R \delta_R$ where

$$\delta_R = \frac{\sigma_n^2}{P_0 N_R} \frac{\Delta_R \sin \phi_{R0}}{2\pi} \ln\left[1 + \frac{P_0 N_R 2\pi}{\sigma_n^2 \Delta_R \sin \phi_{R0}}\right] = \frac{\ln(1 + N_R \gamma)}{N_R \gamma} \qquad (7.30)$$

Here

$$\gamma = \frac{2\pi P_0}{\sigma_n^2 \Delta_R \sin \phi_{R0}} \qquad (7.31)$$

is SNR corresponding to SISO channel. It can be concluded from Equation (7.30) and the fact that $\gamma^{-1} \ln(1 + \gamma)$ is a decreasing function of γ, that per antenna prediction is improved. The major factor in the increased performance is the accumulation of SNR over highly correlated antennas due to a narrow angular spread.

It may appear from Equation (7.31) that an unlimited increase in the number of received antennas allows one to predict the SIMO channel perfectly. However, two important points must be made here

1. As the number N_T of received antennas increases, the angular resolution of the array also increases and a "fine" structure of the cluster can be resolved. In this case approximation of $\sin \phi_T$ by a linear function is no longer valid and the cluster cannot be treated as a small cluster. More discussions are in the following section.

2. While estimation per link improves, the number of links that need to be estimated is also significantly increased. Effects of such an increase on performance of a particular signaling algorithm will be studied separately.

7.2.2.2 Two clusters

Let us now consider a situation where the scattering environment contains two narrow clusters, seen at angles ϕ_{R1} and ϕ_{R2} with angular spreads Δ_1 and Δ_2 respectively. The relative power coming from the first cluster is p_1 and $p_2 = 1 - p_1$ is the relative power received from the second cluster.

Following the same procedure as for a single cluster, we can easily obtain the expression for the cross-spectral matrix $\mathbf{S}(\omega)$:

$$\mathbf{S_{hh}}(\omega) = P_0 N_R \left[\frac{p_1}{\Delta_{R1} \sin \phi_{R1}} \mathbf{aa}^H + \frac{p_2}{\Delta_{R2} \sin \phi_{R2}} \mathbf{bb}^H\right] \qquad (7.32)$$

Appendix 7.4 details the procedure of finding the eigenvalues λ_1 and λ_2 of a matrix presented by $\mathbf{S_{hh}}(\omega)$. In general, these depend on the values of the term $z = \mathbf{b}^H \mathbf{a}$ as well as the weight coefficients.

The simplest case arrives when $z = 0$. In this case the eigenvalues are simply those given by the weights

$$\lambda_1 = N_R P_0 \frac{p_1}{\Delta_{R1} \sin \phi_{R1}} \qquad (7.33)$$

$$\lambda_2 = N_R P_0 \frac{p_2}{\Delta_{R2} \sin \phi_{R2}} \qquad (7.34)$$

and the irreducible error could be calculated to be

$$\delta_R = p_1 \delta_{R1} + p_2 \delta_{R2} \tag{7.35}$$

where δ_k, $k = 1, 2$ is given by

$$\delta_{Rk} = \frac{\sigma_n^2}{P_0 N_R p_k} \frac{\Delta_{Rk} \sin \phi_{Rk0}}{2\pi} \ln \left[1 + \frac{P_0 N_R 2\pi}{\sigma_n^2 \Delta_{Rk} \sin \phi_{Rk0}} \right] = \frac{\ln(1 + N_R \gamma_k)}{N_R \gamma_k} \tag{7.36}$$

and

$$\gamma_k = \frac{2\pi P_0 p_k}{\sigma_n^2 \Delta_{Rk} \sin \phi_{Rk0}} \tag{7.37}$$

Thus, it could be concluded that estimation using multiple receive antennas produces better results per antenna, even in a two cluster situation, due to accumulation of SNR (or, equivalently, by averaging the noise). It can also be seen that the angular spread of each cluster adversely effects the performance of estimator for a fixed AoA.

The effect of non-perfect spatial separation $z \neq 0$ of two clusters can be seen from analysis of Equation (7.77). The effect of such imperfection is equivalent to changing weights p_1 and p_2 and can have both positive and negative effects on estimation quality compared with that of $z = 0$. However, even in this case, an increased number of antennas leads to an improved quality of estimation on per antenna bases.

7.2.3   MISO channel

Let us now turn our attention to a system with a single receive antenna. In this case $N_R = 1$ and $y = 0$ in Equation (7.3) which now can be rewritten in a much simplified form

$$S(x, 0; \omega) = P_0 \int_{-\pi}^{\pi} \exp \left(j 2\pi \frac{x}{\lambda} \sin \phi_T \right)$$

$$\times \frac{p \left(\phi_T, \arccos \frac{\omega}{\omega_D} \right) + p \left(\phi_T, -\arccos \frac{\omega}{\omega_D} \right)}{\sqrt{\omega_D^2 - \omega^2}} d\phi_T \tag{7.38}$$

7.2.3.1   Kronecker model

If the transmit and the receive sides are statistically independent, that is

$$p(\phi_T, \phi_R) = p_T(\phi_T) p_R(\phi_R),$$

then Equation (7.38) simplifies even further to produce the following expression for the spectral matrix $\mathbf{S}(\omega)$

$$\mathbf{S}(\omega) = \frac{P_0}{N_T} \mathbf{R}_T S_0(\omega) \tag{7.39}$$

where $S_0(\omega)$ is the power spectral density corresponding to a SISO channel with unit power and $\mathbf{R}_T = \{R_{Tkl}\}$ is the transmit correlation matrix with elements R_{Tkl} given by

$$R_{Tkl} = \int_{-\pi}^{\pi} p_T(\phi_T) \exp\left[j2\pi \frac{(k-l)d_T}{\lambda} \sin\phi_T \right] d\phi_T \tag{7.40}$$

The normalized MSE of estimation per link now can be calculated in terms of the eigenvalues of the transmit correlation matrix as

$$
\begin{aligned}
\delta_T &= \frac{\sigma_n^2}{P_0} \frac{1}{2\pi} \int_{-\pi}^{\pi} \ln\det\left[\mathbf{I}_{N_T} + \frac{P_0}{\sigma_n^2 N_T} \mathbf{R}_T S_0(\omega) \right] d\omega \\
&= \frac{\sigma_n^2}{P_0} \frac{1}{2\pi} \int_{-\pi}^{\pi} \ln\det\left[\mathbf{I}_{N_T} + \frac{P_0}{\sigma_n^2 N_T} \mathbf{\Lambda}_T S_0(\omega) \right] d\omega \\
&= \frac{\sigma_n^2}{P_0} \sum_{n_t=1}^{N_T} \frac{1}{2\pi} \int_{-\pi}^{\pi} \ln\left[1 + \frac{P_0 \lambda_{n_t}}{\sigma_n^2 N_T} S_0(\omega) \right] d\omega
\end{aligned}
\tag{7.41}
$$

In the case of uncorrelated transmit antennas, the transmit correlation matrix is unity $\mathbf{R}_T = \mathbf{I}$, and all the eigenvalues are equal $\lambda_{n_t} = 1$. Thus, the per link estimation error becomes

$$\delta_T = \frac{\sigma_n^2 N_T}{P_0} \frac{1}{2\pi} \int_{-\pi}^{\pi} \ln\left[1 + \frac{P_0}{\sigma_n^2 N_T} S_0(\omega) \right] d\omega \tag{7.42}$$

This could be recognized as an irreducible estimation error of a SISO link corresponding to the same scattering geometry, but with the power scaled down by the number of antennas as should be expected since each link can be estimated separately with power per antenna reduced: increasing system complexity on the transmit side without increasing transmit power leads to less accurate estimates. This is different from the SIMO link since the harvested power is proportional to the number of receive antennas with the same radiated power.

If scattering geometry is such that matrix $\mathbf{R}_T$ has rank 1, all eigenvalues are zeroes but one: $\lambda_1 = N_T$, $\lambda_{n_t} = 0$ for $2 \leq n_t \leq N_T$. In this case

$$\delta_T = \frac{\sigma_n^2}{P_0} \frac{1}{2\pi} \int_{-\pi}^{\pi} \ln\left[1 + \frac{P_0}{\sigma_n^2} S_0(\omega) \right] d\omega \tag{7.43}$$

that is, the quality of estimation of each link is the same as for the SISO system with the same scattering geometry and available power.

In the intermediate case, the spatial correlation matrix may have $N_{DoF} \leq N_T$ almost equal eigenvalues $\lambda = N_T/N_{DoF}$, with the rest close to zero.[6] In this case the normalized estimation error per link

$$\delta_T = \frac{\sigma_n^2 N_{DoF}}{P_0} \frac{1}{2\pi} \int_{-\pi}^{\pi} \ln\left[1 + \frac{P_0}{N_{DoF}\sigma_n^2} S_0(\omega) \right] d\omega \tag{7.44}$$

---

[6] Careful consideration requires that the SNR level is also taken into account while evaluation N_{DoF}.

Since the quantity $0 < \zeta_T = N_{DoF}/N_T < 0$ could be considered as a sparsity of the channel (including antennas) representation in the angular domain, Equation (7.44) could be considered as a link between sparsity ζ and quality of estimation δ_T: more sparse channels could be estimated better.

7.2.3.2 Single narrow cluster

Let us consider a particular case of a single narrow cluster seen from the transmitter at the angle ϕ_{T0} with angular spread Δ_T, and from the receiver side seen at the angle ϕ_{R0} and the spread Δ_R. The corresponding transmit correlation matrix has approximately

$$N_{DoF} = \lfloor \Delta_T \cos \phi_{T0} N_T d_T \rfloor + 1 \tag{7.45}$$

non-zero eigenvalues, while the received power spectral density is approximately uniform

$$S_0(\omega) = \frac{\Delta_R}{2\pi} \omega_D \cos \phi_{R0}, \ |\omega - \omega_D| \leq \frac{\omega_D \sin \phi_{R0}}{2} \tag{7.46}$$

Therefore, the irreducible error, calculated using Equation (7.44) is given by

$$\delta_T \approx \gamma^{-1} \ln(1 + \gamma) \tag{7.47}$$

where

$$\gamma = \frac{\sigma_n^2}{P_0} \frac{2\pi F_s}{\Delta_T \omega_D \sin \phi_{R0}(\lfloor \Delta_T \cos \phi_{T0} N_T d_T \rfloor + 1)} \tag{7.48}$$

This equation gives a clear view of the effect of different parameters of the MISO system on the prediction quality of an individual link. As expected, the increasing power over noise and reduced bandwidth of the interference improves the quality of estimation. It also transpires that an increase in the number of transmit antennas and the distance between them improves the quality of estimation per individual link.

7.2.3.3 Multiple separable clusters

Let us now consider the situation where there are M narrow scattering clusters located in such a manner that their Doppler spectra do not overlap in the receive antenna. Each cluster is characterized by its relative power $p_m \geq 0$,

$$\sum_{m=1}^{M} p_m = 1 \tag{7.49}$$

mean direction and angular spread on each side, denoted as ϕ_{Tm}, ϕ_{Rm} and $\Delta\phi_{Tm}$, $\Delta\phi_{Rm}$ respectively.

Following the same approach as in the previous section one can easily obtain that the MSE per link is given by

$$\delta_T = \sum_{m=1}^{M} p_m \gamma_m^{-1} \ln(1 + \gamma_m) \tag{7.50}$$

where

$$\gamma_m = p_m \frac{\sigma_n^2}{P_0} \frac{2\pi F_s}{\Delta_T \omega_D \sin \phi_{R0}(\lfloor \Delta_T \cos \phi_{T0} N_T d_T \rfloor + 1)} \tag{7.51}$$

In order to isolate the effect of multiple clusters on the quality of estimation let us assume that each cluster occupies approximately the same band of Doppler frequencies in the received antenna, that is the signal from each individual cluster could be predicted with the same accuracy in one cluster scenario. This is equivalent to the assumption that

$$\gamma_m = p_m \gamma \tag{7.52}$$

where γ is effective SNR if the scattering environment contains only a single cluster. In this case

$$\delta_T = \sum_{m=1}^{M} p_m (p_m \gamma)^{-1} \ln(1 + p_m \gamma) = \gamma^{-1} \sum_{m=1}^{M} \ln(1 + p_m \gamma) > \gamma^{-1} \ln(1 + \gamma) \tag{7.53}$$

Thus, estimation in a multi-cluster MISO environment is more difficult than in a single cluster scenario. The most difficult case arises when all clusters contribute approximately the same amount of energy to the receive antenna, that is in the case $p_m = 1/M$. This should not come as a surprise since the more complex scattering environment corresponds to a greater uniformity of the eigenvalues of the transmit correlation matrix, and therefore it leads to reduced quality of the estimation.

7.2.4   MIMO channel

An analysis of a MIMO scenario could be carried out through the decomposition of the MIMO channel into N_T SIMO channels as follows. Let $\mathbf{R}_T = \mathbf{U}_T \mathbf{\Lambda}_T \mathbf{U}_T^H$ be the spectral decomposition of the transmit correlation matrix. Since the unitary transformation does not change the MSE of estimation, instead of $\mathbf{H}(t)$ one can consider its transformed version

$$\mathbf{H}_u = \mathbf{H} \mathbf{U}_T, \quad \mathbf{h}_u = \left(\mathbf{I}_{N_R} \otimes \mathbf{U}_T^T \right) \mathbf{h} \tag{7.54}$$

It is easy to see, that the transmit correlation matrix $\mathbf{R}_{Tu}$ of the channel $\mathbf{H}_u$ is a diagonal matrix

$$\mathbf{R}_{Tu} = \mathcal{E} \left\{ \mathbf{H}_u^H \mathbf{H}_u \right\} = \mathbf{U}_T^H \mathbf{R}_T \mathbf{U} = \mathbf{\Lambda}_T \tag{7.55}$$

with the diagonal elements coinciding with the eigenvalues of $\mathbf{R}_T$. Therefore, the channel $\mathbf{H}_u$ could be considered as N_T independent SIMO links as shown in Figure 7.5.

Before proceeding further a few important points should be stressed.

1. The m-th "virtual" SIMO channel has a different scale factor $\lambda_m P_0/N_T$ where λ_m is the m-th eigenvalues of the transmit correlation matrix $\mathbf{R}_T$.

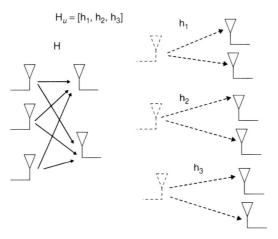

Figure 7.5 Transformation of $N_R \times N_T$ MIMO channel into N_T independent $N_R \times 1$ SIMO channels.

2. The m-th virtual SIMO channel in general has its own AoA PDF $p(\phi_R|m)$ which can be theoretically evaluated from the joint AoA/AoD PDF $p(\phi_T, \phi_R)$ as

$$p_R(\phi_R|m) = \frac{p(\phi_T, \phi_R)}{p_T(\phi_T|m)} \qquad (7.56)$$

where

$$p_T(\phi_T|m) = \int_{-\pi}^{\pi} p_T(\phi_T) p_m(\phi_T) d\phi_T \qquad (7.57)$$

and $p_m(\phi_T) d\phi_T$ is the fraction of power radiated in the direction ϕ_T in m-th eigenmode.

3. For large numbers of transmit antennas the eigenmodes and virtual channels coincide with those defined in (Sayeed 2002), that is they form beams in directions defined by the angles

$$\cos \phi_{Tm} = \frac{m}{N_T} - 1/2, \ m = 1, \cdots, N_T \qquad (7.58)$$

4. In the case of the Kronecker structure of the channel, all "virtual" channels have the same statistical properties in terms of AoA.

5. As the number of antennas increases on either side there are two conflicting trends as shown in Sections 7.2.3 and 7.2.2: an increase in the number of receive antennas improves the quality of estimation per link, while an increased number of transmit antennas decreases it. In addition, if sides are correlated, the eigenmode beamforming on the transmit side may cause some "focusing" on the receive side if the number of antennas is large enough. The net effect depends on the number of antennas on each side, correlation properties on each side, and SNR and correlation between sides.

7.2.4.1 Single narrow cluster

In the case of a single narrow cluster the eigenmodes and eigenvectors correspond to that of modulated DPSS (Sejdic et al. 2008; Slepian 1978). Therefore, for a modest number of transmit antennas, there is only a single non-zero eigenvalue of the transmit correlation matrix[7] $\lambda_{T1} = N_T$, and, therefore, the MIMO system is equivalent to a SIMO system. As a consequence, the results obtained in Section 7.2.2 directly apply.

As the number of antennas increases, so does the angular resolution of the antenna array. For a substantially large number $N_T \geq 1/\Delta_T \sin \phi_{T0}$ the fine structure of the cluster could be resolved and it cannot be modeled as a single cluster. More discussion of this situation is in Section 7.2.4 below.

7.2.4.2 Two narrow clusters

In the case of two narrow separated clusters with weights $p_1 \geq 0$ and $p_2 = 1 - p_1 \geq 0$, the transmit correlation matrix has the following form

$$\mathbf{R}_T = p_1 \mathbf{R}_{T1} + p_2 \mathbf{R}_{T2} = N_T p_1 \mathbf{a}\mathbf{a}^H + N_T p_2 \mathbf{b}\mathbf{b}^H \tag{7.59}$$

where $\mathbf{a}$ and $\mathbf{b}$ are the unit length MDPSS vectors, corresponding to each cluster: $\mathbf{a}^H \mathbf{a} = 1$ and $\mathbf{b}^H \mathbf{b} = 1$. Our next step is to find the eigendecomposition of $\mathbf{R}_T$. In order to do we can utilize results obtained in Appendix 7.4. In this case $z = \mathbf{b}^H \mathbf{a}$ and the required eigenvalues are given by

$$\lambda_{1,2} = N_T \left(p_1 + p_1 q_{1,2} z^* \right) = N_T \left(\frac{1 \pm \sqrt{(p_2 - p_1)^2 + 4p_1 p_2 |z|^2}}{2} \right) \tag{7.60}$$

As expected neither eigenvalue is less than N_T. Therefore, the presence of two clusters decreases the quality of estimation as was shown in Section 7.2.4. The worst case scenario is the case when both eigenvalues are equal, which is possible only for $z = 0$ and $p_1 = p_2 = 1/2$. When the number of transmit antennas grows large, $N_T \gg 1$, it can be seen that $z \approx 0$ and therefore, increasing the number of antennas to this level decreases the quality of estimation.

Thus, in a two cluster scenario, a MIMO channel is equivalent to two SIMO $N_R \times 1$ channels with power

$$P_{1,2} = \frac{P_0}{N_T} \lambda_{1,2} = \frac{1 \pm \sqrt{(p_2 - p_1)^2 + 4p_1 p_2 |z|^2}}{2} P_0 \tag{7.61}$$

The same equation allows us to treat the problem of improved resolution in the case of a single cluster when the number of antennas is increased. In this case the cluster can be thought of as a composition of two non-overlapping clusters with

---

[7] The eigenvectors of this matrix are modulated DPSS as in (Sejdic et al. 2008).

the same weight and angular spread equal to half of the original cluster. In this case one can assume that $p_1 = p_2 = 0.5$ and $|z|$ could be approximated as

$$|z| = |\mathbf{b}^H \mathbf{a}| \approx \left| \frac{1}{N_T} \sum_{n=0}^{N_T-1} \exp \left[j \frac{2\pi n d_T}{\lambda} \frac{\Delta_T}{2} \cos \phi_{T0} \right] \right| = \left| \frac{\sin \left[(N_T - 1)\kappa \right]}{N_T \sin \kappa} \right|$$

(7.62)

where

$$\kappa = \pi (N_T - 1) \Delta_T \frac{d_T}{\lambda} \cos \phi_{T0}$$

(7.63)

The power is split between two clusters as

$$P_1 = (1 - |z|) P_0 / 2 < P_0 \text{ and } P_2 = (1 + |z|) P_0 / 2 \le P_0$$

Thus, as N_T increases even a single cluster with angular spread Δ_T must be treated as two separate clusters. Thus, as the number of transmit antennas increases the number of equivalent "virtual" channels increases.

Our next step is to investigate the effect of the transformation (7.54) on the angular spread seen by the received side. The answer is quite clear in the case of the Kronecker structure of the channel: transformation (7.54) has no effect on the angular characteristics seen by each "virtual" channel. Therefore, the results of Section 7.2.2 can be directly applied to each such channel.

In order to study the effect of the correlation between the receive and the transmit sides let us consider the following joint probability density, corresponding to the two cluster scenario shown in Figure 7.6

$$p(\phi_T, \phi_R) = \sum_{k,l=1}^{2} P_{kl} \, p_{kl}(\phi_T, \phi_R)$$

(7.64)

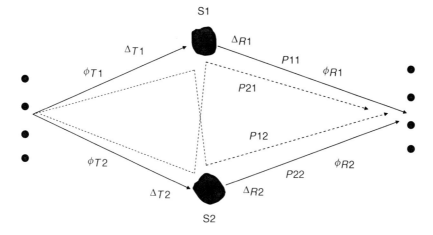

Figure 7.6   Geometry of a two cluster problem with correlation between sides.

where

$$p_{kl} = \frac{1}{\Delta_{Tk}\Delta_{Rl}}, \quad \begin{array}{l} |\phi_T - \phi_{Tk}| \leq \Delta_{Tk}/2 \\ |\phi_R - \phi_{Rl}| \leq \Delta_{Rl}/2 \end{array} \tag{7.65}$$

and coefficients P_{kl} define as follows

$$\begin{array}{ll} P_{11} = \eta\mu & P_{21} = (1-\eta)\alpha\mu \\ P_{12} = \eta(1-\mu) & P_{22} = (1-\eta)(1-\alpha\mu) \end{array} \tag{7.66}$$

Here $0 \leq \eta \leq 1$, $0 \leq \mu \leq 1$, $0 \leq \alpha \leq 1/\mu$. The case $\alpha = 1$ corresponds to the Kronecker structure of this channel.

Each eigenvector $\mathbf{w}_n$ of the transmit correlation matrix $\mathbf{R}_T$ defines its eigen beamforming pattern which redistributes transmitted power among different AoD. It can be seen from Equation (7.60) that for a small value of $|z|$ the eigenmodes almost coincide with the corresponding single modes $\mathbf{a}$ and $\mathbf{b}$ and, therefore, closely resemble modulated DPSS functions. For narrow clusters this also corresponds to approximately Kaiser window (Percival and Walden 1993; van Trees 2002) with a maximum approximately at the direction of the corresponding cluster. Thus, it could be concluded, that for the taper $\mathbf{w}_k$, the power radiated toward the k-th cluster will be enhanced, while the power directed toward the remaining cluster(s) will be suppressed. The enhancement $F_k = \lambda_k \geq 1$ and the suppression $S_k \geq 1$ factors depend on the number of antennas and the angular separation of the clusters. Here λ_k is the eigenvalue of the transmit correlation matrix associated with $\mathbf{w}_k$.

This leads to a redistribution of the power arriving at the receive antenna array. The direction of such redistribution and its impact on the quality of estimation depends on a number of factors and can provide conflicting trends.

One of the trends that can be easily identified is *focusing* at the receive side. Let us consider a limit situation where $\alpha = 0$ and $\mu = 0$. This scenario could describe a real physical environment with a small number of strong scatterers such as stand alone high-rise building where energy comes mainly due to reflection from such scatterers. While in general each receive antenna intercepts from all clusters, the "virtual" channel $\mathbf{H}_u$ will emphasize only a single cluster and suppress all others. Thus, the angular spread on the received side becomes smaller and the quality of estimation better than for the Kronecker channel with the same correlation on the receive and transmit sides.

The quality of suppression depends on the number of transmit antennas and the angular separation of the clusters as seen from the transmit side. For a small number of antennas separation is not very good, especially in the case where clusters are closely located. The improvement from preprocessing is not significant. However, as the number of antennas increases, the separation becomes more effective.

7.3   Summary of infinite horizon prediction results

The effect of a number of antennas on both sides of a MIMO link and scattering geometry has been investigated in this note. The results are compared to the

SISO link with the same scattering geometry by evaluating per link mean square error. There are few conclusions and recommendations that can be made based on this study.

1. The number of transmit antennas N_T decreases the quality of estimation per link due to increased complexity of the link (number of links to be estimated and finer resolution of a larger array) and the fact that the power per transmit antenna decreases. Higher correlation between the transmit antennas increases the quality of estimation due to the reduced number of degrees of freedom to be estimated.

2. In correlated fading, the number of received antennas N_R improves the estimation quality of the link due to averaging of the noise (accumulation of SNR).

3. An increase in the angular width of a scatter decreases the quality of estimation.

4. The number of resolved scatterers decreases the quality of estimation.

5. Transformation $\mathbf{H}_u = \mathbf{H}\mathbf{U}_T$ plays an important role in decomposition of the MIMO channel into independent SIMO channels which can be analyzed separately.

6. Transformation $\mathbf{H}_u = \mathbf{H}\mathbf{U}_T$ may lead to a focusing effect (shrinkage of the spectral spread) in each "virtual" SIMO. The strongest focusing can be observed for so called diagonal channels, that is for the channels with a diagonal coupling matrix (Weichselberger et al. 2005). If the coupling matrix has rank 1, the channel has Kronecker structure and no focusing can be observed.

7. A lower SNR value adversely affects the quality of estimation: mathematically it is described by $\gamma^{-1} \ln(1 + \gamma)$ type function.

7.4 Eigenstructure of two cluster correlation matrix

Let us derive expressions for the eigenvalues and the eigenvectors of the following covariance matrix

$$\mathbf{R}_T = p_1 \mathbf{R}_{T1} + p_2 \mathbf{R}_{T2} = N_T p_1 \mathbf{a}\mathbf{a}^H + N_T p_2 \mathbf{b}\mathbf{b}^H \qquad (7.67)$$

In order to do so let us find the eigenvector $\mathbf{w}$, corresponding to $\mathbf{R}_T$ as a weighted sum[8] $\mathbf{w} = \alpha\mathbf{a} + \beta\mathbf{b}$ of $\mathbf{a}$ and $\mathbf{b}$:

$$\mathbf{R}_T\mathbf{w} = \mathbf{R}_T (\alpha\mathbf{a} + \beta\mathbf{b}) = \lambda\mathbf{w} = \lambda (\alpha\mathbf{a} + \beta\mathbf{b}) \qquad (7.68)$$

---

[8] This is an appropriate approach for the case of two clusters which do not overlap in the dual domain.

Making use of notation $z = \mathbf{b}^H \mathbf{a}$ and the following properties of vector multiplication

$$\mathbf{a}\mathbf{a}^H \mathbf{a} = \mathbf{a}, \quad \mathbf{a}\mathbf{a}^H \mathbf{b} = z^* \mathbf{a} \tag{7.69}$$

$$\mathbf{b}\mathbf{b}^H \mathbf{b} = \mathbf{b}, \quad \mathbf{b}\mathbf{b}^H \mathbf{a} = z\mathbf{b}$$

Equation (7.68) could be recast as a system of two equations

$$\alpha\lambda = N_T p_1 \alpha + N_T p_1 \beta z^* \tag{7.70}$$

$$\beta\lambda = N_T p_2 \alpha z + N_T p_2 \beta \tag{7.71}$$

Multiplying the first equation by β and the second by α and equating the corresponding right-hand sides one obtains an equation which does not contain λ:

$$\alpha\beta p_1 + \beta^2 z^* p_1 = \alpha^2 z p_2 + \alpha\beta p_2 \tag{7.72}$$

If $z = \mathbf{b}^H \mathbf{a} = 0$, that is the eigenvectors corresponding to each cluster are orthogonal, then a system of equations (7.69)–(7.70) has a trivial solution

$$\lambda_1 = N_T p_1, \quad \mathbf{w}_1 = \mathbf{a}, \quad \alpha_1 = 1, \quad \beta_1 = 0 \tag{7.73}$$

$$\lambda_2 = N_T p_2, \quad \mathbf{w}_2 = \mathbf{b}, \quad \alpha_2 = 0, \quad \beta_2 = 1 \tag{7.74}$$

If $z \neq 0$ neither $\alpha = 0$ nor $\beta = 0$. In this case it is valid to set $q = \beta/\alpha$ as a new variable, which converts (7.72) into a simple quadratic equation

$$p_1 z^* q^2 - (p_2 - p_1)q - p_2 z = 0 \tag{7.75}$$

whose solution is

$$q_{1,2} = \frac{p_2 - p_1 \pm \sqrt{(p_2 - p_1)^2 + 4p_1 p_2 |z|^2}}{2p_1 z^*} \tag{7.76}$$

In turn, Equation (7.70) produces the eigenvalues

$$\lambda_{1,2} = N_T \left(p_1 + p_1 q_{1,2} z^* \right) = N_T \left(\frac{1 \pm \sqrt{(p_2 - p_1)^2 + 4p_1 p_2 |z|^2}}{2} \right) \tag{7.77}$$

7.5 Preliminary comments on finite horizon prediction

The question of one step prediction using an infinite amount of past observation was considered above. The infinite amount of data was instrumental in supporting the assumption that the covariance matrices of the MIMO channel under consideration could be assumed to be perfectly known even if estimated from the data. On the other hand, the infinite past allows us to construct band-pass filters with arbitrary good separation between two non-overlapping frequency bands, thus reducing the effect of noise to a minimum.

A finite sample size has two major adverse effects on the performance of practical systems. On the one hand, the estimates of spatial correlation functions become less accurate, especially in terms of eigenvectors. In addition, the finite length of the past data allows only for a reduced resolution between two non-overlapping frequency bands thus not allowing them to achieve separation. As a result, the perceived signal bandwidth widens and the prediction quality declines even for one step prediction.

Since one of the goals we are investigating is the effect of the scattering environment on the limits of prediction, as an intermediate step we can assume that the correlation matrices are still perfectly known and the focus of the effects related to a finite sample size on spectral widening. Later we will make an attempt to include the problem of estimation of the correlation matrix itself.

A general MIMO $N_R \times N_T$ system of ULAs operating in frequency flat fading is considered. It is assumed that the velocity of the vehicular is constant, so is the maximum Doppler shift f_D. N sequential time samples of the channel matrix $\mathbf{H}(n)$ are available at the receiver side. The received signal is received in noise which is spatially and temporally white with variance σ_n^2 per receive dimension.

If $\mathbf{U}_R$ and $\mathbf{U}_T$ are unitary transforms which have passband selectivity in an angular domain then elements of the matrix

$$\mathbf{H}_u(n) = \mathbf{U}_R^H \mathbf{H}(n) \mathbf{U}_T \qquad (7.78)$$

represent the transfer function along virtual paths (Sayeed 2002; Tse and Viswanath 2005). Due to the narrowband (in the angular domain) nature of $\mathbf{U}_R$ and $\mathbf{U}_T$ the spectral content of elements of $\mathbf{H}_u(n)$ could be narrower than the elements of $\mathbf{H}(n)$ itself. The degree of bandwidth reduction depends on the quality of beamformers $\mathbf{U}_R$ and $\mathbf{U}_T$ and, thus the number of antenna elements and their spacing. As the number of antennas increases on both sides of the link it is possible to isolate individual paths, represented by a single entry of the matrix $\mathbf{H}_u$ (Sayeed 2002). If the number of antenna elements and/or their spacing is not sufficiently large, each individual entry of the matrix $\mathbf{H}_u$ is a combination of multiple paths that can be further resolved into individual subpaths only if a sufficient number of time samples is available.

Assume for now that the number of antennas on each side is large enough such that the individual entries of the matrix $\mathbf{H}_u(n)$ are independent. The optimal transform matrices $\mathbf{U}_R$ and $\mathbf{U}_T$ are the Fourier transform matrices (Sayeed 2002). Then each element must be predicted individually. Furthermore, in this case each element can be considered as a single path SISO channel. Prediction of such a channel will be treated in detail in Section 7.6. However, if the number of antennas is relatively small and perfect separation to the individual path is not possible, the elements of the matrix $\mathbf{H}_u$ can be considered as a sum of a few (but closely spaced) individual paths. This will be further treated in Section 7.6.2.

Finally, it worth mentioning that the noise transformation by a unitary matrix $\mathbf{U}_R$ does not change the AWGN nature of the received noise.

CSI quality criteria

It is shown in numerous papers that the MMSE is the ultimate quantity that describes the quality of communications over partially known channels (Misra et al. 2006; Tulino et al. 2005c) and so on. However, it is often possible to relate this parameter to more important communication quality factors such as the target probability of error P_{err} for a given modulation technique, a fraction of the capacity of the channel with perfect CSI, or the target signal to self-interference plus noise ratio (SSINR)

$$\tilde{\gamma} = SSINR = \frac{\sigma_s^2 \sigma_h^2}{\sigma_s^2 \sigma_\varepsilon^2 + \sigma_n^2} = \frac{1 - \delta_\tau \gamma}{1 + \delta_\tau \gamma} \tag{7.79}$$

where $\delta_\tau = \sigma_\varepsilon^2 / \sigma_h^2$ is the normalized error of the channel estimation, also known as the quality of CSI (Misra et al. 2006) and $\gamma = \sigma_h^2 \sigma_s^2 / \sigma_n^2$ is the SNR in the case of a perfectly known channel.

Relations between AoA and the frequency spectrum

If a plane wave with a fluctuating phase arrives to the antenna, moving with velocity v m/s the received signal is perceived as a a bandlimited signal with a central angular frequency

$$\omega_0 = \omega_D \cos \phi = 2\pi \frac{v}{c} f_0 \cos \phi \tag{7.80}$$

Here ϕ is the angle of arrival with respect to the direction of the antenna movement, f_0 is the carrier frequency, and c is the speed of light. Frequency bandwidth $\Delta \Omega$ associated with the fluctuating phase could be alternatively perceived as angular spread of the arriving plane waves (Zatman 1998). If such a spread is relatively narrow, that is $\Delta \omega \ll 2\pi$ can be attributed to a uniform spread $\Delta \phi$ of AoA defined as

$$\Delta \omega = \begin{cases} \omega_D \sin \phi_0 \, \Delta \phi & \text{if } \phi_0 \neq 0 \\ \omega_D \, \Delta \phi^2 / 2 & \text{if } \phi_0 = 0 \end{cases} \tag{7.81}$$

7.6   SISO channel prediction

Understanding SISO channel prediction is an important part of understanding MIMO prediction since, as we show later, it is possible to map the MIMO problem onto a set of loosely correlated SISO problems.

7.6.1   Wiener filter

The exact formal solution in terms of the number of samples and the normalized covariance function $\rho(\tau)$ of the received signal is known in the form of the Wiener filter (Haykin 1989). Using the following notation for the vector of N samples of the normalized covariance function shifted by the prediction horizon τ

$$\mathcal{R}(\tau) = [\rho(\tau), \rho(\tau + T), \cdots, \rho(\tau + (N-1)T)]^T \qquad (7.82)$$

the prediction error δ_τ at the horizon τ, as expressed by the normal equation (Haykin 1989; Papoulis 1991)

$$\delta_\tau = 1 - \gamma \mathcal{R}(\tau)^H (\mathbf{I} + \gamma \mathbf{R})^{-1} \mathcal{R}(\tau) \qquad (7.83)$$

In the simplest case of a single pilot predictor this reduces to

$$\delta_\tau = 1 - \frac{\gamma}{1 + \gamma} |\rho(\tau)|^2 \qquad (7.84)$$

These two expressions are our starting point in analysis of the prediction performance of the Wiener filter.

7.6.2   Single pilot prediction in a two cluster environment

For simplicity we once again consider a case of two sub-bands centered at

$$\omega_1 = \omega_D \cos \phi_{R1} \qquad (7.85)$$

$$\omega_2 = \omega_D \cos \phi_{R2} \qquad (7.86)$$

and bandwidth

$$\Delta \omega_1 = \Delta_{R1} \omega_D \sin \phi_{R1} \qquad (7.87)$$

$$\Delta \omega_2 = \Delta_{R2} \omega_D \sin \phi_{R2} \qquad (7.88)$$

respectively. The relative powers associated with the clusters are $0 \leq P_1 \leq 1$, and $P_2 = 1 - P_1$ respectively. The corresponding covariance function is then given by

$$\rho(\tau) = P_1 \exp(j\omega_1 \tau)\text{sinc}\,(\Delta f_1 \tau) + P_2 \exp(j\omega_2 \tau)\text{sinc}\,(\Delta f_2 \tau) \qquad (7.89)$$

We also assume that $\Delta \omega_1 + \Delta \omega_2 < 2\Delta\Omega = 2|\omega_1 - \omega_2|$. This assures that the sub-bands do not overlap. After a simple algebra, one can obtain the following expression for the squared magnitude of the covariance function

$$|\rho(\tau)|^2 = P_1^2 \text{sinc}^2\,(\Delta f_1 \tau) + P_2^2 \text{sinc}^2\,(\Delta f_2 \tau)$$
$$+ 2 P_1 P_2 \cos(\Delta\Omega\tau)\text{sinc}\,(\Delta\omega_1 \tau)\,\text{sinc}\,(\Delta\omega_2 \tau) \qquad (7.90)$$

Our goal is to investigate how parameters P_1, $\Delta\Omega$ and $\Delta\omega_{1,2}$ affect the prediction error δ. In order to simplify considerations we assume that $\Delta\omega_1 = \Delta\omega_2 = \Delta\omega = 2\pi\Delta f$. Taking this into account and recalling that $P_2 = 1 - P_1$ we can rewrite Equation (7.90) in the following form

$$|\rho(\tau)|^2 = \text{sinc}^2\,(\Delta f \tau) \times \left[P_1^2 + (1 - P_1)^2 + 2P_1(1 - P_1)\cos(\Delta\Omega\tau)\right] \qquad (7.91)$$

One conclusion, consistent with that reached in (Primak 2008a) can be deduced by inspection of (7.91): wider sub-bands $\Delta\omega$ result in a smaller value of $|\rho(\tau)|^2$ and, therefore, in a large variance of estimation error δ. Thus, it is safe to assume that this trend will remain for any number of past samples used for estimation.

The next step is to investigate the influence of the parameter P_1 on the quality of the estimate. It is easy to see that if $P_1 = 0$ or $P_1 = 1$, that is in the case of just a single cluster, the value of δ is defined only by spectral width $\Delta\omega$ of the corresponding cluster:

$$\delta_\tau = 1 - \text{sinc}^2\left(\Delta f \tau\right) \frac{\gamma}{1 + \gamma} \tag{7.92}$$

In order to find a possible extremum[9] let us consider a derivative of (7.91) with respect to P_1:

$$\frac{\partial}{\partial P_1}|\rho(\tau)|^2 = -2\text{sinc}^2\left(\Delta f \tau\right)\left[(1 - 2P_1)(1 - \cos\phi)\right] = 0 \tag{7.93}$$

which, for $\phi \neq 2n\pi$, has a unique solution $P_1 = 1/2$. It is easy to see that this value corresponds to the minimum of $|\rho(\tau)|^2$ and, therefore, the worst estimate.[10] The corresponding value of δ can be easily found to be

$$\delta_\tau = 1 - \text{sinc}^2\left(\Delta f \tau\right) \frac{\gamma}{1 + \gamma} \cos^2 \frac{\Delta\Omega\tau}{2} \tag{7.94}$$

Since we have assumed that $\Delta\Omega \gg \Delta\omega$, it is important to note that the term $\cos^2(\Delta\Omega\tau/2)$ varies faster than the term $\text{sinc}^2(\Delta f \tau)$ and, therefore, dominates the estimation error for the short prediction horizon $\tau \leq 1/\Delta\Omega$. In this case

$$\delta_\tau = 1 - \frac{\gamma}{1 + \gamma}\text{sinc}^2\left(\Delta f \tau\right)\cos^2\frac{\Delta\Omega\tau}{2} \approx 1 - \frac{\gamma}{1 + \gamma}\left(\frac{\Delta\omega^2}{3} + \frac{\Delta\Omega^2}{4}\right)\tau^2$$

$$= 1 - \frac{\gamma}{1 + \gamma}\frac{\Delta\omega_{eq}^2}{3} \approx 1 - \frac{\gamma}{1 + \gamma}\text{sinc}^2\left(\sqrt{3\Delta\Omega^2/4 + \Delta\omega^2}\right) \tag{7.95}$$

that is the prediction error δ is roughly the same as for a signal with the bandwidth

$$\Delta\omega_{eq} = \sqrt{3\Delta\Omega^2/4 + \Delta\omega^2} \approx 0.87\Delta\Omega \sim \Delta\Omega\tau \tag{7.96}$$

In other words, for a prediction based on a single pilot, the process appears as having bandwidth roughly equal to $\Delta\Omega$. Thus, for the estimation based on a single pilot and a short horizon the finite bandwidths $\Delta\omega$ of the signal from a single cluster could be neglected.

This is new phenomena which was not predicted in (Primak 2008a) for the infinite amount of samples in the past. This is of course due to a pure spectral resolution of a predictive filter with a single pole. At this point we would just speculate that as the number of samples in the past increases, a better resolution could be achieved and dominance of the term $\Delta\Omega$ will be reduced in favor of $\Delta\omega$.

---

[9] Expression (7.91) is a quadratic polynomial in P_1 and therefore may have no more than one extremum on the interval $[0; 1]$.

[10] The same result has been obtained in (Primak 2008a) in the case of an infinite number of samples in the past.

7.6.3 Single cluster prediction with multiple past samples

In this section we consider prediction in a single cluster environment, described by AoA ϕ_{R0} and angular spread Δ_R. The normalized covariance matrix of such channel is a simple modulated sinc function

$$\rho(\tau) = \exp\left(j2\pi\omega_D\cos\phi_{R0}\right)\,\text{sinc}\left(\Delta_R\omega_D\cos\phi_{R0}\tau\right) \tag{7.97}$$

In this case the signal covariance matrix $\mathbf{R}$ could be represented in terms of MDPSS (Percival and Walden 1993; Sejdic et al. 2008)

$$\mathbf{R} \approx \mathbf{U}^H \begin{bmatrix} \frac{N}{M}\mathbf{I}_M & \mathbf{0} \\ \mathbf{0}^H & \mathbf{0}_{N-M} \end{bmatrix} \mathbf{U} \tag{7.98}$$

Here

$$M = \lfloor \Delta_R N f_D \sin\phi_R / F_s \rfloor + 1 \tag{7.99}$$

is a number of degrees of freedom provided by the cluster, $\mathbf{0}$ is $M \times (N - M)$ zero matrix and $\mathbf{0}_{N-M}$ is $(N - M) \times (N - M)$ zero matrix. The matrix of the eigenvectors $\mathbf{U} = [\mathbf{U}_1 \mathbf{U}_0]$ consists of MDPSS rows with $\mathbf{U}_1$ being $M \times N$ matrix corresponding to non-zero eigenvalues and $\mathbf{U}_0$ is $(N - M) \times N$ matrix of eigenvectors corresponding to zero eigenvalues.

Furthermore, using notation $\mathcal{R}_1(\tau) = \mathbf{U}_1\mathcal{R}(\tau)$ one can rewrite error Equation (7.83) as

$$\delta_\tau = 1 - \gamma\mathcal{R}(\tau)^H\left(\mathbf{I} + \gamma\mathbf{R}\right)^{-1}\mathcal{R}(\tau)$$

$$= 1 - \frac{\gamma}{1 + \frac{N}{M}\gamma}\mathcal{R}_1(\tau)^H\mathcal{R}_1(\tau) = 1 - \frac{\gamma}{1 + \frac{N}{M}\gamma}||\mathcal{R}_1(\tau)||_F^2 \tag{7.100}$$

This is a natural generalization of the one sample prediction formula (7.84).

For a modest number of available samples in the past $M = 1$ since $\Delta_R N f_D \sin\phi_R / F_s < 1$. In this case $\mathbf{U}_1 = \mathbf{w}$ is a row vector approximately corresponding to the Kaiser window of length N (Percival and Walden 1993). In this case

$$\mathcal{R}_1(\tau)^H\mathcal{R}_1(\tau) = \sum_{n=0}^{N-1} w(n)^2|\rho(n)|^2 \tag{7.101}$$

where

$$\rho(n) = \rho(\tau + nT_s) \tag{7.102}$$

It is well known that the first DPSS is a symmetric function with respect to its peak at $n = N/2$. The function $|\rho(n)|^2$ can be approximated by its MacLaurent series around $t = \tau + N/2T_s$ as

$$|\rho(n)| = |\rho(\tau + NT_s/2)|^2 + s_1(n - N/2) + \frac{1}{2}s_2(n - N/2)^2 + \cdots \tag{7.103}$$

where

$$s_i = \frac{d^i}{dt^i} |\rho(t)|^2 \Big|_{t=\tau+NT_s/2} \tag{7.104}$$

Using this expansion in Equation (7.101), using symmetry of $\mathbf{w}$ and recalling that $\mathbf{w}^H \mathbf{w} = 1$ one can obtain the following approximation for the normalized prediction error

$$\delta_\tau = 1 - \frac{N\gamma}{1+\gamma N} |\rho(\tau + NT_s/2)|^2 - \frac{1}{2} \frac{N\gamma}{1+\gamma N} s_2 \sigma_w^2 \tag{7.105}$$

where

$$\sigma_w^2 = \sum_{n=0}^{N-1} |w(n)|^2 (n - N/2)^2 \tag{7.106}$$

is the time spread of the taper $\mathbf{w}$. The following conclusions can be reached through inspection of Equation (7.105):

- The increase in the number of past samples used for prediction increases effective SNR of the process from γ to $N\gamma$. This could be attributed to averaging. The largest gain from the increasing number of taps in the past is achieved for low SNR if the number of taps is small.

- At the same time using larger number of samples increases the prediction horizon from τ to $\tau + NT_s/2$.

- Most of the increase in performance is achieved due to the increase in SNR through accumulation over the few past samples that are strongly correlated.

- The strongest improvement is observed at $\tau = 0$ with the quality inadvertently decreasing as the prediction horizon increases.

It must be noted that a significant increase in N beyond the limits prescribed by Equation (7.99) would render the assumption of $M = 1$ invalid and Equation (7.105) could not be applied.

Results of numerical simulations of a single cluster scenario are discussed in Section 7.9.1 and shown in Figures 7.7 and 7.8.

7.6.4 Two cluster prediction with multiple past samples

In order to focus on the effect of the increased number of taps in the past on the quality of prediction in the case of the two cluster scattering environment, we assume for simplicity that both clusters have equal power $P_1 = P_2 = 1/2$ and identical angular spread $\Delta_{R1} = \Delta_{R2} = \Delta_R$. As was shown in Section 7.6.2 the angular spread of the cluster does not contribute significantly do the quality of prediction in comparison to the angle between clusters. Therefore, at first we assume $\Delta_R = 0$ for simplicity. The clusters are seen at the angles ϕ_{R1} and $\phi_{R2} \neq \pm\phi_{R1}$ respectively.

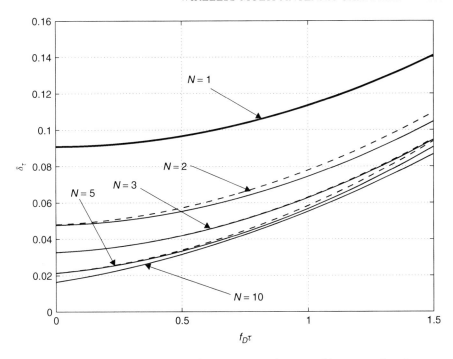

Figure 7.7   Prediction horizon of the Wiener predictor and its approximation as a function of the number N of past taps. Mean AoA of the signal from a single cluster is $\phi_{R0} = 30°$, $\Delta_R = 10°$, SNR 10 dB.

Let us define the following vectors of unit norm and length N.

$$\mathbf{q}_i = \frac{1}{\sqrt{N}} \left[1, a_i, a_i^2, \cdots, a_i^{N-1} \right], \ i = 1, 2 \qquad (7.107)$$

where

$$a_i = \exp \left[-j\omega_D T_s \cos \phi_{Ri} \right] \qquad (7.108)$$

It will be seen from the considerations below that the quantity

$$z = \mathbf{q}_2^H \mathbf{q}_1 = \exp \left[j\omega_D T_s N \left(\cos \phi_2 - \cos \phi_1 \right) / 2 \right]$$

$$\times \frac{\sin \left[\omega_D T_s N \left(\cos \phi_2 - \cos \phi_1 \right) / 2 \right]}{N \sin \left[\omega_D T_s \left(\cos \phi_2 - \cos \phi_1 \right) / 2 \right]} = |z| \exp(j\psi_z) \qquad (7.109)$$

Since we consider relatively slow fading $\omega_F T_s \ll 1$ for a small number of taps or very closely spaced clusters $|\cos \phi_2 - \cos \phi_1| \ll 1$ the quantity z is close to unity: $|z| \approx 1$. However, if N increases significantly, the value of z approaches zero even for signals spaced closely in an angular domain.

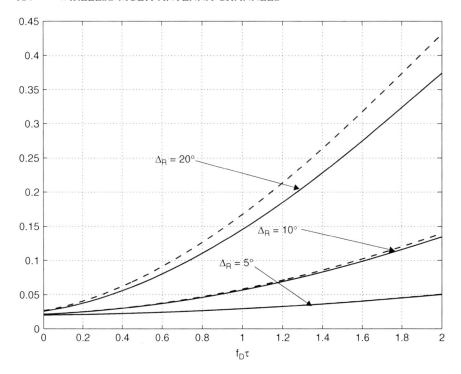

Figure 7.8   Prediction horizon of the Wiener predictor and its approximation as a function of the number received angular spread Δ_R. The number of taps $N = 5$. Mean AoA of the signal from a single cluster is $\phi_{R0} = 30°$, SNR 10 dB.

The covariance function, corresponding to two infinitely narrow clusters is then simply expressed in terms of $\mathbf{q}_i$ as

$$\mathbf{R} = \frac{N}{2}\mathbf{q}_1\mathbf{q}_1^H + \frac{N}{2}\mathbf{q}_2\mathbf{q}_2^H = [\mathbf{q}_1 \ \mathbf{q}_2]\begin{bmatrix} P_1N & 0 \\ 0 & P_2N \end{bmatrix}\begin{bmatrix} \mathbf{q}_1^H \\ \mathbf{q}_2^H \end{bmatrix} \qquad (7.110)$$

and

$$\mathcal{R}(\tau) = a_1^{\tau/T_s} P_1\mathbf{q}_1 + a_2^{\tau/T_s} P_2\mathbf{q}_2 \qquad (7.111)$$

Unless $z = 0$, expression (7.110) does not represent the SVD decomposition of $\mathbf{R}$. However, using the fact that the matrix $\mathbf{R}$ has rank 2 and $\mathbf{q}_1 \neq \mathbf{q}_2$ one can represent the eigenvectors $\mathbf{u}_i$ of $\mathbf{R}$ as a linear combination of $\mathbf{q}_1$ and $\mathbf{q}_2$, that is $\mathbf{u}_i = \alpha_i(\mathbf{q}_1 + \beta_i\mathbf{q}_2)$, $i = 1, 2$. Here real coefficient $\alpha_i > 0$ must be chosen in such a way that $\mathbf{u}^H\mathbf{u} = 1$. It is easy to see that the α_i must be chosen as

$$\alpha = \frac{1}{\sqrt{1 + |\beta|^2 + \beta^* z + \beta z}} \qquad (7.112)$$

Furthermore, simple calculations show that in the case of $P_1 = P_2 = 1/2$ the following equations must be satisfied

$$1 + \beta z^* = 2\lambda \tag{7.113}$$

$$z + \beta = 2\lambda\beta \tag{7.114}$$

which the solutions

$$\beta_{1,2} = \pm\exp(j\psi_z) \tag{7.115}$$

$$\lambda_{1,2} = \frac{1 \pm |z|}{2} \tag{7.116}$$

$$\alpha_{1,2} = \frac{1}{\sqrt{2(1 \pm |z|)}} \tag{7.117}$$

Alternatively, inversion of matrix $\gamma\mathbf{I} + \mathbf{R}$ can be accomplished using the so-called Woodbury's identity as discussed in Appendix 7.19. In this case

$$\left(\gamma^{-1}\mathbf{I} + \mathbf{R}\right)^{-1} = \gamma\left[\mathbf{I} - \frac{1}{d}\mathbf{F}\right] \tag{7.118}$$

where

$$\mathbf{F} = \left(1 + \frac{1}{N\gamma P_2}\right)\mathbf{q}_1\mathbf{q}_1^H - z^*\mathbf{q}_1\mathbf{q}_2^H - z\,\mathbf{q}_2\mathbf{q}_1^H + \left(1 + \frac{1}{N\gamma P_1}\right)\mathbf{q}_2\mathbf{q}_2^H \tag{7.119}$$

and

$$d = \left(1 + \frac{1}{N\gamma P_1}\right)\left(1 + \frac{1}{N\gamma P_2}\right) - |z|^2 \tag{7.120}$$

It is interesting to note, that the number N of taps in the past enters into the expression for the normalized variance of error only in two ways: through parameter z reflecting the decrease in $|z|$ with N and by increasing SNR by a factor of N. When number N of taps in the past is significant

$$N > \frac{2}{f_D T_s \left(\cos\phi_{R2} - \cos\phi_{R1}\right)} \tag{7.121}$$

the parameter z is small $z \approx 0$. In this case the normalized prediction error is given by

$$\delta_\tau = P_1\left(1 - \frac{N\gamma P_1}{1 + N\gamma P_1}\right) + P_2\left(1 - \frac{N\gamma P_2}{1 + N\gamma P_2}\right) \tag{7.122}$$

It is easy to recognize by inspection that the error δ_τ could be considered as a weighted sum of errors of prediction of each component separately with effective SNR $P_i\gamma N$: $\delta_\tau = P_1\delta_1 + P_2\delta_2$.

Thus, for large N (that is for small z) with an increase in N not only does SNR accumulate to produce a better estimate but the angular resolution also increases leading to a reduction of the spectral width of the signal and improving the overall

SNR. However, for small N angular resolution is very poor and the improvement is achieved solely on the increase in SNR.

It follows from considerations above that there are two different modes of operation of the Wiener filter depending on the number of taps and the length of the observation interval. If the length of the observation interval $N f_D T_s \ll 1/ |\cos \phi_{R1} - \cos \phi_{R2}|$ is such that signals corresponding to two clusters cannot be resolved in the angular domain, the effect of having N samples translates into the noise reduction by accumulating SNR $\gamma_{eff} = N\gamma$. However, clusters remain unresolved and the prediction horizon is determined by the net spectral width of the signal $\omega = \omega_D |\cos \phi_{R1} - \cos \phi_{R2}| \gg \omega_1, \omega_2$.

If a significantly larger observation interval is available, two clusters could be effectively separated in the angular domain and predicted separately. The effective SNR accumulated on each such cluster is scaled by the fraction of the net power of such a cluster and is less than the SNR accumulated over all samples. However, the spectral width of a resolved cluster could be significantly less than that of an unresolved cluster, thus increasing the prediction horizon. Some simulation results are discussed in Section 7.9.2.

It is reasonable to assume that accumulation of SNR is effective only on the interval of correlation of the signal. Therefore, it is expected that for a single cluster, the accumulation of SNR comes from $N_{eff} = \tau_{corr}/T_s$ samples. Thus, although a faster sampling rate does not increase angular resolution, it allows us to accumulate higher effective SNR.

All that said above can be summarized into the following rule of thumb. Let us consider a signal with correlation interval τ_{corr} sampled at the rate $F_s = 1/T_s$. If N samples of such a process are available the following approximation for the prediction error can be constructed:

1. if $N T_s \leq \tau_{corr}$ the fine structure of the signal cannot be considered. Wiener filtering performs an accumulation of SNR over all N samples. The predicted error is approximated by the following expression

$$\delta_\tau = 1 - \frac{N\gamma}{1 + N\gamma} |\rho(\tau)|^2 \qquad (7.123)$$

   where $\rho(\tau)$ represents the covariance function of the unresolved process

2. if the observation is long, that is $N T_s \gg \tau_{corr}$ there are $N_c = N T_s/\tau_{corr}$ independent samples that can be used to resolve the fine structure of the scattering environment. At the same time, SNR is accumulated over $N_{eff} = \tau_{corr}/T_s$. The expression for the error in prediction can be approximated as

$$\delta_\tau = \sum_{n=1}^{N_c} P_n \delta_{\tau n} \qquad (7.124)$$

   where P_n is the fraction of power associated with the n-th resolved mode (cluster) with covariance function $\rho_n(\tau)$ and

$$\delta_{\tau n} = 1 - \frac{P_n N_{eff} \gamma}{1 + P_n N_{eff} \gamma} |\rho_n(\tau)|^2 \qquad (7.125)$$

Since the resolution of a cluster results in a narrower spectral spread of such a cluster, the prediction horizon increases. On the opposite trend, the effective SNR of each cluster is reduced. This conclusion is consistant with one drawn in (Primak 2008a).

It is quite clear that the most difficult prediction is for a situation where there are many clusters distributed in such a way that corresponding power spectral density is uniform.[11]

While the considerations below shed some light on relationship between complexity, its structure, and the number of taps in the past used for a SISO channel prediction, an important characteristic is left out: how does SNR affect the number of identifiable modes? An intuitive answer to this is as follows: for a given number N of past taps the number of modes is equal to the number of singular values of the matrix $\mathbf{R}$ which exceed the noise level γ^{-1}, that is

$$|z| < 1 - \frac{2}{\gamma} \qquad (7.126)$$

This question will be further discussed in Section 7.7.

7.6.5   Role of oversampling

It follows from Equation (7.109) that the larger number of samples decreases with correlation between two harmonic vectors $\mathbf{q}_1$ and $\mathbf{q}_2$. The larger number of samples can be achieved through oversampling by zero-padding in the frequency domain (Papoulis 1991) since it is known that the signal is band-limited. It can be easily seen from Equation (7.109) that oversampling does not change the correlation between the harmonics since it depends on the overall duration of the available sample.

An increased number of samples obtained through zero-padding resampling also does not change the effect of accumulating SNR. Indeed, since zero-padding is used, samples of the noise become colored rather than white, making averaging much less efficient. Only new samples with independent noise can improve the performance of a predictor.

Thus, while it is beneficial to take samples at the higher rate by applying a high sampling rate with independent noise samples, obtaining such samples through zero-padding has no benefit. In pilot assisted estimation and prediction schemes, additional samples with independent noise can be obtained by either inserting additional pilots or by using recovered data symbols as pilots. Since such samples improve accumulation of SNR, data assisted prediction is useful mainly in low SNR regimes.

---

[11] This does not correspond to the uniform distribution of AoA as used in the principle of maximum entropy (Muller 2002). However, as we will see in the SIMO case, for a rectangular, or virtual array the maximum entropy distribution much more closely resembles the uniform one than in the SISO case. This is due to a better resolution of a rectangular array in both broadside and fire-end directions.

7.7 What is the narrowband signal for a rectangular array?

The material in this section is an extension of the method presented in (Zatman 1998) and reused in (van Trees 2002). The main goal is to define a rule on which a process can be judged a narrowband. Equivalently, it is related to the question of the array resolution in the angular domain which has a direct implication on the predictability of the process under consideration. For simplicity we restrict ourselves to the case of $2 - D$ geometry.

Let us consider a rectangular, uniformly spaced array of identical sensors with sides parallel to the x and y axes respectively. The number of elements and their spacing are L_x, L_y, d_x, and d_y respectively. Let two plane waves from two separate point sources seen at the azimuth ϕ_1 and ϕ_2 arrive at the array. The noise-free response of the rectangular array to each of these signals of unit power can be described by the following function of two variables

$$\mathbf{r}(m_x, m_y) = \exp\left(j2\pi \frac{m_x d_x \sin\phi + m_y d_x \cos\phi}{\lambda} \right) \tag{7.127}$$

The correlation $\mathbf{r}_s(m_{x2} - m_{x1}, m_{y2} - m_{y1})$ between any two elements (m_{x1}, m_{y1}) and (m_{x2}, m_{y2}) has Kronecker structure

$$\mathbf{r}_s(m_{x2} - m_{x1}, m_{y2} - m_{y1}) = \exp\left(j2\pi \frac{(m_{x2} - m_{x1})d_x \sin\phi}{\lambda} \right)$$

$$\times \exp\left(j2\pi \frac{(m_{y2} - m_{y1})d_y \cos\phi}{\lambda} \right) = a_x(\phi) \cdot a_y(\phi) \tag{7.128}$$

Therefore the full correlation matrix can be represented as

$$\mathbf{R}(\phi) = \mathbf{a}(\phi)\mathbf{a}^H(\phi) = \left(\mathbf{a}_x(\phi) \otimes \mathbf{a}_y(\phi)\right)\left(\mathbf{a}_x(\phi) \otimes \mathbf{a}_y(\phi)\right)^H$$

$$= \mathbf{a}_x(\phi)\mathbf{a}_x(\phi)^H \otimes \mathbf{a}_y(\phi)\mathbf{a}_y(\phi)^H = \mathbf{R}_x \otimes \mathbf{R}_y \tag{7.129}$$

where

$$\mathbf{a}_x(\phi) = \left[1, \ z_x, \ z_x^2, \ \cdots, \ z_x^{L_x-1}\right]^H \tag{7.130}$$

$$\mathbf{a}_y(\phi) = \left[1, \ z_y, \ z_y^2, \ \cdots, \ z_y^{L_y-1}\right]^H \tag{7.131}$$

and

$$z_x = \exp\left(j2\pi \frac{d_x}{\lambda} \sin\phi \right) \tag{7.132}$$

$$z_y = \exp\left(j2\pi \frac{d_y}{\lambda} \cos\phi \right). \tag{7.133}$$

are the response vectors from x and y directions respectively.

For two sources, the noise-free covariance matrix is a weighted sum of covariance matrices corresponding to each direction, that is

$$\mathbf{R}_s = \sigma_1^2 \mathbf{R}(\phi_1) + \sigma_2^2 \mathbf{R}(\phi_2) \tag{7.134}$$

where $\mathbf{R}(\phi_1)$ and $\mathbf{R}(\phi_2)$ are given by Equation (7.129) and σ_1^2 and σ_2^2 represent the power of waves coming from corresponding directions. In order to facilitate subspace separation of sources one has to find the eigenvectors $\mathbf{v}_k$ and corresponding eigenvalues λ_k, $k = 1, 2$ of $\mathbf{R}_s$. Proper separation could be achieved if both eigenvalues $\lambda_1 > \lambda_2 > \sigma_n^2$ exceed the noise level σ_n^2.

Following the procedure suggested in (Zatman 1998), one can represent the eigenvector $\mathbf{v}$ as a weighted sum of the eigenvectors $\mathbf{a}(\phi_1)$ and $\mathbf{a}(\phi_2)$ of $\mathbf{R}(\phi_1)$ and $\mathbf{R}(\phi_2)$ respectively

$$\mathbf{v} = \alpha \mathbf{a}(\phi_1) + \beta \mathbf{a}(\phi_2) \tag{7.135}$$

Thus

$$\lambda \mathbf{v} = \mathbf{R}_s \lambda = L_x L_y \left[\sigma_1^2 \alpha \mathbf{a}(\phi_1) + \sigma_1^2 \beta \psi \mathbf{a}(\phi_1) \right.$$
$$\left. + \sigma_2^2 \alpha \psi^* \mathbf{a}(\phi_2) + \sigma_2^2 \beta \mathbf{a}(\phi_2) \right] = \lambda \alpha \mathbf{a}(\phi_1) + \lambda \beta \mathbf{a}(\phi_2) \tag{7.136}$$

The quantity $\psi = \mathbf{a}(\phi_1)\mathbf{a}^H(\phi_2)/L_x L_y$ is the cosine of the angle between $L = L_x L_y$ dimensional vectors $\mathbf{a}(\phi_1)$ and $\mathbf{a}(\phi_2)$. It is found in (Zatman 1998) that

$$\lambda_{1,2} = \frac{1}{L} \left(\sigma_1^2 + \sigma_2^2 \right) \left[1 \pm \sqrt{1 - \frac{4\sigma_1^2 \sigma_2^2 \left(1 - |\psi|^2 \right)}{\left(\sigma_1^2 + \sigma_2^2 \right)^2}} \right] \tag{7.137}$$

which is further simplified for the case of equal power sources $2\sigma_1^2 = 2\sigma_2^2 = \gamma \sigma_n^2$

$$\lambda_{1,2} = \frac{L\gamma}{2} \sigma_n^2 \left(1 \pm |\psi| \right) \tag{7.138}$$

Here γ is the signal-to-noise ratio. As was mentioned earlier, two point sources can be distinguished if both eigenvalues exceed the variance of noise, that is if

$$\frac{L\gamma}{2} \left(1 - |\psi| \right) \geq 1 \tag{7.139}$$

Equation (7.139) defines a condition on the minimum angle $|\psi|$ between vectors $\mathbf{a}(\phi_1)$ and $\mathbf{a}(\phi_2)$ so that sources can be seen as separate

$$|\psi|_{\min} = 1 - \frac{2}{\gamma L} \tag{7.140}$$

Finally, we need to relate the value of $|\psi|$ with ϕ_1 and ϕ_2. This can be accomplished as follows

$$L\psi = \mathbf{a}^H(\phi_1)\mathbf{a}(\phi_2) = \mathbf{a}_x^H(\phi_1) \otimes \mathbf{a}_y^H(\phi_1) \cdot \mathbf{a}_x(\phi_2) \otimes \mathbf{a}_y(\phi_2)$$
$$= \left(\mathbf{a}_x^H(\phi_1)\mathbf{a}_x(\phi_2) \right) \otimes \left(\mathbf{a}_y^H(\phi_1)\mathbf{a}_y(\phi_2) \right) \tag{7.141}$$

Since products on both sides of the $\otimes$ signs are scalars, this Kronecker product is simply reduced to multiplication.

$$|\psi| = \left| \frac{\sin\left(L_x \pi \frac{d_x}{\lambda} \Delta \cos\phi\right)}{L_x \sin\left(\pi \frac{d_x}{\lambda} \Delta \cos\phi\right)} \frac{\sin\left(L_y \pi \frac{d_y}{\lambda} \Delta \sin\phi\right)}{L_y \sin\left(\pi \frac{d_y}{\lambda} \Delta \sin\phi\right)} \right| \qquad (7.142)$$

This quantity can be easily calculated directly for a small Δ. Indeed, using series expansion of ϕ with respect to Δ

$$|\psi| = 1 - \left[\left(L_x^2 - 1\right) d_x^2 \cos^2\phi + \left(L_y^2 - 1\right) d_y^2 \sin^2\phi\right] \frac{\pi^2}{6\lambda^2} \Delta^2 \qquad (7.143)$$

Comparing (7.143) with (7.140) one obtains the following expression for the minimum separation of point sources which can be observed as separate in noisy environments in the direction ϕ

$$\Delta(\phi) = \frac{2}{\pi} \sqrt{\frac{3\lambda^2}{\gamma L_x L_y \left[\left(L_x^2 - 1\right) d_x^2 \cos^2\phi + \left(L_y^2 - 1\right) d_y^2 \sin^2\phi\right]}} \qquad (7.144)$$

In the case of a virtual rectangular array $L_y = N_R$ is the number of transmit antennas, $L_x = N$ is the number of time samples available, $d_y = d_R$ is the distance between antennas, and $d_x = vT_s = f_D T_s \lambda$.

For a given value of SNR γ one can split the angular domain into bins $\Delta(\phi_k)$ corresponding to minimum angular distances for cluster separation. Within each bin the signal can be treated as uniformly distributed in an angle with.

7.8   Prediction using the UIU model

In this section we consider the prediction of a MIMO UIU channel (Tulino et al. 2005b) with constant spatial correlation properties, that is represented in the following form

$$\mathbf{H}(t) = \mathbf{U}_R \mathbf{H}_I(t) \mathbf{U}_R^H \qquad (7.145)$$

Equivalently, after vectorization,

$$\mathbf{h}(t) = \text{vec}\,\mathbf{H}(t) = \mathbf{U}_T^T \otimes \mathbf{U}_R \,\text{vec}\,\mathbf{H}_I(t) = \mathbf{U}\mathbf{h}_I(t) \qquad (7.146)$$

where the matrix $\mathbf{U} = \mathbf{U}_T^T \otimes \mathbf{U}_R$ is again unitary. The UIU model implies that all entries $h_k(t)$, $1 \le k \le N_T N_R$, of $\mathbf{h}_I(t)$ are independent between themselves. Therefore, prediction can be made component-wise. If the Wiener filtering is used on each $h_k(t)$, the resulting estimate is also optimal for $\mathbf{h}_I(t)$ and, therefore, $\mathbf{H}(t)$. Of course, the optimal estimation can also be obtained from correlation properties of $\mathbf{h}(t)$ directly. However, such a process may not be component-wise and, thus, more computationally complex. In the following we investigate sub-optimal element-wise prediction in MIMO channels.

Let $\mathbf{R}_I(\tau) = \mathcal{E}\left\{\mathbf{h}_I(t+\tau)\mathbf{h}_I^H(t)\right\}$ be the covariance matrix of the process $\mathbf{h}_I(t)$. Under UIU assumption the matrix $\mathbf{R}_I(\tau)$ is a diagonal, that is

$$\mathbf{R}_I(\tau) = \begin{bmatrix} \sigma_1^2\rho_1(\tau) & 0 & \cdots & 0 \\ 0 & \sigma_2^2\rho_2(\tau) & \cdots & 0 \\ \cdots & \cdots & \cdots\cdots \\ 0 & 0 & \cdots & \sigma_{N_T N_R}^2\rho_1(\tau) \end{bmatrix} \qquad (7.147)$$

We call the process $\mathbf{R}_I(\tau)$ *separable* in space and time if $\rho_k(\tau) = \rho(\tau)$ for all $1 \le k \le N_T N_R$, that is

$$\mathbf{R}_I(\tau) = \mathbf{D}\rho(\tau) \qquad (7.148)$$

where $\mathbf{D} = diag\{\sigma_k^2\}$ is a diagonal matrix of variances of components of $\mathbf{h}_I(t)$. Diagonal elements of these matrices also coincide with corresponding elements of the Weichselberger's coupling matrix $\mathbf{\Omega}$ (Weichselberger et al. 2006).

The covariance matrix $\mathbf{R}_\mathbf{h}(\tau)$ of the observable process $\mathbf{h}(t)$ can be represented as

$$\mathbf{R}_\mathbf{h}(\tau) = \mathcal{E}\left\{\mathbf{h}(t+\tau)\mathbf{h}^H(t)\right\} = \mathbf{U}\mathbf{R}_I(\tau)\mathbf{U}^H \qquad (7.149)$$

The latter could be further specialized to

$$\mathbf{R}_\mathbf{h}(\tau) = \mathbf{U}\mathbf{D}\rho(\tau)\mathbf{U}^H\rho(\tau) = \mathbf{R}_S\rho(\tau) \qquad (7.150)$$

if there is separation of correlation in time and space. Here $\mathbf{R}_S$ is the full spatial covariance function of the MIMO channel.

Finally, let $\hat{\mathbf{h}}_I(t)$ be an estimate of $\mathbf{h}_I(t)$ and $\boldsymbol{\varepsilon}_I(t) = \mathbf{h}_I(t) - \hat{\mathbf{h}}_I(t)$ be the corresponding estimation error. Then, the error $\boldsymbol{\varepsilon}_h(t)$ in estimation of $\mathbf{h}(t)$ is simply given by

$$\boldsymbol{\varepsilon}_h(t) = \mathbf{U}\boldsymbol{\varepsilon}_I(t) \qquad (7.151)$$

The variance matrices of estimation errors are also simply related to each other through the following unitary transformation

$$\mathbf{\Upsilon}_\mathbf{h} = \mathcal{E}\left\{\boldsymbol{\varepsilon}_h(t)\boldsymbol{\varepsilon}_h^H(t)\right\} = \mathbf{U}\mathbf{\Upsilon}_I\mathbf{U}^H \qquad (7.152)$$

and

$$||\mathbf{\Upsilon}_\mathbf{h}||_F^2 = tr\left(\mathbf{\Upsilon}_\mathbf{h}\mathbf{\Upsilon}_\mathbf{h}^H\right) = tr\left(\mathbf{U}\mathbf{\Upsilon}_I\mathbf{U}^H\right) = tr\left(\mathbf{\Upsilon}_I\mathbf{U}^H\mathbf{U}\right) = ||\mathbf{\Upsilon}_I||_F^2 \qquad (7.153)$$

7.8.1 Separable covariance matrix

For a scalar complex Gaussian process $\xi(t)$ with known normalized covariance function $\rho(\tau)$, the optimal predictor is well known (Papoulis 1991). For simplicity we consider predictions based only on a current measurement. The optimal predictor and the variance of error is then given by

$$\hat{\xi}(\tau) = \rho(\tau)\xi(0) \qquad (7.154)$$

$$\sigma_\varepsilon^2(\tau) = (1 - |\rho(\tau)|^2)\sigma_\xi^2 \qquad (7.155)$$

In the case of element-wise prediction of the vector $\mathbf{h}(t)$ one can easily obtain that vectors of variances of prediction errors have the following form

$$\sigma_\varepsilon^2 = (1 - |\rho(\tau)|^2) \text{ vec diag } \mathbf{R}_S \tag{7.156}$$

with overall estimation error for the whole channel

$$\sigma_\varepsilon^2(\tau) = (1 - |\rho(\tau)|^2) \text{ tr } \mathbf{R}_S \tag{7.157}$$

Now let us turn to estimation based on the unitary transform $\mathbf{h}_I(t)$ of $\mathbf{h}(t)$. It is easy to see that in this case the total prediction error is given by

$$\sigma_{\varepsilon,U}(\tau) = (1 - |\rho(\tau)|^2) \text{ tr } \mathbf{D} = (1 - |\rho(\tau)|^2) \text{ tr } \mathbf{R}_S = \sigma_\varepsilon^2(\tau) \tag{7.158}$$

Thus, it can be seen that in case of separable space-time covariance function the overall prediction error remains the same. It is important to note at this stage that we did **not** take into account measurement of noise.

7.8.2   1 × 2 unseparable example

Let us consider a simple case of a vector with two elements. This could correspond to a scenario when Alamouti coding is used. To simplify calculations even further we assume that both components have equal and unitary variance. In general, we assume that the spatial correlation coefficient $\rho_S = |\rho_S| \exp(j\psi_S)$ is known. Thus the observable process is given by

$$\mathbf{h}(t) = \begin{bmatrix} 1 & 0 \\ \rho_S & \sqrt{1 - |\rho_S|^2} \end{bmatrix} \mathbf{h}_I(t) \tag{7.159}$$

Here

$$\mathcal{E}\left\{\mathbf{h}_I(t + \tau)\mathbf{h}_I(t)^H\right\} = \begin{bmatrix} \rho_1(\tau) & 0 \\ 0 & \rho_2(\tau) \end{bmatrix} \tag{7.160}$$

and $\rho_1(0) = \rho_2(0) = 1$. Using Equations (7.159) and (7.160) the covariance function of the observable process $\mathbf{h}(t)$ can be found to be

$$\begin{aligned} \mathbf{R_h}(\tau) &= \mathcal{E}\left\{\mathbf{h}(t + \tau)\mathbf{h}(t)^H\right\} \\ &= \begin{bmatrix} 1 & 0 \\ \rho_S & \sqrt{1 - |\rho_S|^2} \end{bmatrix} \begin{bmatrix} \rho_1(\tau) & 0 \\ 0 & \rho_2(\tau) \end{bmatrix} \begin{bmatrix} 1 & \rho_S^* \\ 0 & \sqrt{1 - |\rho_S|^2} \end{bmatrix} \\ &= \begin{bmatrix} \rho_1(\tau) & \rho_1(\tau)\rho_S^* \\ \rho_1(\tau)\rho_S & \rho_1(\tau)|\rho_S|^2 + \rho_2(\tau)\left(1 - |\rho_S|^2\right) \end{bmatrix} \end{aligned} \tag{7.161}$$

It can be easily seen that

$$\mathbf{R_h}(0) = \begin{bmatrix} 1 & \rho_S^* \\ \rho_S & 1 \end{bmatrix} = \mathbf{R}_S \tag{7.162}$$

that is it poses a proper spatial correlation.

If prediction is made on each component of $\mathbf{h}(t)$ separately, the corresponding variance of the prediction error is a sum of variances of each component and, therefore, is given by

$$\sigma_{\varepsilon h}^2 = (1 - |\rho_1(\tau)|^2) + (1 - |\rho_1(\tau)|\rho_S|^2 + \rho_2(\tau)\left(1 - |\rho_S|^2\right)|^2) \tag{7.163}$$

At the same time, if the prediction is made based on the process $\mathbf{h}_l(t)$, the corresponding error is

$$\sigma_{\varepsilon l}^2 = (1 - |\rho_1(\tau)|^2)(1 + |\rho_S|^2) + (1 - |\rho_2(\tau)|^2)(1 - |\rho_S|^2) \tag{7.164}$$

After simple algebra we obtain that the difference between the estimation errors is given by

$$\Delta\sigma_\varepsilon^2 = \sigma_{\varepsilon h}^2 - \sigma_{\varepsilon l}^2 = |\rho_S|^2 \left(1 - |\rho_S|^2\right) |\rho_1(\tau) - \rho_2(\tau)| \geq 0 \tag{7.165}$$

As expected, the sub-optimal estimation error $\sigma_{\varepsilon h}^2$ is no less than $\sigma_{\varepsilon l}^2$ which is optimal for the UIU model. It can also be seen that equality is reached if both channels are uncorrelated $\rho_s = 0$, perfectly correlated $|\rho_S| = 1$, or have identical covariance functions. All this is consistent with the previous discussions.

7.8.3 Large number of antennas: no noise

If the number of transmit and receive ULA elements is large, it has been shown (Sayeed 2002) that the matrices $\mathbf{U}_R$ and $\mathbf{U}_T$ are simply DFT matrices. In this case the elements of the virtual MIMO channel $\mathbf{H}_l(t) = \mathbf{U}_R^H \mathbf{H}(t)\mathbf{U}_T$ represent narrow (in angular domain) beams leaving the transmit antennas at the spatial angle (AoD) ϕ_k and arriving at a spatial angle (AoA) ψ_l. Angles are measured with respect to the broadside of the corresponding arrays. The energy σ_{kl}^2 associated with such virtual channels can be easily calculated from the bi-directional power delay profile $P(\phi, \psi)$ as

$$\sigma_{kl}^2 = \int_{\phi_k^-}^{\phi_k^+} \int_{\psi_l^-}^{\psi_l^+} P(\phi, \psi)d\phi d\psi \approx \Delta\phi_k \Delta\psi_l P(\phi_k, \psi_l) \tag{7.166}$$

where $\Delta\phi_l = \phi_l^+ - \phi_l^-$ and $\Delta\psi_l = \psi_l^+ - \psi_l^-$ are angular spreads on each side of the link. The corresponding time covariance function can be approximated by a sinc type function (Salz and Winters 1994)

$$\rho_{kl}(\tau) = \exp\left[j2\pi f_D\tau \cos\hat{\psi}_l\right] \mathrm{sinc}\left(\Delta\psi_l f_D\tau \sin\hat{\psi}_l\right) \tag{7.167}$$

or, in general, with an exponent with slowly decaying magnitude $D_l(\tau)$

$$\rho_{kl}(\tau) = D_l\left(\Delta\psi_l f_D\tau \sin\hat{\psi}_l\right)\exp\left[j2\pi f_D\tau \cos\hat{\psi}_l\right] \tag{7.168}$$

Here $\hat{\psi} = \psi_l - \alpha_v$, where α_v is the angle between the velocity vector and the broadside of the array.

As the number of receive antennas N_T increases the angular resolution increases as $1/N_T$, that is

$$\Delta \psi_l \sim \frac{1}{N_T} \tag{7.169}$$

This, in turn, causes function $D_l(\tau)$ to vary more slowly and reduces the prediction error for a given prediction horizon τ. For a prediction based on a single snapshot of the channel

$$\sigma_{I,kl}^2(\tau) = \left(1 - |\rho_{kl}(\tau)|^2\right) \sigma_{kl}^2 = \left(1 - D_l^2(\tau)\right) \sigma_{kl}^2 \tag{7.170}$$

For sufficiently large N_T, or equivalently, for sufficiently small $\Delta \psi_l$, one can approximate the magnitude function $D_l(\tau)$ by a second order polynomial

$$1 - D_l(\tau) \approx a^2 \Delta^2 \psi_l \tau^2 \sim \frac{1}{N_R^2} \tag{7.171}$$

Thus, the prediction error in a single element of the vector $\mathbf{h}_I(t)$ rapidly approaches zero. Equivalently, the prediction horizon for a fixed prediction error increases. However, with the increase in number of antennas, the number of elements to be predicted also increases. At the same time the actual number of significant modes depends on the type of the bi-directional power angular spectrum $P(\phi, \psi)$.

If $P(\phi, \psi)$ is composed of L discrete (specular) components

$$P(\phi, \psi) = \sum_{l=1}^{L} P_l \delta(\phi - \phi_l) \delta(\psi - \psi_l) \tag{7.172}$$

the prediction error could be eventually driven to zero even based on a single time snapshot. This is compatible with the case considered in (Svantesson and Swindlehurst 2006).

The situation is remarkably different if $P(\phi, \psi)$ occupies a finite angular range in a continuous manner as in the case of distributed scatterers or clusters. For example, assuming that AoA and AoD are independent and uniform in the band Δ_ϕ and Δ_ψ, one would expect the number of significant virtual rays be proportional to $\Delta_\phi \Delta_\psi N_T N_R$. Therefore, the total error accumulates as N_T/N_R, that is it may remain constant if the aspect ratio of the MIMO system remains constant as the number of antennas increases.

7.8.4 Large number of antennas: estimation in noise

In the previous section it was assumed that $\mathbf{h}(0)$, or, equivalently $\mathbf{h}I(0)$ were known precisely. In practice, however, these values must be estimated from noisy measurements by means of pilot signals. This question is considered in detail in (Abdel-Samad et al. 2006) and we only state the result here (for now): The estimate $\hat{\mathbf{H}}$ of the channel $\mathbf{H}$ is such that the error matrix $\Delta \mathbf{H} = \mathbf{H} - \hat{\mathbf{H}}$ has independent equally distributed components with variance

$$\sigma_\varepsilon^2 = \gamma_p^{-1} N_T \tag{7.173}$$

where γ_p is the pilot SNR. Since the unitary transformation preserves independence and equal variance, it can be said that the variance of error in estimation of elements of $\mathbf{h}_l(0)$ is also given by Equation (7.173). Therefore, the prediction error of each virtual path, based on a single snapshot, can be written as

$$\sigma_{I,kl}^2(\tau) = \left(1 - |\rho_{kl}(\tau)|^2 \frac{\gamma_{kl}}{1+\gamma_{kl}}\right)\sigma_{kl}^2$$

$$= \left(1 - D_l^2(\tau)\frac{\gamma_{kl}}{1+\gamma_{kl}}\right)\sigma_{kl}^2 \qquad (7.174)$$

where

$$\gamma_{kl} = \frac{\sigma_{kl}^2}{\sigma_\varepsilon^2} = \frac{\sigma_{kl}^2}{N_T}\gamma_p \qquad (7.175)$$

is the virtual mode SNR. Once again using approximation (7.171) one can obtain the following approximation of the prediction error of a single virtual path

$$\sigma_{I,kl}^2(\tau) \approx \frac{1}{1+\gamma_{kl}} + a^2\Delta^2\psi_l\tau^2\frac{\gamma_{kl}}{1+\gamma_{kl}} \sim \frac{1}{1+\gamma_{kl}} + \frac{b^2\Delta^2\psi_l\tau^2}{N_R^2}\frac{\gamma_{kl}}{1+\gamma_{kl}} \qquad (7.176)$$

This error will not approach zero for any shape of the bi-directional angular spectrum. In fact, if the number of transmit antennas increases too much, this will lead to a very low mode SNR γ_{kl}, and thus to a significant prediction error.

7.8.5 Effects of the number of antennas, scattering geometry, and observation time on the quality of prediction

7.8.5.1 Link parameters

In this section we would like to discuss how different parameters of the MIMO link and propagation scenario may affect the quality of prediction of the MIMO process in terms of the squared Frobenious norm $\|\Delta\mathbf{H}\|_F^2$ of the matrix prediction error. It is assumed that the link has ULA antennas and the scattering environment is described by a bi-directional power angle spectrum (PAS) $Pp(\phi_T, \phi_R)$. There are N_T transmit antennas separated by distance d_T and N_R receive antennas separated by distance d_R. The vehicular moves with velocity v, therefore the maximum Doppler shift if $f_D = v/cf_0$. Samples of the channel are taken at the rate F_s samples per second. N time samples $\mathbf{H}(n)$, $n = 0, \cdots, N-1$ in the past are available for prediction. Due to a limited amount of data it is assumed that only one sided covariance matrices $\mathbf{R}_T = \mathcal{E}\{\mathbf{H}^H\mathbf{H}\}$ and $\mathbf{R}_R = \mathcal{E}\{\mathbf{H}\mathbf{H}^H\}$ could be properly estimated and used.

The main principle behind evaluation of the quality and horizon of prediction is the fact that *for a bandlimited process in noise the horizon of prediction is inversely proportional to a frequency band occupied by the signal*. Since the frequency spread for frequency flat fading is due to the angular spread in angle of arrival to the mobile terminal, this is equivalent to the fact that *the prediction horizon is in inverse relation to the angular spread of the signal at the receive terminal*.

7.8.5.2 Decoupling on the transmit side

Let $\mathbf{U}_T \mathbf{\Lambda}_T \mathbf{U}_T^H$ be SVD decomposition of the transmit correlation matrix. The unitary transformation

$$\mathbf{H}_1(n) = \mathbf{H}(n)\mathbf{U}_T \qquad (7.177)$$

allows us to convert a problem of prediction of $N_R \times N_T$ matrix MIMO problem into N_T independent $N_R \times 1$ SIMO problems. Indeed

$$\mathcal{E}\{\mathbf{H}_1^H \mathbf{H}_1\} = \mathbf{U}_T^H \mathcal{E}\{\mathbf{H}^H \mathbf{H}\}\mathbf{U}_T = \mathbf{\Lambda}_T = \text{diag}\{\lambda_{T,n_t}\} \qquad (7.178)$$

The last equation states that the columns $\mathbf{h}_{1,n_t}$, $n_t = 1, \cdots, N_T$ of the matrix $\mathbf{H}_1$ are independent. Therefore, the prediction could be performed column-wise as stated above. The errors of prediction $\sigma_{n_t}^2$ of each column could be then aggregated into the total error

$$||\Delta\mathbf{H}||_F^2 = \sum_{n_t=1}^{N_T} \lambda_{T,n_t} ||\Delta\mathbf{h}_{1,n_t}||_F^2 \qquad (7.179)$$

The transformation (7.177) can be considered as transmit eigen beamforming. Therefore, it creates N_T distinct angularly separated areas which are illuminated by a single transmit mode. This also may lead to some contraction of AoA on the receiving side due to the fact that for a given mode only a fraction of the total scattering region is now illuminated. If the joint distribution of AoA/AOD factorizes (Kronecker model), that is $p(\phi_T, \phi_R) = p_T(\phi_T)p_R(\phi_R)$ the choice of the mode on the transmit side does not affect the distribution of AoA on the receive side, the width of the receive spectrum remains the same for each SIMO channel. The only effect of the transformation (7.177) is to reduce the power allocated to each sub-channel. The normalized prediction error will be the same for each SIMO sub-channel since they all will have the same statistical properties, defined solely by distribution of AoA. The total error will become

$$||\Delta\mathbf{H}||_F^2 = ||\Delta\mathbf{h}_1||_F^2 \sum_{n_t=1}^{N_T} \lambda_{T,n_t} = ||\Delta\mathbf{h}_1||_F^2 \text{tr}\,\mathbf{R}_T \qquad (7.180)$$

Thus, we can conclude that for the Kronecker model the number of transmit antennas does not affect the overall quality of estimation.

If the scattering environment deviates from the Kronecker structure, the effect of the number of transmit antennas could be different. Indeed, for each transmit mode the conditional distribution of AoA $p_{R|n_t}(\phi_R|n_t)$ is no more spread than the unconditional AoA distribution $p_R(\phi_R)$ due to the fact that only a fraction of the scattering volume is illuminated by each transmit mode. Therefore, prediction error in each SIMO sub-channel will be smaller. The larger the number of transmit antennas, the better angular isolation between different modes can be achieved. At the same time the angular spread of each mode will be reduced, further constricting the illuminated scattering volume and the possible range of AoA for a given mode.

Therefore, it could be expected that the increase in the number of N_T will lead to reduction of the prediction error.

It is interesting to note that this conclusion is quite consistent with the maximum entropy principle (Muller 2002). Indeed, for two random variables with fixed marginal distribution, the joint distribution achieving maximum entropy corresponds to the independent case. Thus, the Kronecker model corresponds to the case of the maximum entropy and is the hardest to predict among the models with fixed one side correlations.

7.8.5.3 Prediction of SIMO sub-channel

At this stage we confine our considerations to a single SIMO (virtual) sub-channel with distribution of AoA given by a conditional distribution $p_{R|n_t}(\phi_T | n_t)$. Since all virtual channels are independent we can drop index n_t to shorten the notation.

One possible view of the prediction process based on N past time samples is to form a rectangular virtual array comprising time samples in one direction and spatial samples in another direction. The electrical size of such an array is $(N_R - 1)d_T \times (N - 1)f_D/F_s$. The resolution of such an array in the $u = \cos\phi_T$ domain (van Trees 2002) is defined by the extent of the diagonal

$$D^2 = (N_R - 1)^2 d_T^2 + (N - 1)^2 \frac{f_D^2}{F_s^2} \qquad (7.181)$$

Since all time samples have been used to form a virtual array, the prediction could be thought of as a prediction using a current sample only. Due to an limited angular resolution of the array, roughly

$$\Delta u \approx \frac{2}{D} \qquad (7.182)$$

At this juncture we may approach the problem in at least two ways. On one hand, for each virtual path $1 \leq n_t \leq N_T$ departing from the transmitter one can form the received covariance matrix $\mathbf{R}_{R,n_t}$

$$\mathbf{R}_{R,n_t} = \mathcal{E}\left\{\mathbf{h}_{n_t} \mathbf{h}_{n_t}^H\right\} \qquad (7.183)$$

Eigenvectors of corresponding SVD decomposition

$$\mathbf{R}_{R,n_t} = \mathbf{U}_{R,n_t} \Lambda_{\mathbf{R,n_t}} \mathbf{U}_{R,n_t}^H \qquad (7.184)$$

will further decompose each SIMO channel into N_R independent SISO channels. Overall, an arbitrary MIMO channel will be decomposed into $NT N_R$ independent SISO channels. This decoupling is precise in contrast to the Weichselberger UIU model. As in the Weichselberger method, parameters of each covariance matrix are estimated by averaging over time, receive and transmit antennas, in contrast to the direct estimation of the full $NTNR \times NTNR$ covariance matrix where each element is obtained by averaging only over time. The number of elements to be estimated is in this case $N_T N_R^2 + N_T^2$.

Stacking columns of the matrix $\mathbf{H}_1$ one can obtain the following model of the spatial coloring

$$\mathbf{h}_1 = \begin{bmatrix} \mathbf{R}_{R1} & \mathbf{0} & \cdots & \mathbf{0} \\ \mathbf{0} & \mathbf{R}_{R2} & \cdots & \mathbf{0} \\ \cdots & \cdots & \cdots & \cdots \\ \mathbf{0} & \mathbf{0} & \cdots & \mathbf{R}_{RN_T} \end{bmatrix} \boldsymbol{\xi} = \mathbf{R}_1 \boldsymbol{\xi} \tag{7.185}$$

Finally, one obtains

$$\mathbf{h} = \mathbf{U}_T \otimes \mathbf{I}_{N_R} \mathbf{R}_1 \boldsymbol{\xi} = \mathbf{R}_{\mathbf{h}}^{1/2} \boldsymbol{\xi} \tag{7.186}$$

that is, it does produce the same result as the estimation of the full correlation matrix.

7.9   Numerical simulations

7.9.1   SISO channel single cluster

In this section we illustrate the performance of the Wiener filter on the example with the parameters (Zemen and Mecklenbrauker 2005). Let us assume that the scatterers are concentrated around the transmit antenna only. In this case the assumption of a single narrow scatterer is justified if the distance between the antennas is much greater than the distance between the transmit antenna and major scattering centers. In this case $F_c = 2\,\text{GHz}$, $v = 19.4\,\text{m/s}$ (or $70\,\text{km/h}$). The bit duration is $T_s = 20\,\mu s$, the block size is $M = 256$. Following (Zemen and Mecklenbrauker 2005) we choose 5 uniformly spaced pilot symbols per block. Results of the simulations are shown in Figure 7.7 and 7.8. The Figures show good agreement between the approximation (7.105) for the range of interest and the exact results. Figure 7.8 indicates clearly that the narrower clusters make prediction easier.

7.9.2   Two cluster prediction

The same system parameters are used as in Section 7.9.1 above. However, the scattering environment is now represented by two clusters rather than one. The goal of the simulation is to verify some conclusions reached in Section 7.6.2. The results of prediction in the two cluster environment with $\phi_{R1} = 30°, \phi_{R1} = 35°$ and $P_1 = 0.3$ are shown in Figure 7.9. As could be seen for a small number of past taps, Equation (7.123) provides a reasonable fit, indicating that no clusters are resolved but SNR is accumulated. As the number of taps increases approximately to $N = 50$ a better spatial resolution is achieved and it is now Equation (7.124) that provides a better fit.

A similar scenario is shown in Figure 7.10. However, in this case the clusters are separated by a wider angle. For small N this results in a smaller prediction horizon since the spectral width of such a cluster is wider. However, as N increases, the approach to the solution provided by Equation (7.124) is much faster. This is consistent with the conclusions presented in Section 7.6.2 since a smaller resolution (i.e., a smaller number of independent samples) is required to resolve these clusters.

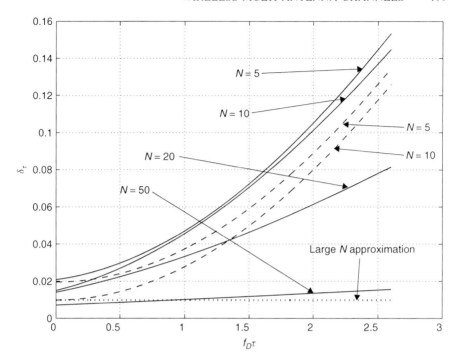

Figure 7.9   Prediction horizon of the Wiener predictor and its approximation as a function of the number N of past taps in the case of two clusters: $\phi_{RI} = 30°, \phi_{RI} = 35°, \Delta_R = 1°$, SNR 10 dB. Solid lines correspond to the exact Wiener filter solution. Dashed line is an approximation using Equation (7.123), while the approximation using Equation (7.124) is shown as a dotted line.

7.10   Wiener estimator

The Wiener estimator and its prediction error for the process with the normalized covariance function $\rho(\tau)$ and N taps can be obtained as

$$\mathbf{w}^H(\tau) = \frac{1}{\sqrt{E_p}} \mathcal{R}^H(\tau) \left(\mathcal{R} + \gamma_p^{-1}\mathbf{I}\right)^{-1} \tag{7.187}$$

$$\delta_\tau = \frac{\sigma_\varepsilon^2(\tau)}{\sigma_h^2} = 1 - \mathcal{R}^H(\tau)\left(\mathcal{R} + \gamma_p^{-1}\mathbf{I}\right)^{-1}\mathcal{R}(\tau) \tag{7.188}$$

Here $\gamma_p = E_p\sigma_p^2/\sigma_n^2$ is the pilot SNR;

$$\mathcal{R} = \begin{bmatrix} 1 & \rho(\tau_{p(N_T-n_t)} - \tau_{p(N_T-n_t+1)}) & \cdots & \rho(\tau_{N_T-n_t} - \tau_{pN_T}) \\ \rho(\tau_{\tau_{p(N_T-n_t+1)-p(N_T-n_t)}}) & 1 & \cdots & \rho(\tau_{N_T-n_t} - \tau_{pN_T}) \\ \cdots & \cdots & \cdots \cdots & \\ \rho(\tau_{pN_T} - \tau_{N_T-n_t}) & \rho(\tau_{pN_T} - \tau_{N_T-n_t+1}) & \cdots & 1 \end{bmatrix}$$

$$\tag{7.189}$$

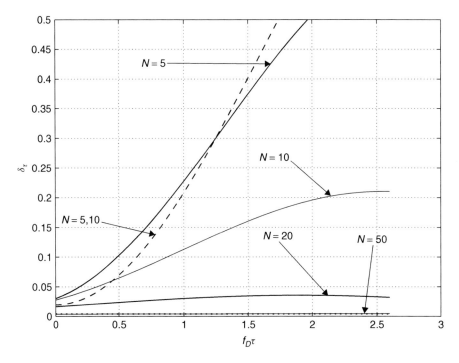

Figure 7.10   Prediction horizon of the Wiener predictor and its approximation as a function of the number N of past taps in the case of two clusters: $\phi_{R1} = 30°, \phi_{R1} = 45°, \Delta_R = 1°$, *SNR 10 dB. Solid lines correspond to the exact Wiener filter solution. Dashed line is an approximation using Equation (7.123) while the dotted line is an approximation using Equation (7.124).*

is the normalized covariance matrix at the last n_t pilot locations and $\mathcal{R}(\tau)$ is defined as

$$
\mathcal{R}(\tau) = \frac{1}{\sigma_h^2} \mathcal{E} \left\{ h(\tau) \begin{bmatrix} h^*(-\tau_D + \tau_{p(N_T - n_t)}) \\ h^*(-\tau_D + \tau_{p^*(N_T - n_t + 1)}) \\ \dots \\ h(-\tau_D + \tau_{p(N_T)}) \end{bmatrix} \right\}
$$

$$
= \begin{bmatrix} \rho(\tau + \tau_D - \tau_{p(N_T - n_t)}) \\ \rho(\tau + \tau_D - \tau_{p^*(N_T - n_t + 1)}) \\ \dots \\ \rho(\tau + \tau_D - \tau_{p(N_T)}) \end{bmatrix} \tag{7.190}
$$

If only one training symbol, located at τ_{p1} in the frame, equations for the estimation filter, the estimate, and the variance of the error becomes

$$
w^*(\tau) = \frac{1}{\sqrt{E_p}} \frac{\rho(\tau + \tau_D - \tau_{p1})}{1 + \gamma_p^{-1}} \tag{7.191}
$$

$$\hat{h}(\tau) = \frac{r_{p1}}{\sqrt{E_p}} \frac{\rho(\tau + \tau_D - \tau_{p1})}{1 + \gamma_p^{-1}} \tag{7.192}$$

$$\delta_\tau = 1 - \frac{|\rho(\tau + \tau_D - \tau_{p1})|^2}{1 + \gamma_p^{-1}} \tag{7.193}$$

Finally, the instantaneous effective SNR $\gamma_{eff}(\tau)$ is given by

$$\gamma_{eff}(\tau) = \frac{E_s |\hat{h}(\tau)|^2}{\sigma_\varepsilon^2 E_s + \sigma_n^2}$$

$$= \frac{|\rho(\tau + \tau_D - \tau_{p1})|^2}{\left[\left(1 + \gamma_p^{-1}\right)\left(1 + \gamma_s^{-1}\right) - |\rho(\tau + \tau_D - \tau_{p1})|^2\right] E_p \left(1 + \gamma_p^{-1}\right)} \frac{|r_{p1}|^2}{} \tag{7.194}$$

It is worth mentioning that once a frame is transmitted and its parameters are estimated at the receiver, the quality of the estimates at the receiver is better than the quality of the prediction based on the outdated pilot tones. This is due to the fact that the receiver uses the most recent pilots. Thus, on average, the predicted effective SNR is a lower bound for the effective SNR at the receiver, thus insuring the desired BEP.

It also follows from Equation (7.194) that the optimal place for placement of the pilot, from the transmitter perspective is in the very end of the frame. In this case the estimates are the least outdated. Of course, an assignment based only on the prediction without considering conditions at the receiver are conservative and do not implement an optimal strategy. Joint consideration, similar to that in (Falahati et al. 2005) is needed.

7.11   Approximation of the Wiener filter

While an exact expression for the Wiener filter (7.187) allows for efficient numerical simulation as long as the condition number of the covariance matrix is not too large, it is beneficial to obtain an approximate but simple expression which allows for analysis of the prediction error.

For practical purposes we are interested only in a region where MMSE is relatively small, say $\varepsilon^2 \le 0.2$. This means that we are considering relatively short prediction horizons (in terms of $f_D T$), and, therefore, the covariance function $\rho(\tau)$ can be well approximated by its Taylor series for the non-negative[12] $\tau \ge 0$ with respect to the origin.

$$\rho(\tau) \approx 1 + \rho'(0)\tau + 0.5\rho''(0)\tau^2 = 1 + \mu a_1 \tau - \mu^2 a_2 \tau^2 \tag{7.195}$$

where $\mu = f_D T_s$ is a small parameter. This representation allows us to use perturbation theory for approximating the behavior of the Wiener filter and of the

---

[12] This assumption restricts us only to a prediction problem. However, it allows us to avoid the fact that the covariance function may not have a continuous derivative at $\tau = 0$.

prediction error. Let us further assume that N past samples of the signal are used to predict the channel at the horizon $\tau_0 = m_0 T_s$.

7.11.1   Zero order approximation

In this case it is assumed that all elements of the matrix $\mathcal{R}$ are equal to one and the elements of the vector $\mathcal{R}$ are all equal to $\rho(\tau_0)$, that is

$$\mathcal{R} = \mathbf{1}\mathbf{1}^H, \quad \mathcal{R}(\tau_0) = \rho(\tau_0)\mathbf{1} \tag{7.196}$$

where $\mathbf{1} = [1, 1, \cdots, 1]^T$ is a column vector of size $N \times 1$.

Using the Sherman–Morrison form of the Woodbury formula

$$\left(\mathbf{A} + \mathbf{u}\mathbf{v}^H\right)^{-1} = \mathbf{A}^{-1} - \frac{\mathbf{A}^{-1}\mathbf{u}\mathbf{v}^H\mathbf{A}^{-1}}{1 + \mathbf{v}^H\mathbf{A}^{-1}\mathbf{u}} \tag{7.197}$$

(van Trees 2002) one can easily obtain the expression for the inverse of the matrix $\mathcal{R} + \gamma_d^{-1}\mathbf{I}$ as follows

$$\left(\mathcal{R} + \gamma_d^{-1}\mathbf{I}\right)^{-1} = \gamma_d\mathbf{I} - \frac{\gamma_d^2}{1 + \gamma_d N}\mathbf{1}\mathbf{1}^H \tag{7.198}$$

Furthermore, the equation for the filter coefficients $\mathbf{w}(\tau_0)$ and the normalized estimation error δ_{τ_0} will simply become

$$\mathbf{w}^H(\tau_0) \approx \mathcal{R}(\tau_0)^H \left(\mathcal{R} + \gamma_d^{-1}\mathbf{I}\right)^{-1}$$

$$\approx \rho^*(\tau_0)\mathbf{1}^H \left(\gamma_d\mathbf{I} - \frac{\gamma_d^2}{1 + \gamma_d N}\mathbf{1}\mathbf{1}^H\right) = \rho^*(\tau_0)\frac{\gamma_d}{1 + \gamma_d N}\mathbf{1}^H \tag{7.199}$$

and

$$\delta_{\tau_0}^0 \approx 1 - |\rho(\tau_0)|^2 \frac{\gamma_d N}{1 + \gamma_d N} \tag{7.200}$$

respectively. These results can be simply interpreted as follows. On the rough scale, the Wiener filter performs averaging of the available channel measurements and their scaling to take into account variance of the additive noise. Then prediction is made assuming the first order prediction model, that is prediction is based only on the current (but averaged) sample.

7.11.2   Perturbation solution

The next step is to take into account the perturbation of the solution (7.199) and (7.200) caused by variation of the correlation function in time. In other words we need to consider higher order terms in expansion (Equation 7.195). If the bandwidth occupied by the covariance function is F and we consider N samples of the covariance function taken every T seconds then the rank of the covariance matrix is approximately equal to $\lfloor 2FT \rfloor + 1$. For modest N this is still approximately one. Therefore, only the first eigenvalue and corresponding eigenvector in spectral decomposition of $\mathbf{R} = \mathbf{U}\mathbf{\Lambda}\mathbf{U}^H$ should be taken into account. Since for the normalized covariance matrix $\mathbf{R}$ has the trace equal to the dimension of the matrix $\text{trace}(\mathbf{R}) = N$, the largest eigenvalue is also equal to N: $\lambda_1 \approx N$. Therefore

$$\mathbf{R} \approx N\mathbf{u}_1\mathbf{u}_1^H \tag{7.201}$$

where $\mathbf{u}_1$ is the unity norm eigenvector corresponding to the eigenvalue λ_1 and $\mathbf{u}_1^H\mathbf{u}_1 = 1$. Application of the Woodbury equality (7.197) now leads to the following expression for inverse of the matrix $\mathbf{R} + \gamma_p^{-1}\mathbf{I}$:

$$\left(\mathbf{R} + \gamma_p^{-1}\mathbf{I}\right)^{-1} \approx \gamma_d\mathbf{I} - \frac{\gamma_d^2 N}{1 + \gamma_d N}\mathbf{u}_1\mathbf{u}_1^H \tag{7.202}$$

For convenience, and without loss of generality, we assume that $N = 2M + 1$ is an odd number. In this case the vector $\mathcal{R}(\tau_i)$ could be approximated as

$$\mathcal{R}(\tau_0) \approx \rho(\tau_0 + MT_s)\left(1 + \rho'(\tau_0 + MT_s)\mathbf{x_1}\right) \tag{7.203}$$

where $\rho'(\tau_0 + MT_s)$ is the slope of the covariance function as $\tau = \tau_0 + MT_s$ and $\mathbf{x_2} = [-M, -M+1, \cdots, M-1, M]^T$.

Once again, for moderate values of N the eigenvector $\mathbf{u}_1$ can be approximated by the Kaiser window (Percival and Walden 1993) or by a Gaussian window. Therefore, it could be represented as

$$\mathbf{u}_1 \approx \frac{1}{\sqrt{N}}\left(1 - \beta^2\mathbf{x}_2\right) \tag{7.204}$$

where β is some coefficient and $\mathbf{x}_2 = \mathbf{x}_1 \odot \mathbf{x}_1$. Therefore, if we are looking for a correction term of order $\mathbf{x}_1$ only, we can use $\mathbf{u}_1 \approx 1/\sqrt{N}$, while for higher order approximation a correction term of the second order must be used. Since $\mathbf{1}^H\mathbf{x} = \mathbf{x}^H\mathbf{1} = 0$ it is easy to see that the first order correction terms vanishes, that is

$$\delta_{\tau_0}^1 \approx 1 - |\rho(\tau_0 + MT_s)|^2\frac{\gamma_d N}{1 + \gamma_d N} \tag{7.205}$$

which is only a slight modification of the Equation (7.200).

7.12 Element-wise prediction of separable process

Let us assume that a zero mean complex vector Gaussian process $\mathbf{x}(t)$ is described by the following covariance matrix function

$$\mathbf{R}(\tau) = \mathcal{E}\left\{\mathbf{x}(t + \tau)\mathbf{x}^H(t)\right\} = \mathbf{R}_s\rho(\tau) \tag{7.206}$$

where $\rho(\tau)$ is a normalized covariance function; $\rho(0) = 1$. The optimal element-wise predictor for this process can be written as

$$\hat{\mathbf{x}}(\tau) = \mathbf{W}(\tau)\mathbf{x}(0) = \text{diag}\{\mathbf{R}_s\}\rho(\tau)\mathbf{x}(0) \tag{7.207}$$

with corresponding total MSE equal to

$$\sigma_\varepsilon^2 = \left(1 - |\rho(\tau)|^2\right) \text{trace } \mathbf{R}_s \tag{7.208}$$

Let $\mathbf{R}_s = \mathbf{U}\mathbf{\Lambda}_s\mathbf{U}^H$ be SVD decomposition of the the matrix $\mathbf{R}_s$. Consider vector $\mathbf{y}(t) = \mathbf{U}^H\mathbf{x}(t)$. Its covariance function is simply

$$\mathbf{R}_y(\tau) = \mathcal{E}\{\mathbf{y}(t + \tau)\mathbf{y}(t)\} = \mathbf{U}^H\mathbf{R}_s\mathbf{U}\rho(tau) = \mathbf{\Lambda}_s\rho(\tau) \tag{7.209}$$

Therefore, the optimal element-wise prediction strategy is

$$\hat{\mathbf{y}}(\tau) = \mathbf{\Lambda}_s\rho(\tau)\mathbf{y}(0) \tag{7.210}$$

with the total MSE error

$$\sigma_{\varepsilon\mathbf{y}}^2 = \left(1 - |\rho(\tau)|^2\right) \text{trace } \mathbf{\Lambda}_s \tag{7.211}$$

7.13   Effect of prediction and estimation errors on capacity calculations

The capacity of fading channels is a well-studied topic in communications theory under the assumption that perfect channel state information (CSI) is available at the receiver and possibly at the transmitter (Cover and Thomas 2002; Goldsmith and Varaiya 1997). For such ideal conditions, achievable data rates are bounded by the capacity and depend only on the first order statistics of the channel. In reality, however, the CSI has to be estimated from data sent by the transmitter and is therefore inherently inaccurate due to receiver noise, non-zero fading rates and a limited amount of available measurements. Therefore, the achievable rates are limited in practice not only by noise, but also by channel estimation errors, which also depend on the fading rate.

The question of finding maximal achievable rates and optimal design of systems employing pilot symbol assisted estimation has been studied for quite a while with initial results obtained in (Cavers 1991). Most of the authors have focused on optimizing pilot placement based on the desired bit error rate for fixed rate transmission (Zhang and Ottersten 2003) or block adaptive modulation (Cai and Giannakis 2005) and assume that the same quality of estimates are available at all signal-to-noise ratio (SNR) values and all data symbol positions (Cai and Giannakis 2005; Øien et al. 2004). A new approach to solve this problem is proposed in (Abou-Faycal et al. 2005; Medard 2000) by finding a lower bound on the capacity of the channel with imperfect CSI. However, only the case of CSI at the receiver and a first-order Markov channel model was considered.

The main contributions of this section are as follows. Analytical expressions for the achievable data rates are derived for a number of pilot-based estimators based on prior knowledge of the channel covariance function. An optimal power control scheme for a channel with imperfect CSI is then proposed and its performance is evaluated. The optimal frame size is also studied for synthetic channels with varying angular spreads.

The rest of this chapter is organized as follows. Section 7.14 introduces the system model and analytical parameters related to the channel estimation quality that are used in this section. In Section 7.15, the optimal transmitter power allocation strategy for the case of imperfect CSI is developed and the corresponding optimal achievable rates are derived. Performance examples are presented in Section 7.16, and conclusions are drawn in Section 7.17.

7.14 Channel estimation and effective SNR

7.14.1 System model

The frequency flat fading channel is modeled to be a complex zero-mean circularly symmetric Gaussian random process $h(t)$ with covariance function $R(\tau) = \sigma_h^2 \rho(\tau)$, $\rho(0) = 1$. The received signal $r(t)$ is given by

$$r(t) = h(t)s(t) + \xi(t), \quad t = nT_s, \; n \in \mathcal{N}^+. \tag{7.212}$$

Here, T_s is the bit duration and $s(nT_s)$ represents the signal transmitted during the n-th time interval. The additive white Gaussian noise (AWGN) $\xi(t)$ has the variance σ_n^2 and is independent of both $h(t)$ and $s(t)$. The transmitted signal has the variance $\sigma_s^2 = \gamma \sigma_n^2$, where γ denotes the data SNR. The time-varying channel is estimated at the receiver side using pilots periodically inserted into the transmitted data stream: one pilot per frame consisting of N symbols. The energy of the pilot is $E_p = \gamma_p T_s \sigma_n^2$, where the quantity γ_p represents the pilot SNR.

7.14.2 Estimation error

While it is common to distinguish between interpolation, estimation, and prediction based on the location of the symbols of interest (Papoulis 1991), in the following we use the term estimation for all three cases without loss of generality. The minimum mean-squared error (MMSE) estimator for the model assumed above is linear (Papoulis 1991)

$$\hat{h}(\tau) = \sum_{l=-L_1}^{L_2} \alpha_l(\tau) \left[h(lNT_s)\sqrt{\frac{E_p}{T_s}} + \xi(lNT_s) \right] \tag{7.213}$$

Here, $\tau = nT_s$, $n \in [1, N-1]$, L_1, and L_2 are the number of pilots in the past and future which can be used for estimation; $\alpha_l(\tau)$ are estimation filter coefficients. The notation $P(L_1, L_2)$ is used to denote an optimal estimation scheme which uses L_1 and L_2 past and future pilots, respectively. The weights $\alpha(\tau)$ for every prediction horizon τ can be found using the normal equation (see equation (14-65) on page 500 in (Papoulis 1991)). Using the principle of orthogonality of the estimation error and the observation (Papoulis 1991), we can obtain expressions for the normalized variance δ_τ of the estimation error $\varepsilon(\tau) = h(\tau) - \hat{h}(\tau)$ and the variance $\sigma_{\hat{h}}^2(\tau)$ of the estimate $\hat{h}(\tau)$

$$\delta_\tau = 1 - \sigma_{\hat{h}}^2(\tau)/\sigma_h^2 = \sigma_\varepsilon^2(\tau)/\sigma_h^2, \tag{7.214}$$

where $\sigma_\varepsilon^2(\tau)$ is the variance of the estimation error. The quantity δ_τ is called the CSI quality by some authors (Misra et al. 2006). Analytical expressions can be easily obtained for the following estimation schemes: $P(0, 0)$ (causal single pilot), $P(1, 0)$ (causal two pilots), and $P(0, 1)$ (non-causal two pilots). Moreover, we can show that for the $P(0, 0)$ scheme

$$\delta_\tau = 1 - \frac{\gamma_p}{1 + \gamma_p}|\rho(\tau)|^2, \tag{7.215}$$

while for the $P(0, 1)$ and $P(1, 0)$ schemes one has

$$\delta_\tau = 1 - \frac{\left(1 + \gamma_p^{-1}\right)\left(|\rho(\tau)|^2 + |\rho(s)|^2\right) - 2\Re\rho^*(T)\rho(\tau)\rho(s)}{\left(1 + \gamma_p^{-1}\right)^2 - |\rho(\tau)|^2}. \tag{7.216}$$

Here, $\Re(z)$ is the real part of z, $T = NT_s$, $s = T - \tau$ holds for the $P(0, 1)$ scheme, and $s = T + \tau$ holds for the $P(1, 0)$ scheme. It is important to note that δ_τ is a function of the pilot SNR γ_p, which was not taken into account in earlier publications (Yoo and Goldsmith 2006).

The best possible estimator takes into account all past and future samples, that is $L_1 = L_2 = \infty$. While impractical, this asymptotic case provides a lower bound on the estimation error that can be related to the properties of the fading signal. Due to overlap of the noise and signal power spectral densities, estimates that are based on even infinitely many samples in the past and future are subject to error. The normalized MMSE error δ is given by the (van Trees 2001)

$$\delta = \frac{1}{\sigma_s^2}\frac{1}{2\pi}\int_{-\pi}^{\pi}\ln\left[\sigma_n^2 + S(\omega)\right]d\omega = \frac{\sigma_n^2}{\sigma_s^2}\frac{1}{2\pi}\int_{-\pi}^{\pi}\ln\left[1 + \frac{S(\omega)}{\sigma_n^2}\right]d\omega. \tag{7.217}$$

If a signal has power spectral density (PSD) $S(\omega)$ composed of K spectrally separable segments with uniform power density

$$S(\omega) = \sigma_s^2\sum_{k=1}^{K}\frac{2\pi P_k}{\Delta\omega_k}\Pi(\omega_k, \Delta\omega_k) \tag{7.218}$$

then the corresponding error can be easily computed as

$$\delta = \frac{\sigma_n^2}{\sigma_s^2}\frac{1}{2\pi}\int_{-\pi}^{\pi}\ln\left[1 + \frac{\sigma_s^2}{\sigma_n^2}\sum_{k=1}^{K}\frac{2\pi P_k}{\Delta\omega_k}\Pi(\omega_k, \Delta\omega_k)\right]d\omega$$

$$= \sum_{k=1}^{K}P_k\frac{\Delta\omega_k}{2\pi\gamma P_k}\ln\left[1 + \gamma\frac{2\pi P_k}{\Delta\omega_k}\right]d\omega = \sum_{k=1}^{K}P_k\gamma_k^{-1}\ln(1 + \gamma_k). \tag{7.219}$$

Here $\Pi(\omega_k, \Delta\omega_k) = 1$ for $|\omega - \omega_k| \leq \Delta\omega_k/2$ and 0 otherwise, and $0 \leq P_k \leq 1$ is the fraction of the total signal power concentrated in the sub-band $[\omega_k - \Delta\omega_k/2; \omega_k + \Delta\omega_k/2]$.

If all the power is concentrated in a single sub-band of width $0 \le \Delta\omega_1 \le 2\pi$, $P_1 = 1$, then the MMSE error becomes

$$\delta_1(\Delta\omega_1, \gamma) = \frac{\Delta\omega_1}{2\pi\gamma} \ln\left(1 + \frac{2\pi\gamma}{\Delta\omega_1}\right). \tag{7.220}$$

Simple calculus shows that for a fixed SNR $\gamma > 0$, the error δ_1 in (7.220) is an increasing function of the bandwidth $\Delta\omega_1$ with minimum $\delta_{min} = \delta_1(0, \gamma) = 0$ and maximum $\delta_{max} = \delta_1(2\pi, \gamma) = \gamma^{-1}\ln(1 + \gamma)$. Thus, regardless of the SNR, if the process has zero bandwidth, it could be estimated without any error for any non-zero SNR. Otherwise, a residual estimation error is unavoidable and it increases with bandwidth due to the signal/noise overlap.

Since working with long frames is unrealistic (Abou-Faycal et al. 2005), it is safe to assume that the normalized covariance function is evaluated only for small lags and, therefore, can be approximated by a few terms of its Taylor series with respect to a small parameter $f_D T_S$:

$$\rho_n = \rho(nT_s) \approx 1 + \rho'(0)nf_D T_s + 0.5\rho''(0)n^2(f_D T_s)^2. \tag{7.221}$$

Here f_D is the maximum Doppler spread encountered in the channel. This can be easily calculated for analytical models of the covariance function or from measurements.

For realistic measurements, values of $\rho'(0) = ja_1$ and $\rho''(0) = -a_2^2/2$ can be estimated from the power spectral density $S(\omega)$ by

$$\rho'(0) = \frac{j}{2\pi}\int_{-2\pi f_D}^{2\pi f_D} \omega S(\omega)d\omega, \quad \rho''(0) = -\frac{1}{2\pi}\int_{-2\pi f_D}^{2\pi f_D} \omega^2 S(\omega)d\omega. \tag{7.222}$$

Note that PSD estimates are often available at the receiver. Keeping in mind that $S(\omega)$ is a real function, one can obtain the following simple approximation for $|\rho(nT_S)|^2$

$$|\rho(nT_s)|^2 \approx 1 + (a_1^2 + a_2^2)n^2 f_D^2 T_s^2. \tag{7.223}$$

7.14.3 Effective SNR

An equivalent channel model, which takes into account the channel estimation error, can be expressed as a known channel $\hat{h}(\tau)$ plus an aggregate noise $\eta(\tau)$ model in the following form

$$r(\tau) = \hat{h}(\tau)s(\tau) + \varepsilon(\tau)s(\tau) + \xi(\tau) = \hat{h}(\tau)s(\tau) + \eta(\tau). \tag{7.224}$$

Here, $\eta(\tau)$ represents the effects of AWGN and self-interference $\varepsilon(\tau)s(\tau)$. In general, $\eta(\tau)$ is not Gaussian distributed and is correlated with $s(t)$ and $h(t)$. However, it is shown in (Hassibi and Hochwald 2003; Medard 2000), that if $\eta(t)$ is assumed to be an AWGN process with variance $\sigma_\eta^2 = \sigma_\varepsilon^2\sigma_s^2 + \sigma_n^2$, it represents the worst case scenario in terms of capacity and probability of error. The resulting "effective"

SNR, or signal to self-interference plus noise ratio (SSINR), $\gamma_{eff}(\tau)$, can then be expressed as

$$\gamma_{eff}(\tau) = \frac{\sigma_{\hat{h}}^2(\tau)\sigma_s^2}{\sigma_\varepsilon^2(\tau)\sigma_s^2 + \sigma_n^2} = \frac{(1 - \delta_\tau)\gamma}{1 + \delta_\tau \gamma} \leq \gamma, \quad \delta_\tau \leq 1. \tag{7.225}$$

Here, we made use of (7.214) and the fact that $\gamma = \sigma_h^2 \sigma_s^2 / \sigma_n^2$ is the channel SNR. It can be seen that the transformation (7.225) is a one-to-one monotonic function with the inverse given by

$$\gamma = \frac{\gamma_{eff}(\tau)}{1 - \delta_\tau \left(1 + \gamma_{eff}(\tau)\right)}. \tag{7.226}$$

As the SNR γ approaches infinity, the effective SNR approaches a finite limit $\gamma_{eff\,\infty}(\tau) = 1/\delta_\tau - 1$. This causes a well-known error floor phenomenon (Proakis 2001) for a given modulation scheme. It also limits achievable data rates in channels with pilot assisted channel estimation.

It can be seen from (7.225) that the effective SNR, and therefore the achievable data rates, vary with position of the symbol in the data block, in other words the block fading assumption is often invalid. The authors of (Abou-Faycal et al. 2005) treat such channels as an array of $N - 1$ partial channels, obtained by sampling the original channel at the rate $1/NT_s$ and shifted by one symbol with respect to each other with each partial channel having a fixed effective SNR. The optimal communication strategy over such channels involves optimizations over each partial channel. The overall achievable rate is a sum of rates over all partial channels.

7.15 Achievable rates in Rayleigh channels with partial CSI

7.15.1 No CSI at the transmitter

Assuming that the noise $\eta(\tau)$ in (7.224) is AWGN, we can obtain a lower bound $C(\tau)$ for the achievable data rates in each partial channel without power control (i.e., no CSI at the transmitter)

$$C(\tau) = \int_0^\infty \ln\left(\frac{1 + \gamma}{1 + \delta_\tau \gamma}\right) p(\gamma) d\gamma = \exp\left(\frac{1}{\bar{\gamma}}\right) \mathrm{Ei}\left(-\frac{1}{\bar{\gamma}}\right)$$

$$- \exp\left(\frac{1}{\delta_\tau \bar{\gamma}}\right) \mathrm{Ei}\left(-\frac{1}{\delta_\tau \bar{\gamma}}\right) = C_\infty(\bar{\gamma}) - C_\infty(\delta_\tau \bar{\gamma}). \tag{7.227}$$

Here $\mathrm{Ei}(x)$ is the exponential integral function (Abramowitz and Stegun 1965), $p(\gamma) = 1/\bar{\gamma} \exp(-\gamma/\bar{\gamma})$ is the probability density function (PDF) of the instantaneous SNR with average $\bar{\gamma}$. Thus, the lower bound is the difference between the capacity of a perfectly known Rayleigh fading channel with average SNR $\bar{\gamma}$ and the capacity of a perfectly known Rayleigh fading channel with average SNR $\bar{\gamma}\delta_\tau(\bar{\gamma})$.

The explicit dependence of δ_τ on $\bar{\gamma}$ emphasizes the fact that these two quantities cannot be chosen independently to produce artificially loose bounds.

7.15.2   Partial CSI at the transmitter

If the transmitter power is adaptive, the optimal allocation of power $\Psi(\gamma)$ can be found as a strategy which maximizes the following integral

$$C_{opt} = \max \int_0^\infty \ln \left(1 + \frac{\Psi(\gamma)}{\hat{\gamma}} \gamma_{eff} \right) p(\gamma) d\gamma \tag{7.228}$$

subject to the average power constraint

$$\int_0^\infty \Psi(\gamma) p(\gamma) d\gamma = \bar{\gamma} . \tag{7.229}$$

Changing variables according to (7.225), the optimization becomes

$$C_{opt} = \max \int_0^\infty \ln \left(1 + \frac{F(\gamma_{eff})}{\hat{\gamma}} \gamma_{eff} \right) p_e(\gamma_{eff}) d\gamma_{eff} \tag{7.230}$$

subject to the average power constraint

$$\int_0^\infty F(\gamma_{eff}) p_e(\gamma_{eff}) d\gamma_{eff} = \hat{\gamma}, \quad F(\gamma_{eff}) = \Psi \left(\frac{(1 - \delta_\tau)\gamma}{1 + \delta_\tau \gamma} \right), \tag{7.231}$$

where $p_e(\gamma_{eff})$ is the PDF of the effective SNR. Here, the solution is given by the water-filling algorithm (Goldsmith and Varaiya 1997)

$$\frac{F(\gamma_{eff})}{\bar{\gamma}} = \begin{cases} \frac{1}{\gamma_0} - \frac{1}{\gamma_{eff}} & \text{if } \gamma_{eff} > \gamma_0 \\ 0 & \text{if } \gamma_{eff} < \gamma_0 \end{cases} \tag{7.232}$$

where γ_0 is the solution of

$$\int_{\gamma_0}^\infty \left(\frac{1}{\gamma_0} - \frac{1}{\gamma_{eff}} \right) p_e(\gamma_{eff}) d\gamma_{eff} = 1 . \tag{7.233}$$

Changing the variable again according to (7.225) and (7.233), we find that the optimal power allocation strategy is

$$\frac{\Psi(\gamma)}{\bar{\gamma}} = \begin{cases} \frac{1}{\gamma_0} - \frac{\delta_\tau}{1 - \delta_\tau} - \frac{1}{(1 - \delta)\gamma} & \text{if } \gamma > \frac{\gamma_0}{1 - \delta_\tau(1 + \gamma_0)} \\ 0 & \text{if } \gamma < \frac{\gamma_0}{1 - \delta_\tau(1 + \gamma_0)} \end{cases} \tag{7.234}$$

where γ_0 is a positive solution of the following equation

$$\bar{\gamma} \left[1 - \delta_\tau(1 + \gamma_0) \right] \exp \left(-\frac{\gamma_0}{[1 - \delta_\tau(1 + \gamma_0)] \bar{\gamma}} \right)$$

$$- \gamma_0 \text{Ei} \left(\frac{\gamma_0}{1 - \delta_\tau(1 + \gamma_0)\bar{\gamma}} \right) = \bar{\gamma} \gamma_0 (1 - \delta_\tau) . \tag{7.235}$$

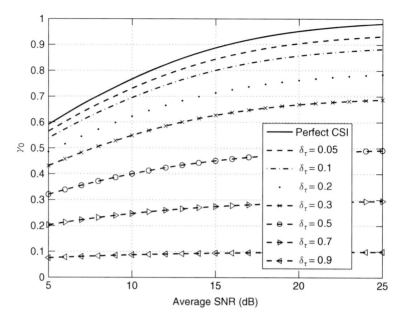

Figure 7.11   Optimal cut off SNR values for different values of δ_τ.

The corresponding optimal achievable rates are then

$$C_{opt} = \mathrm{Ei}\left(\frac{\gamma_0}{[1 - \delta_\tau(1 + \gamma_0)]\,\bar{\gamma}}\right)$$

$$- \exp\left(\frac{1}{\bar{\gamma}\delta_\tau}\right)\mathrm{Ei}\left(\frac{1 - \delta_\tau}{[1 - \delta_\tau(1 + \gamma_0)]\,\delta_\tau\bar{\gamma}}\right). \qquad (7.236)$$

The results of the numerical evaluation of C_{opt} with fixed δ_τ and different average SNR are presented in Figure 7.11. Let us emphasize here that we consider individual partial channels. While the average SNR remains the same for each partial channel, the variance of the estimation error varies from one partial channel to another. The same value of δ_τ may correspond to a different partial channel. It can be seen that the level of estimation error significantly affects the achievable rates, especially for high average SNRs. The saturation of the capacity as a function of average SNRs is due to self-interference induced by the error in estimating the channel. Such errors may result from a combination of two factors: the finite SNR of the pilot signal and the fading rate. The quality of different estimation schemes are considered in Figure 7.12. As for the perfect CSI case, the power adaptation provides some gain at lower average SNRs and almost no gain for average SNRs higher than 10 dB.

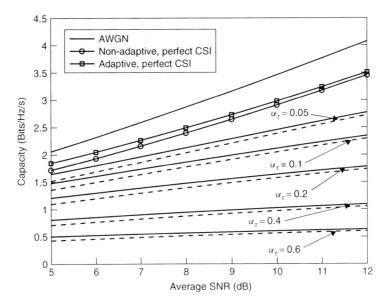

Figure 7.12   Improvement achieved through power adaptation. Solid lines: adaptive power allocation; dashed lines: constant power allocation.

7.15.3   Optimization of the frame length

The total achievable rates C_T for the channel can be obtained by summing the achievable rates of each partial channel, that is,

$$C_T = N^{-1} \sum_{n=1}^{N-1} C(nT_s). \tag{7.237}$$

The factor $1/N$ is included to take into account that each partial channel occupies only a fraction of the frame. Since for a frame of length 2 only half of the transmission time is used for data transmission, an increase in the frame length will result in a better utilization of the channel. However, longer frames lead to larger estimation errors, thus decreasing the achievable rates. The balance between these trends is achieved at some optimal frame length N_{opt}, which can be found numerically. A few examples are included in Section 7.16.

7.16   Examples

7.16.1 $P(0, 0)$ Estimation

In order to analytically evaluate the effects of pilot SNR and finite correlation intervals on the achievable rates, we consider the simplest case of $P(0, 0)$ estimation.

Each partial channel can be treated as a separate channel with perfect CSI and an effective SNR given by (7.225). Since perfect CSI is available, some sort of power adaptation can be exercised over each partial channel. At the same time, since different partial channels experience different average equivalent noise conditions, one may consider the optimal distribution of power among the partial channels.

Without power adaptation, the achievable rates are given by Equation (7.227), where, according to (7.215),

$$\delta_\tau = 1 - |\rho(\tau)|^2 \frac{\gamma_p}{1 + \gamma_p} . \tag{7.238}$$

The case of perfect CSI is recovered by setting $\gamma_p = \infty$ and $\rho = 1$, yielding the well-known expression for the capacity of a Rayleigh channel with perfect CSI (Goldsmith and Varaiya 1997) at the receiver:

$$C_\infty = \exp\left(\frac{1}{\bar{\gamma}}\right) \mathrm{Ei}\left(-\frac{1}{\bar{\gamma}}\right) . \tag{7.239}$$

The effect of the channel correlation alone can be traced by setting $\gamma_p = \infty$ and $b = 1 - |\rho(\tau)|^2$, thus resulting in the following expression

$$C_\rho = C_\infty - \exp\left(\frac{1}{\left(1 - |\rho(\tau)|^2\right)\bar{\gamma}}\right) \mathrm{Ei}\left(-\frac{1}{\left(1 - |\rho(\tau)|^2\right)\bar{\gamma}}\right). \tag{7.240}$$

As expected, $C_\rho \to 0$ as $\rho \to 0$. The effect of imperfect estimation in very slow fading can be obtained by setting $\rho = 1$, resulting in

$$C_{\gamma_p} = C_\infty - \exp\left(\frac{1 + \gamma_p}{\bar{\gamma}}\right) \mathrm{Ei}\left(-\frac{1 + \gamma_p}{\bar{\gamma}}\right) . \tag{7.241}$$

The negative terms in (7.240) and (7.241) represent a decrease in achievable rates due to imperfect correlation and due to the finite power of the pilot energy, respectively. The results are shown in Figure 7.13. This figure allows us to track the impact of two factors on the achievable rates. As expected, a constant, perfectly known channel provides the maximum capacity. A fading, but perfectly known channel results in only slightly smaller rates for all SNRs. Furthermore, if the channel is unknown and being estimated but does not vary in time ($f_D = 0$), additional degradation of the achievable rates can be observed. This degradation is almost the same for all levels of SNR, approximately 1 bit/sec/Hz. However, even if a fading rate $f_D T_s = 0.04$ is taken into account, it can be seen from Figure 7.13 that in a high SNR region the achievable rate saturates at a level significantly lower than that predicted by the zero rate fading model. This can be explained by the fact that at a high SNR level the estimation error is negligible compared to the error due to time selectivity ($f_D T_s > 0$). Thus, further increase in the SNR does not improve the net estimation error and the performance saturates. In contrast, for low SNRs, the estimation error is dominated by noise and the performance is only marginally lower than that of zero rate fading.

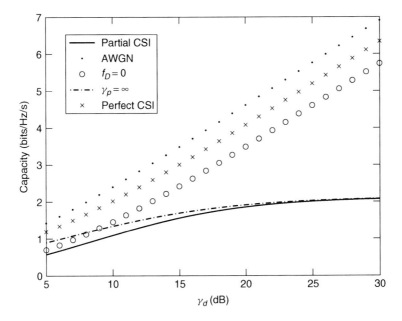

Figure 7.13   Channel capacities in different scenarios. $f_D T_s = 0.04$.

Figure 7.14 demonstrates the dependence of the optimal frame length on the average SNR and fading rate. A more pronounced maximum corresponds to a faster fading scenario. It may appear strange that for the same value of $f_D T_s$ a lower SNR will result in longer optimal frames. However, for low SNRs, the effective SNR is dominated by the noise and not by self-interference: the optimum frame length is then defined by a balance between the loss in accuracy of the estimate and the utilization of the power, which increases with the frame duration. In contrast, for a high average SNR, the effective SNR is dominated by self-interference, which increases with the frame size: the balance between power utilization and the increased interference is achieved for shorter frames. Similar conclusions can be reached for other estimation schemes.

7.16.2   Effect of non-uniform scattering

In this section we assume that the scattering environment forming the channel is such that scattering appears uniformly from only a single cluster, seen at the azimuth angle ϕ_0 ($\phi_0 = 0$ corresponds to the direction of the mobile movement) and angular spread Δ. The squared magnitude of the covariance function of the fading signal is then (Loyka 2005)

$$|\rho(\tau)|^2 = \text{sinc}^2\left(\Delta f_D \tau \cos \phi_0\right) \approx 1 - 0.33 \,\Delta^2 \pi^2 f_D^2 \tau^2 \cos \phi_0^2. \qquad (7.242)$$

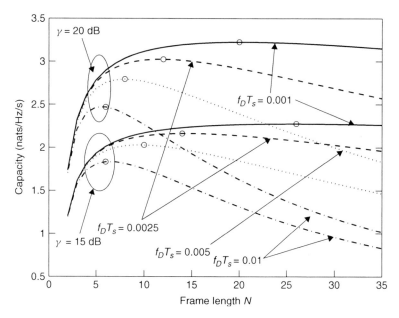

Figure 7.14   Optimal frame size for two values of average SNR, $\gamma_p = \gamma$ as function of fading rate and the frame length.

Using (7.242), one can obtain insight into how a cluster's angular spread and orientation affects the achievable rates. A smaller spread Δ results in a slower decay of $|\rho(\tau)|^2$ and in a slow increase in estimation error. As a result, the quality of CSI, δ_τ, is better for a narrower cluster and longer frames could be used: a wider angular spread corresponds to a wider fading bandwidth making the channel less predictable (Papoulis 1991). Similarly, orientation of the cluster plays a role in changing δ_τ since the increase in ϕ_0 from 0 to 90° leads to a slower decay of $|\rho(\tau)|$ and a slower growth of δ_τ. Optimal frame length designs for different angular spreads Δ and $P(0, 0)$ and $P(1, 0)$ estimators are shown in Figure 7.15. It can be seen that while the use of an additional pilot increases the overall quality of estimation and achievable rates, the optimal frame length is less sensitive to the estimation scheme compared to the rate of decay of the fading covariance function. A situation with multiple clusters and measurements could be similarly analyzed.

7.17   Conclusions

The achievable data rates of systems with pilot based channel estimates have been analyzed. Expressions for the maximum achievable data rates were derived in terms of the fading channel estimation quality. It has been shown how the classical water-filling algorithm should be modified for application in scenarios with imperfect CSI at the transmitter. It has been found that the gain achieved through power

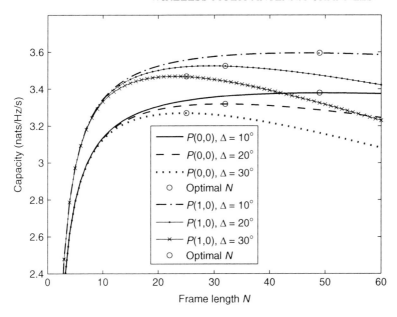

Figure 7.15 Achievable rates for a single scattering cluster with different angular spread Δ and for different estimators ($\bar{\gamma} = 15$ dB, $f_D T_s = 0.004$, and $\phi_0 = 30°$).

adaptation is more pronounced at low SNRs and for good estimates, but overall it is not significant. The impact of the estimation scheme on achievable data rates has been studied for physical channel models with varying angular spreads. It has been observed that for causal estimates, the optimal frame sizes are relatively short and the optimum is well pronounced. In the case of non-causal estimation, the optimum block size is at least twice as long and the corresponding maxima are relatively flat.

7.18   Appendix A: Szegö summation formula

Let one desire to evaluate the following sum

$$S_n = \sum_{n=0}^{N-1} f(n) \tag{7.243}$$

for a large N. This can be accomplished using the following approximation of a sum by an integral

$$S_n = N \sum_{n=0}^{N-1} f(n) \frac{1}{N} = N \sum_{n=0}^{N-1} f\left(N\frac{n}{N}\right) \frac{1}{N} \approx N \int_0^1 f(Nx)dx \tag{7.244}$$

7.19 Appendix B: matrix inversion lemma

The so-called **matrix inversion lemma** (van Trees 2002) is a useful tool in analyzing behavior of the Wiener filter in the time domain. In the most general form it reads

$$(\mathbf{A} + \mathbf{BCD})^{-1} = \mathbf{A}^{-1} - \mathbf{A}^{-1}\mathbf{B}\left(\mathbf{DA}^{-1}\mathbf{B} + \mathbf{C}^{-1}\right)^{-1}\mathbf{DA}^{-1} \tag{7.245}$$

Here $\mathbf{A}$ and $\mathbf{C}$ are invertible square matrices of size $N \times N$ and $M \times M$ respectively, $\mathbf{B}$ is $N \times M$ matrix and $\mathbf{D}$ is $M \times N$ matrix.

If $\mathbf{B} = \mathbf{D}^H = \mathbf{w}$ is a $N \times 1$ vector, the identity (7.245) becomes the Woodbury's identity (also known as the Sherman–Morrison form)

$$\left(\mathbf{A} + \mathbf{ww}^H\right)^{-1} = \mathbf{A}^{-1} - \frac{\mathbf{A}^{-1}\mathbf{ww}^H\mathbf{A}^{-1}}{1 + \mathbf{w}^H\mathbf{A}^{-1}\mathbf{w}} \tag{7.246}$$

If $\mathbf{A} = \gamma^{-1}\mathbf{I}$ and $\mathbf{B}^H = \mathbf{D}$ the identity (7.245) is reduced to the following simplified form (van Trees 2002)

$$\left(\gamma^{-1}\mathbf{I} + \mathbf{U}^H\mathbf{CU}\right)^{-1} = \gamma\left[\mathbf{I} - \mathbf{U}^H\left(\mathbf{UU}^H + \gamma^{-1}\mathbf{C}^{-1}\right)^{-1}\mathbf{U}\right] \tag{7.247}$$

In particular, let us consider application of 7.247 to the prediction of the sum two exponents signal. In this case the covariance matrix could be represented as

$$\mathbf{R} = [\mathbf{q}_1\mathbf{q}_2]\begin{bmatrix} NP_1 & 0 \\ 0 & NP_2 \end{bmatrix}\begin{bmatrix} \mathbf{q}_1^H \\ \mathbf{q}_2^H \end{bmatrix} \tag{7.248}$$

that is $\mathbf{V} = \mathbf{U}^H = [\mathbf{q}_1\mathbf{q}_2]$ and $\mathbf{C} = \begin{bmatrix} NP_1 & 0 \\ 0 & NP_2 \end{bmatrix}$ in Equation (7.247). Taking into account that $\mathbf{q}_2^H\mathbf{q}_1 = z$ and $\mathbf{q}_1^H\mathbf{q}_2 = z^*$ one can derive the following step-by-step equations for calculation of $\left(\gamma^{-1}\mathbf{I} + \mathbf{U}^H\mathbf{CU}\right)^{-1}$:

$$\mathbf{VV}^H = [\mathbf{q}_1\mathbf{q}_2]\begin{bmatrix} \mathbf{q}_1^H \\ \mathbf{q}_2^H \end{bmatrix} = \begin{bmatrix} 1 & z^* \\ z & 1 \end{bmatrix} \tag{7.249}$$

$$\mathbf{VV}^H + \gamma^{-1}\mathbf{C}^{-1} = \begin{bmatrix} 1 + \frac{1}{N\gamma P_1} & z^* \\ z & 1 + \frac{1}{N\gamma P_2} \end{bmatrix} \tag{7.250}$$

$$\left(\mathbf{VV}^H + \gamma^{-1}\mathbf{C}^{-1}\right)^{-1} = \frac{1}{\Delta}\begin{bmatrix} 1 + \frac{1}{N\gamma P_2} & -z^* \\ -z & 1 + \frac{1}{N\gamma P_1} \end{bmatrix} \tag{7.251}$$

where

$$\Delta = \left(1 + \frac{1}{N\gamma P_1}\right)\left(1 + \frac{1}{N\gamma P_2}\right) - |z|^2 \tag{7.252}$$

Finally

$$\left(\gamma^{-1}\mathbf{I} + \mathbf{R}\right)^{-1} = \gamma\left[\mathbf{I} - \frac{1}{\Delta}\mathbf{F}\right] \tag{7.253}$$

where

$$\mathbf{F} = \left(1 + \frac{1}{N\gamma P_2}\right)\mathbf{q}_1\mathbf{q}_1^H - z^*\mathbf{q}_1\mathbf{q}_2^H - z\mathbf{q}_2\mathbf{q}_1^H + \left(1 + \frac{1}{N\gamma P_1}\right)\mathbf{q}_2\mathbf{q}_2^H \tag{7.254}$$

8

Coding, modulation, and signaling over multiple channels

8.1 Signal constellations and their characteristics

We begin by considering construction of orthogonal space-time block codes (OSTBC). It is shown in (Gharavi-Alkhansari and Gershman 2005) that construction and analysis of code performance can be accomplished in two steps:

- Design of the orthogonal code and choice of modulation constellation. These two steps can be accomplished independently.

- Given the choice of constellation, a particular performance measure, such as pair-wise probability of error, can be derived.

In this section we will focus on the aspect of the design related to choice of proper constellations, mainly based on two-dimensional (complex) signals.

In order to provide a fair comparison between different constellations, in this chapter we will use the following terminology:

- M is the number of signals in an assembly.

- A coefficient

$$\alpha = \frac{d_{min}}{2\sqrt{E}} \tag{8.1}$$

Wireless Multi-Antenna Channels: Modeling and Simulation, First Edition.
Serguei L. Primak and Valeri Kontorovich.
© 2012 John Wiley & Sons, Ltd. Published 2012 by John Wiley & Sons, Ltd.

describes sensitivity of a chosen constellation to noise. Here $E = P_s T_s / \log_2 M$, P_s is the average power of a signal from the assembly, T_s is the symbol duration.

•

$$\gamma = \log_2 \frac{M}{N} \qquad (8.2)$$

is normalized transmission speed. Here $N = 2 F_s T_s$ is the base of the constellation, F_s is the bandwidth occupied by a symbol, and T_s is the symbol duration.

It is assumed in (8.1) that $P_s < \infty$ is limited, however the same results will be valid if one puts limitation on the peak power $P_{s,p} < \infty$ instead and

$$E = \frac{P_{s,p} T_s}{\log_2 M} \qquad (8.3)$$

If the signal assembly is composed of $M = 2^k$ signals where k is an integer, then

$$\gamma = \log_2 \frac{2^k}{2 F_s T_s} = \frac{R}{F_s} \qquad (8.4)$$

where R is the information rate of the constellation. If $N = 2$, that is $F_s T_s = 1$, then the constellation under consideration is a set of narrowband signals. This will be assumed for the most part of the discussions below.

According to the theory of signal detection based on the criteria of minimum average error (Proakis 2001), the decision regions of the complex plane are separated by equidistant lines between two nearest constellation points (Voronoi diagram). In this case, optimization of the constellation is reduced to optimizing packing of identical spheres (circles) centered at the constellation points (Foshini et al. 1974). These constellation points are for a regular grid[1] (mesh) allowing for the densest packing. If $M \gg 1$ such a system of dense packing is optimal in the mean error probability sense. Examples of such constellations are shown in Figure 8.1 and the numerical characteristics of such constellations are presented in Table 8.1.

In order to compare two-dimensional ($N = 2$) and multi-dimensional ($N > 2$) constellations, we confine ourselves to the case of $N = 2$ and $N = 4$. As an example

Table 8.1   Table of energy parameters of different modulations.

M	8	8	16	32
$\frac{d_{min}}{E}$	0.945	0.945	0.667	0.46
γ	1.5	1.5	2	2.5

---

[1] Triangular, rectangular, etc.

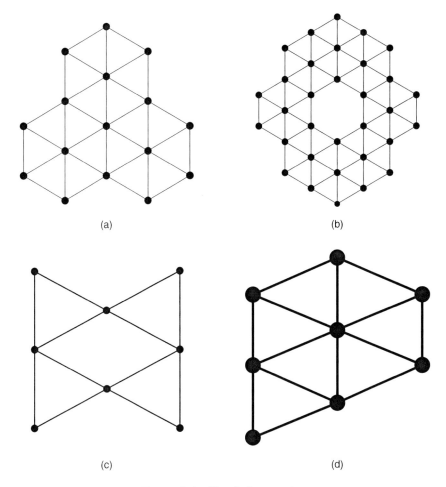

(a)   (b)

(c)   (d)

Figure 8.1   Simulation results.

of such signals let us consider simplex, bi-simplex, and biorthogonal constellations[2]
using the concept of dimension of signal space N_y.

The orthogonal system of signals is formed by such constellation points that
coincide with the vector basis of signal space (Foshini et al. 1974; Proakis 2001).
It is possible to show that in this case

$$\alpha = \sqrt{\log_2 N}, \ \ \gamma = \frac{\log_2 N}{N} \ d_{min} = \sqrt{2E} \tag{8.5}$$

The orthogonal signals are equidistant by construction with $\alpha = \sqrt{2E}$ but are not
optimal since the constellation points are located only along the orthogonal basis
vectors and most of the signal space appears to be unoccupied.

---

[2] Signals of permutation modulation, suggested in (Slepian 1965) are not considered here.

In a bi-orthogonal system of signals one of $M - 1$ signals is negative $s_1 = -s_0$ to a given signal s_0 and the remaining $M - 2$ are mutually orthogonal. In this case

$$d_{min} = \frac{2\sqrt{E}}{N}, \quad \gamma = \frac{\log_2 2N}{N} \tag{8.6}$$

Simplex signals are located at the edges of a right simplex (equilateral triangle in the case of $N = 2$) with the coordinate origin located at the center of symmetry. In this case

$$d_{min} = \sqrt{2\frac{N+1}{N}E} \tag{8.7}$$

Let us note that in the case of orthogonal signals $d_{min} = \sqrt{2E}$, and thus, the gain in energy of simplex constellations is $(N + 1)/N$ in order to obtain the same probability of error (i.e., by keeping the same d_{min}). As a result

$$\alpha = \sqrt{(N+1)\log_2\frac{N+1}{2N}}, \quad \gamma = \frac{\log_2(N+1)}{N} \tag{8.8}$$

Bi-simplex signals are formed in the very same manner as bi-orthogonal by adding the inverted version of each simplex constellation point. In this case, if the dimension of the signal space is N, then $M = 2(N + 1)$ and

$$d_{min} = \sqrt{2\frac{N-1}{N}E}, \quad \alpha = \sqrt{\frac{N-1}{2N}\log_2 2(N+1)}, \quad \gamma = \frac{\log_2 2(N+1)}{N} \tag{8.9}$$

Corresponding parameters for $N = 4$ are summarized in Table 8.2.

It is important to mention that the metrics (8.1) and (8.2) are not the only metrics possible. For example, the authors of (Su and Xia 2004) suggest the idea of "diversity product" for OSTBC in the following form

$$\xi = \frac{1}{2\sqrt{k}}\min\{d_1, d_2, \cdots, d_k\} \tag{8.10}$$

where k is the number of symbols in OSTBC, and d_i, $i = 1, \cdots, k$ is the minimum distance $d_{min,k}$ used for constructing OSTBC symbols. Assuming that identical constellations are used, one obtains that $d_{min,k} = d_{min} = const$ and

$$\xi = \frac{1}{2\sqrt{k}}d_{min} \tag{8.11}$$

Table 8.2   Table of energy parameters of different modulations.

Parameter	Simplex	Bi-simplex	Bi-orthogonal
M	5	10	10
$\frac{d_{min}}{\sqrt{E}}$	1.57	1.23	0.5
γ	$\frac{\log_2 5}{4}$	$\frac{1+\log_2 5}{4}$	0.75

In other words, for a fixed k and identical constellations this metric is identical to (8.1).

For two-dimensional signals $N = 2$, $M = 2^p$, $E = 1$

$$d_{min} = \frac{2\alpha}{p}, \; \xi = \frac{\alpha}{p\sqrt{k}} \tag{8.12}$$

For multidimensional signals $N > 2$ it is easy to show that (8.11) becomes

$$\xi = \begin{cases} \frac{1}{2}\sqrt{2(N+1)}Nk & \text{for simplex signals} \\ \frac{1}{\sqrt{kN}} & \text{for bi-orthogonal signals} \end{cases} \tag{8.13}$$

Furthermore, it is shown in (Su and Xia 2004) that if g is a time of transmission of quasi-OSTBC, then

$$\xi \leq \frac{1}{2}\sqrt{\frac{\gamma}{2g}} \tag{8.14}$$

Thus, metrics (8.1) and (8.2), introduced above are convenient for evaluating both the diversity product and application to non-OSTBC.

It is interesting to conduct a comparative analysis of Tables 8.1 and 8.2 keeping in mind a compromise between coefficients α and γ.

1. For two-dimensional signals the increase in value of M from 8 to 32 is equivalent to a loss in power by approximately 3 dB. However, the normalized rate γ increases by the same factor.

2. Simplex and bio-orthogonal signals with the same value of M lead to a similar trade-off, that is for relatively small N multidimensional signals do not provide significant gain in terms of α and γ compared to two-dimensional signals. However if, for some reason, it is necessary that $d_{min}/\sqrt{E}$ and γ be close to unity, then, according to Table 8.2, one would have to turn attention to bi-simplex and bi-orthogonal signals.

3. It is important to take into account that as the number of transmit and receive antennas increases, both OSTBC and quasi-OSTBC could achieve a very low probability of errors by exploiting diversity. In this case, limitations in their spectral efficiency are due to the relatively small information rate, which, in part, could be compensated for by choosing higher constellations M. Therefore, it seems important to introduce an additional parameter, known as energy efficiency (Foshini et al. 1974)

$$\beta = \frac{R}{P_s/N_0} \tag{8.15}$$

where P_s is the mean power of the signal and N_0 is the noise spectral density. In contrast to the parameter γ, which defines the frequency effectiveness of modulation, β defines its energy efficiency. In other words β shows the rate of increase in the information rate with the increase in SNR.

Table 8.3   Comparison of frequency and energy efficiency of some constellations.

M	16	64 Fig. 8.1
Fig. 8.1	$\gamma = 2.0,\ \beta = -13.7\,\text{dB}$	$\gamma = 3,\ \beta = -18.1\,\text{dB}$
16-PSK	$\gamma = 20,\ \beta = -18.0\,\text{dB}$	

Calculation of β is very complicated, since R itself depends on P_s/N_0 and the probability of error. Therefore, we limit our considerations to only a few particular cases: the constellations shown in Figure 8.1 and 16-PSK having the same $M = 16$ and $\gamma = 2$. This is summarized in Table 8.3. It can be seen that the coefficient β for constellations in Figure 8.1 is higher than for 16-PSK by 4.3 dB. On the other hand, comparing constellations in Figure 8.1, $M = 16$ with Figure 8.1, $M = 64$ which have similar $\beta \approx -18\,dB$, it is seen that 16-PSK has a frequency efficiency 1.5 times that of constellation in Figure 8.1.

Based on the brief considerations above, based on the parameters α, β and γ one can conclude the following:

- Starting from $M = 16$ two dimensional signals of amplitude-phase modulation (APM) of the type shown in Figure 8.1 provide better effectiveness compared to signals formed by phase modulation and this advantage increases with increase of M (Proakis 2001), (Foshini et al. 1974).

- Advantages of multidimensional signals $N > 2$ show with an increase of N, (see Table 8.2). However, this is achieved partly through complication of the system implementation.

- The biggest advantage is provided by APM modulation with $M \gg 1$ based on a triangular grid (see (Foshini et al. 1974; Su and Xia 1981). However, they are much more difficult to implement in practice than those based on the rectangular grid.

- These recommendations justify the choice of APM with $M = 32, 64, 128, 256$.

8.2   Performance of OSTBC in generalized Gaussian channels and hardening effect

8.2.1   Introduction

It is well known (Paulraj et al. 2003) that the use of OSTBC in a MIMO channel provides the following benefits:

- OSTBC allows realization of the full diversity provided by a MIMO channel. Indeed, assuming a slow flat fading, known CSI at the receiver, and independent individual sub-channels, the order of diversity is $N_R N_T$, that is

$$P_{err} \sim \bar{\gamma}^{-N_R N_T}, \ \bar{\gamma} \to \infty \tag{8.16}$$

where $\bar{\gamma}$ is the average SNR on the individual link.

- Coding and detection are relatively simple.

However, OSTBC have a relatively low code rate, especially for high SNR. As a result, they do not allow us to approach the capacity of the MIMO link when power conditions are favorable. There is also no general procedure for designing such codes for an arbitrary number of transmit and receive antennas.

Since the use of OSTBC converts a MIMO channel into a set of independent SIMO channels, decoding of OSTBC and its performance evaluation is very similar to analysis of diversity combining.

8.2.2 Channel representation

Let us consider the MIMO channel formed by N_T transmit and N_R receive antennas. In this case the received vector at the output of the MIMO channel could be described in the following way (Larsson and Stoica 2003; Paulraj et al. 2003):

$$\mathbf{z}(t) = \int_{-\infty}^{\infty} \mathbf{H}(t, \tau)\mathbf{s}(t - \tau)d\tau + \mathbf{n}(t) \tag{8.17}$$

where $\mathbf{s}(t - \tau)$ is the transmitted signal vector of size $N_T \times 1$, $\mathbf{z}(t)$ is the received vector of size $N_R \times 1$, $\mathbf{n}(t)$ is a $N_R \times 1$ vector of additive, white, Gaussian noises with known power spectrum density N_0. Finally $\mathbf{H}(t, \tau)$ is a time-varying complex impulse response matrix of size $N_R \times N_T$. Each complex element $h_{n_T n_R}(t, \tau)$ of $\mathbf{H}(t, \tau)$ can be modeled as a complex Gaussian process in both variables, that is:

$$h_{n_T n_R}(t, \tau) = x_{n_T n_R}(t, \tau) + j y_{n_T n_R}(t, \tau) \tag{8.18}$$

$\tau \in [0 : \tau_{max}]$, $t \in [0 : pT_s]$. Here $x_{n_T n_R}(t, \tau)$ and $y_{n_T n_R}(t, \tau)$ are real and imaginary Gaussian components, described by mean $m_{x.n_T.n_T}$, $m_{y.n_T.n_T}$ and variances $\sigma_{x.n_T.n_T}^2$, $\sigma_{y.n_T.n_T}^2$. It is important that we do not make any assumptions about the equality of $\sigma_{x.n_T.n_T}^2$, $\sigma_{y.n_T.n_T}^2$. In other words we do not require that $h_{n_T n_R}(t, \tau)$ is a proper Gaussian process (Schreier and Scharf 2003a). Here $1 \leq n_T \leq N_T$, $1 \leq n_R \leq N_R$; τ_{max} is the maximum delay spread, T_s is the symbol interval and $p \geq \min(N_T, N_R)$ is the length of the OSTBC. It is common to convert continuous delay τ in $\mathbf{H}(t, \tau)$ into a set of tap delays τ_l, $1 \leq l \leq L$ and then convert the frequency selective problem to an equivalent frequency flat problem of size $N_R \times N_T L$ (Giannakis et al. 2007). Furthermore, using channel organization in both space and time (Alcocer et al. 2005a; Fechtel 1993; Sayeed 2002) the channel could be represented by a matrix with $Q = N_R N_T L$ independent components, or artificial eigenmodes (trajectories) (Alcocer et al. 2005a; Kontorovich 2007; Sayeed 2002). Therefore, after using pre- and post-processing, the problem of transmission of OSTBC over a fading frequency selective channel can be transformed to the following matrix low-pass equivalent form

$$\mathbf{Z}(t) = \mathbf{H}_Q(t)\mathbf{S}(t) + \mathbf{N}(t), \ t = nT_s \tag{8.19}$$

where $\mathbf{H}_Q(t) = \left\{ h_q(t) = h_{(l-1)(N_R)+n_R,n_T}(t) \right\}$ is the augmented channel matrix of size $N_R L \times N_T$, $\mathbf{S}(t)$ is $p \times N_T$ block of a OSTBC, $\mathbf{Z}(t)$ is $N_R L \times p$ received matrix, and $\mathbf{N}(t)$ is $N_R L \times p$ matrix of AWGN.

If the orthogonalization is ideal, then all elements of the extended matrix $\mathbf{H}$ are independent. In practice both virtual channel representation (Sayeed 2002) and "universal basis orthogonalization" (Alcocer et al. 2005a) produce slightly correlated components. However, this correlation could be neglected, especially for a large number of antennas (Sayeed 2002).

In general, the elements of matrix $\mathbf{H}_Q$ are non-Gaussian (Sayeed 2002), since the orthogonalization transformation tends to reduce the angular spread of each virtual path h_{l,n_t,n_r}. However, it is possible to model such components as a four-parametric distribution (Simon and Alouini 2000) as described above. It is important to mention that in contrast to a common assumption about the spherical symmetry of $h_{l,n_t,n_r} = x_{l,n_t,n_r} + jy_{l,n_t,n_r}$ we do not assume this and consider a more general case. Therefore, the in-phase x_{l,n_t,n_r} and the quadrature x_{l,n_t,n_r} components must be treated individually.

In order to simplify notations, in the future we would use a single index q to represent the triplet $\{l, n_T, n_R\}$, that is $x_q = x_{l,n_T,n_R}$. Furthermore $m_{xl} = E\{x_l\}$, $m_{yl} = E\{y_l\}$, $\sigma_{xl}^2 = E\{(x_l - m_{xl})^2\} \geq 0$, and $\sigma_{yl}^2 = E\{(y_l - m_{yl})^2\} \geq 0$ are the means and the variances of the corresponding variables. In addition, it is beneficial to define the following auxiliary normalization parameters

$$q_q = \frac{m_{xq}^2 + m_{yq}^2}{\sigma_{xq}^2 + \sigma_{yq}^2}, \ \beta_q = \frac{\sigma_{xq}^2}{\sigma_{yq}^2}, \ \phi_q = \arctan \frac{m_{yq}}{m_{xq}}$$

$$\Omega_1 = m_{xq}^2 + m_{yq}^2 + \sigma_{xq}^2 + \sigma_{yq}^2, \ E_{S_q} = P_{S_q} T_S, \ \gamma_{q0} = \Omega_q^2 \frac{E_{S_q}}{N_0} \tag{8.20}$$

Here P_{S_q} is the average signal power in the q-th artificial trajectory. It is important to note at this point that the distribution of the channel instantaneous power $|h_q|^2$ is now defined by four parameters m_{xq}, m_{yq}, σ_{xq}, and σ_{yq}, or, equivalently by q_q, β_q, Ω_q, and ϕ_q. This gives rise to the term "four-parametric distribution" as used in (Simon and Alouini 2000).

It is well known from the theory of the OSTBC that application of such codes leads to decomposition of a $N_R \times N_T$ MIMO system to a N_T independent SIMO arms of size $Q = N_R \cdot N_T \cdot L$ with each symbol decoded separately (Agrawal et al. 2002b; Giannakis et al. 2007) by a single arm. The optimal detection in each SIMO system is achieved by performing maximal ratio combining (MRC). This scheme is also known as a $2 - D$ RAKE receiver (Luo et al. 2002). In this case each transmitted symbol experiences the same channel, and, therefore, the average symbol error probability is equal to that calculated based on a single symbol detection. The equivalent SNR γ at the output of the $2 - D$ RAKE receiver can be calculated in terms of the equivalent SNR of each

$$\gamma = \frac{1}{N_T} \sum_{n_T=1}^{N_T} \sum_{n_R=1}^{N_R} \sum_{l=1}^{L} \gamma_{l,n_T,n_R} = \frac{1}{N_T} \sum_{q=1}^{Q} \gamma_q \qquad (8.21)$$

with the term N_T^{-1} accounting for the division of power at the transmitter side.

In order to evaluate the probability of error in detection of each symbol from a chosen constellation, one has first to consider a channel without fading, and then average the results over the distribution of SNR γ. In addition, in order to consider practical systems, one has to take into account errors caused by prediction and estimation of channel parameters, usually unknown on either side of the communication link (Kontorovich and Primak 2008).

8.2.3 Probability of error

One of the main advantages of OSTBC is decomposition of the decoding algorithm into N_T statistically independent SIMO channels (Paulraj et al. 2003) with $Q = N_T N_R L$ receive branches. Each symbol is detected independently, while an optimal detection strategy of a single symbol is MRC combining over all Q branches (Agrawal et al. 2002b; Kontorovich et al. 2008; Paulraj et al. 2003). As is shown above, the average SNR at the output of the MRC combiner is given by (8.21).

Calculating an average probability of symbol error can be achieved by averaging the symbol error probabilities in a fixed channel with AWGN over distribution of the SNR. In the following we assume that fading in all brunches is independent and is described by the generalized Gaussian model, described in Section 2.2.

Since the case of $M = 2^k$ is most interesting in practice, we can capitalize on considerations of Section 8.2.1. A number of exact results are known for M-QAM (Beaulieu 2006) and M-PSK (Okunev 1997). The expression for QAM has the following form

$$P_{0,MQAM} = 2 \left(1 - \frac{1}{M_I}\right) Q(A_I) + 2 \left(1 - \frac{1}{M_Q}\right) Q(A_Q)$$
$$- \frac{1}{4} \left(1 - \frac{1}{M_I}\right) \left(1 - \frac{1}{M_Q}\right) Q(A_Q)Q(A_I) \qquad (8.22)$$

Here the $M = M_I M_Q$ QAM constellation is considered with a $M_I \times M_Q$ grid. In other words the in-phase component is M_I-ary PAM while the quadrature component is M_Q-PAM. As usual $Q(\bullet)$ is the Q-function. Furthermore, $A_I = d_I/\sigma_n$ and $A_Q = d_Q/sigma_n$ are decision distances normalized by a standard deviation of noise.

For M-PSK one can provide upper and lower bounds in the following form

$$Q\left(\sqrt{2\gamma} \sin \frac{\pi}{2^k}\right) \leq P_{0,MPSK} \leq 2Q\left(\sqrt{2\gamma} \sin \frac{\pi}{2^k}\right) \left(1 - \frac{1}{2}Q\left[\sqrt{2\gamma} \sin \frac{\pi}{2^k}\right]\right)$$
$$(8.23)$$

Other methods of finding bounds are known (Simon and Alouini 2000). It can be seen that expressions (8.22)–(8.23) are poorly suited to further averaging over the distribution of fading: even if it is possible to obtain an analytical expression it will be too complicated for analysis.

Fortunately, for the case of $M \gg 1$ there are a few simple approximations for P_0 (Proakis 2001).

$$P_{0,MQAM} \leq \frac{1}{M} \sum_{n=1,n\neq m}^{M} \sum_{m=1}^{M} Q(\alpha_{n,m}^2 \gamma) \tag{8.24}$$

where $\gamma = \sqrt{P_s T_s / N_0}$, and $\alpha_{n,m}^2 = 0.25 d_{n,m}/E_s$, where $d_{n,m}$ is distance between n-th and m-th constellation points. Using bounds on Q function this could be further simplified to produce

$$P_{0,MQAM} \leq \frac{1}{M} \sum_{n=1,n\neq m}^{M} \sum_{m=1}^{M} \frac{\sqrt{2N_0}}{\sqrt{2\pi}} \exp\left(-\frac{d_{m,n}^2}{4N_0}\right) \tag{8.25}$$

Both of these equations are based on the upper bound and are valid for the case of small P_0, or, equivalently, of large SNR.

Similar consideration can be applied to the case of M-QAM. In the case of $M_I = M_Q = \sqrt{M}$ and $A_I = A_Q = h = \sqrt{\bar{\gamma}}$ one can easily obtain for $h \gg 1$ that

$$P_{0,MQAM} \leq 4 \left(1 - \frac{1}{\sqrt{M}}\right) Q(\sqrt{\bar{\gamma}}) \tag{8.26}$$

Similarly, for $M - PSK$

$$P_{0,MPSK} \leq 2Q\left(\sqrt{2\bar{\gamma}} \sin\frac{\pi}{2^k}\right) \tag{8.27}$$

Therefore, the question of performance evaluation can be reduced to averaging of $Q(\psi\bar{\gamma})$, with $\psi = \alpha_{m,n}, 1, \sqrt{2\bar{\gamma}} \sin\frac{\pi}{2^k}$.

Taking into consideration that the instantaneous SNR is a quadratic function of $\lambda = [\mathbf{x}^T, \mathbf{y}^T]^T$, one has to average the following quantity $\alpha Q(\psi \lambda^T \Gamma \lambda)$ Here again $\mathbf{x}$ and $\mathbf{y}$ are vectors in-phase and quadrature components of the channel transfer function, and $\Gamma = \text{diag}\{\bar{\gamma}_1, \bar{\gamma}_2, ..., \bar{\gamma}_Q,\}$ is a matrix of virtual path's average SNRs.

Using exponential bound $Q(x) \leq exp(-x^2/2)$ for the $Q(x)$ at high SNR averaging of $\alpha Q(\psi \lambda^T \Gamma \lambda)$ results in the following expression

$$P_0 = \int_{-\infty}^{\infty} \int_{-\infty}^{\infty} \exp\left[-\sum_{q=1}^{Q} \frac{\psi}{N_T}\bar{\gamma}_q \sqrt{x_q^2 + y_q^2}\right] \prod_{q=1}^{Q} p_{x_q}(x_q) p_{y_q}(y_q) dx_q dy_q \tag{8.28}$$

$$= C_Q^{2Q-1} \prod_{q=1}^{Q} \frac{\pi N_T^2}{\psi^2 \bar{\gamma}_i} p_{x_q}(0) p_{y_q}(0) \tag{8.29}$$

After taking into account the particular shape of the distribution of the in-phase and quadrature components of the channel response, one can obtain, after some algebra

$$P_0 \le \alpha C_Q^{2Q-1} \prod_{i=1}^{Q} \frac{\exp\left[q_i^2(1+\beta_i^2)\left(\cos^2\phi_i + \beta_i^2\cos^2\phi_i\right)\right]}{2\beta_i/(1+\beta_i^2)(1+q_i^2)} \frac{1}{2\prod_{i=1}^{Q}\bar{\gamma}\psi^2} \quad (8.30)$$

It is important to note that Equation (8.30) is not valid for one-sided Gaussian fading in each branch. In this case, equations should be rewritten to produce

$$P_0 \le \frac{\alpha\Gamma\left(\frac{Q+1}{2}\right)}{Q\Gamma(Q/2)\sqrt{\pi}} \prod_{i=1}^{Q} \left(\bar{\gamma}_i\psi^2\right)^{-1/2} \quad (8.31)$$

Relations (8.30)–(8.31) describe the dynamics of variation of P_0 as Q increases in the case of small P_0 (or high SNR). Absorbing all parameters of the channel in a constant one obtains

$$P_0 \le A \left(\prod_{i=1}^{Q} \gamma_i\right)^{-1} \quad (8.32)$$

Since Equations (8.30)–(8.31) were obtained by assuming that the receiver has perfect CSI about the channel, bounds derived in (8.30)–(8.31) can only be considered "lower" upper bounds.

In order to account for practical errors in channel evaluation one has to consider the following issues. Let us assume that estimates of unknown parameters of the channels (CSIR) $\hat{x}_i$, $\hat{y}_i$ are obtained by use of pilot signals and the quality of such estimation is high in each received branch. In this case it is possible to use notion of additional "system" or equivalent estimation noise (Almustafa et al. 2009; Hassibi and Hochwald 2003). A simple approximation could be

$$N_{syst}^{(i)} \sim N_0 \frac{E_i}{E_{P_i}} \quad (8.33)$$

where E_{P_i} is the energy of a pilot signal in i-th diversity brunch and E_i is the energy of the OSTBC symbol in the same diversity brunch. In this case the effective SNR (which accounts for estimation noise) can be written as

$$\hat{\gamma}_i = \frac{P_{S_i} T_s}{N_0 + N_{syst}^{(i)}} \le \gamma_i \quad (8.34)$$

Therefore, the same Equations (8.30)–(8.31) can be used to account for estimation errors.

8.2.4   Hardening effect

It has been known for a while (Klovski 1982b; Proakis 2001) that given target P_0 power gain with respect to an AWGN channel using more diversity brunches slows down as the number of branches Q increases. The advantage of diversity reception constitutes itself in the form of two factors. First of all is the accumulation of SNR by properly adding signals in different branches, thus reducing noise variance. However, the second effect of reduction of variance of the channel variation often has a predominant effect on gain obtained through the diversity reception. As Q increases to infinity, the gain achieved in channel variance reduction becoming smaller and saturates at a certain level. Therefore, adding additional branches may not be very beneficial. It is clear from physical considerations that saturation is fastest when the branches are homogeneous and is much slower for non-homogeneous branches. The goal of this section is to provide a quantitative description of this process, called channel hardening (Hochwald et al. 2004).

One possible approach was suggested by (Hochwald et al. 2004) and is based on measuring the fluctuating component of instantaneous channel capacity in channels with OSTBC. This approach is sound from an information-theoretical point of view.

However, it is more practical to quantify this effect by measuring power gain between the channel with diversity and a constant AWGN channel, achieved by adding an additional diversity branch. It also allows us to find a value of Q_{lim} which allows us to obtain most of the practical gain achievable through diversity, and beyond which further increase in Q is not reasonable.

Let us consider this in more detail. Let η be the power gain of the MRC receiver with Q diversity branches in the generalized Gaussian channel with given parameters and fixed probability of error P_0 in a high SNR regime (i.e., $\gamma_i \gg 1$). Using Equations (8.30), (8.31) one can easily obtain

$$
\eta_Q = \left[\frac{\prod_{i=2}^{Q} \delta_i^2 \gamma_i^2}{C_{2Q-1}^Q} N_T \right] \frac{1}{P_0^{(1-1/Q)}}
\tag{8.35}
$$

where $\delta_i^2 = \bar{\gamma}_i / \bar{\gamma}_1$ describes non-homogeneity of the diversity branches.

A similar equation for a one-sided Gaussian channel becomes

$$
\eta_Q = \frac{1}{\pi^2 P_0^{2(1-1/Q)}} \left(\frac{Q \sqrt{\pi} \Gamma(Q/2)}{\Gamma(Q/2+1/2)} \prod_{i=2}^{Q} \bar{\gamma}_i \gamma_i^2 \right)
\tag{8.36}
$$

The Table 8.4 and Figures 8.2–8.4 illustrates the power gains for $P_0 = 10^{-4}$ and various parameters δ_i^2, q, β^2, and Q assuming homogeneous diversity branches. It can be seen that the higher the average SNR is, the less gain diversity provides. At the same time the more inhomogeneous channels are, the less gain can be obtained from diversity.

The example above is concerned with independent fading in all branches. In the other extreme case all branches are correlated and

$$
\eta_Q = \frac{1}{N_T} \sum_{i=1}^{Q} \bar{\gamma}_i \gamma_i^2
\tag{8.37}
$$

Table 8.4   Channel hardening for different values of the parameters.

Q	$\delta = 0.5$		$\delta^2 = 1$		$\delta^2 = 0.1$	
	$q^2 = 10,$ $\beta^2 = 1$	$q^2 = 0,$ $\beta^2 = 1$	$q^2 = 10,$ $\beta^2 = 1$	$q^2 = 0,$ $\beta^2 = 1$	$q^2 = 10,$ $\beta^2 = 1$	$q^2 = 0,$ $\beta^2 = 1$
1	0	0	0	0	0	0
2	13.13	31.57	16.63	33.08	9.63	28.08
3	18.5	43.14	20.51	45.16	13.81	38.48
4	20.87	48.43	23.13	50.69	15.75	43.19
5	22.4	51.72	24.81	54.14	16.81	46.14

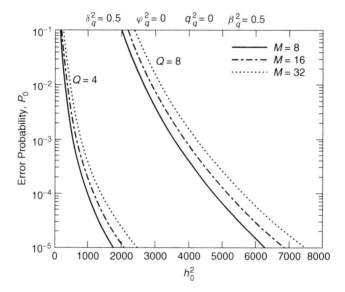

Figure 8.2   Simulation results.

In order to estimate the hardening effect one has to consider ratio

$$\frac{\eta_{Q+1}}{\eta_Q} \leq 1 \tag{8.38}$$

and find such a value of Q_{lim} such that the inequality becomes equality. It follows from (8.35) that for $Q \gg 1$ and $\delta^2 < 1$

$$Q_{lim} \geq \lfloor \frac{\log_1 0 P_0}{2 \log_1 0 \delta_i^2} \rfloor \tag{8.39}$$

For example, if $P_0 = 10^{-4}$, $\delta_i^2 = 0.5$ then $Q_{lim} \approx 6$. In homegenous diversity branches this number is even smaller.

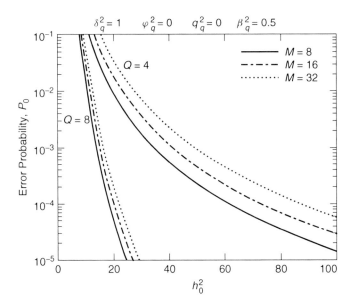

Figure 8.3   Simulation results.

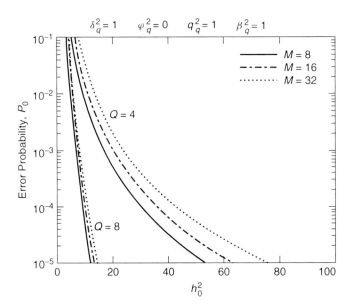

Figure 8.4   Simulation results.

8.3 Differential time-space modulation (DTSM) and an effective solution for the non-coherent MIMO channel

8.3.1 Introduction to DTSM

In the previous section, a coherent reception of OSTBC was considered, thus emphasizing the importance of CSI at the receiver. However, accurate estimation of CSI in MIMO channel is not always feasible, especially in cases of high power. In this case the situation is similar to those of SISO channels: coherent reception is possible only if fading is very slow. Otherwise, in order to avoid direct channel estimation, non-coherent differential demodulation (such as DPSK) can be used (Okunev 1997; Simon and Alouini 2000).

Differential time-space modulation (DTSM) can be considered a generalization of ideas of DPSK in SISO channels to MIMO channels. Let us consider DTSM in some more detail. Following (Hughes 2003) let us assume that there is a constellation χ with unit energy and fading in a MIMO channel that could be described by a block-fading model. In other words, the value of the channel transfer function $\mathbf{H}(t)$ does not change during the transmission time T_s of the symbol.

Let us further assume that a symbol $C_{i,k} \in \chi_j$ is transmitted from the $n_{t,i}$-th antenna during the interval $(k-1)T_s \le t \le kT_s$. If a $N_T \times n$ code matrix $\mathbf{C}_m$ is transmitted over the duration of n symbols, one can write it in the form

$$\mathbf{C}_m = \begin{bmatrix} C_{1,1}(m) & \cdots & C_{1,n}(m) \\ \cdots & \cdots \cdots \\ C_{N_T,1}(m) & \cdots & C_{N_T,n}(m) \end{bmatrix}, \ 1 \le m \le M \tag{8.40}$$

assuming that up to M such matrices could be constructed. It could be further assumed (Hughes 2003) that matrices $\mathbf{C}(m)$ are constructed in such a way that they are proportional to a unitary matrix, that is

$$\mathbf{C}(m)\mathbf{C}(m)^H = n\mathbf{I}_{N_T}, \ 1 \le m \le M \tag{8.41}$$

Then the set $\mathcal{G} = \{\mathbf{G}(m)\}$ of these matrices forms a so-called a unitary code (Hughes 2003). Unitary codes can be further used to construct unitary group codes in the following manner. Let χ be a given constellation and χ be an $n \times n$ matrix with elements belonging to χ. From all possible $n \times n$ matrices $\mathbf{G}(m)$ let us choose a group g such that $\mathbf{C}(m) \in \chi$ and $\mathbf{G}_m\mathbf{G}_m^H = \mathbf{G}_m^H\mathbf{G}_m = n\mathbf{I}_n$. This is generalization of Slepian's group codes to the MIMO case. In order to provide a reference point in an unknown channel one also must generate a $N_T \times n$ matrix $\mathbf{D}$ such that $\mathbf{DG} \in \chi^{N_T \times n}$.

In the initial stage of a transmission a matrix $\mathbf{D}$ is sent, that is $\mathbf{X}_0 = \mathbf{D}$. For the further data symbol at k-th time slot the transmitted symbol is defined as

$$\mathbf{X}_k = \mathbf{X}_{k-1}\mathbf{G}_k, \ 1 \le k \le \tag{8.42}$$

where K is the maximum number of symbols sent before $\mathbf{D}$ is recent again.[3] In this formulation the unitary group code is nothing else but a generalization of DPSK extended to MIMO channels. This becomes transparent if one considers the so-called cyclic group codes suggested in (Hughes 2003)

$$g = \left\{ \mathbf{I}_n, \ \mathbf{G}_0, \ \mathbf{G}_0^2, \cdots, \ \mathbf{G}_0^{M-1} \right\} \tag{8.43}$$

where

$$\mathbf{G}_0 = \begin{bmatrix} \exp j \frac{2\pi k_1}{M} & \cdots & 0 \\ \cdots & \cdots & \cdots \\ 0 & \cdots & \exp j \frac{2\pi k_{N_T}}{M} \end{bmatrix} \tag{8.44}$$

In this case the unitary group code is formed from PSK symbols and during each symbol interval only one non-zero symbol is send, in other words only one antenna is active.

In order to detect such a symbol, passed through a MIMO channel

$$\mathbf{Y} = \mathbf{HC} + \mathbf{N} \tag{8.45}$$

where $\mathbf{N}$ is $N_R \times n$ matrix of AWGN, one can deploy the so-called autocovariance receiver, based on the following simple rule (Hughes 2003; Okunev 1997)

$$\mathbf{S} = \mathrm{diag}\mathbf{Y}_k^H \mathbf{Y}_{k-1} = \mathrm{diag}\left\{ S_{nr,nr} \right\} \tag{8.46}$$

Expression (8.46) indicates that the cyclic group code DTSM is parallel coding without redundancy. As a result, the decoding procedure must also be parallel and autocovariance based, separately for each sent symbol.[4] Furthermore, taking into account the fact that the parallel coding is without redundancy, the probability of the symbol error P_0 completely describes the performance of such codes.

In the following we confine ourselves to considering the transformation of (8.46) which is suggested in (Hughes 2003). If $\mathbf{C}_m$ is an MPSK constellation with unity energy, then each $\mathbf{C}_m$ is in g and algorithm (8.46) can be rewritten as

$$\hat{g} = \arg \max_{g \in \mathbf{C}_M} \Re \left\{ \sum_{i=1}^{N_T} g^{k_i} S_{ii} \right\} \tag{8.47}$$

8.3.2   Performance of autocorrelation receiver of DSTM in generalized Gaussian channels

It is well known (Hughes 2003; Okunev 1997) that the autocorrelation receiver is a non-coherent algorithm with a loss of 3 dB with respect to optimal ML decoding.

---

[3] Both the receiver and the transmitter are aware of K and the initial time interval.

[4] It is proved in (Hughes 2003) that each full rank unitary group code is equivalent to a cyclic group code containing $M = 2^k$ code combinations from M-PSK constellations ($n = N_T$). Thus, the results obtained below are quite general.

While symbol k_i is being transmitted over the interval $(k_i - 1)T_s \le t \le k_i T_s$ over a MIMO channel with independent links, the received signal in j-th diversity branch is given by

$$y_{k_i}^{(j)}(t) = \mu^{(j)} s_{k_i}(t) + n \qquad (8.48)$$

8.3.3 Comments on MIMO channel model

The band-pass equivalent of a frequency-flat MIMO channel can be fully characterized by the complex matrix of the impulse responses $\mathbf{H}(t)$ of the size $N_R \times N_T$ with independent entries. Here N_T is the number of transmit antennas and N_R is the number of the receive antennas. Therefore, the input-output relation for the MIMO channel can be written in the following form (Paulraj et al. 2003)

$$\mathbf{z}(t) = \mathbf{H}(t)\mathbf{s}(t) + \mathbf{n}(t) \qquad (8.49)$$

where $\mathbf{s}(t)$ is a $N_T \times 1$ signal vector, $\mathbf{z}(t)$ is $N_R \times 1$ receive vector and $\mathbf{n}(t)$ is a $N_T \times 1$ vector of Gaussian noises with the covariance matrix

$$\mathcal{E}\left\{\mathbf{n}(t)\mathbf{n}(t)^H\right\} = \sigma_n^2 \mathbf{I}_{N_R} \qquad (8.50)$$

We also assume block-fading model (Paulraj et al. 2003).

Each statistically independent complex element $h_{kl}(t)$ of the matrix $\mathbf{H}(t)$ in Equation (8.49) can be represented through its in-phase and quadrature component as

$$h_{kl}(t) = x_{kl}(t) + jy_{kl}(t), \ 0 \le k \le N_T, \ 0 \le l \le N_R \qquad (8.51)$$

The generalized Gaussian model (Klovski 1982b; Simon and Alouini 2000) considered hereafter is characterized by the Gaussian distribution of the quadrature components with arbitrary means m_x and m_y, and variances σ_x^2 and σ_y^2 respectively. As a special case this model includes Rayleigh, Rice, and Hoyt distribution as well as a good approximation of the Nakagami distribution (Klovski 1982b; Simon and Alouini 2000).

For future development (particularly that of Section 8.3.5) we will need a set of auxiliary parameters, obtained from m_x, m_y, σ_x^2 and σ_y^2 for each component of $h_{kl}(t)$ of $\mathbf{H}(t)$

$$q^2 = \frac{m_x^2 + m_y^2}{\sigma_x^2 + \sigma_y^2}, \beta = \frac{\sigma_x^2}{\sigma_y^2} \qquad (8.52)$$

$$\Omega = m_x^2 + m_y^2 + \sigma_x^2 + \sigma_y^2, \ \phi_p = \arctan\frac{m_y}{m_x} \qquad (8.53)$$

8.3.4 Differential space-time modulation

The differential space-time modulation (DSTM) (Hochwald et al. 2004; Hughes 2000, 2003) is a generalization of the well known approach of DPSK

(Okunev 1997; Simon and Alouini 2000) and generalizes differential modulations to the MIMO case. We are following the basic ideas of (Hughes 2000, 2003).

Let $\mathcal{C}$ be a signal constellation with unitary energy. We also assume that the fading could be well described by the so-called block-fading model (Paulraj et al. 2003), that is, its value does not change over the interval of $nT_s \leq T$ seconds. Here T_s is the symbol duration (interval), n is the number of symbols in a single codeword, and T is the fading coherence time. Let $c_{n_t,k} \in \mathcal{C}$ be a constellation point selected at the n_t-th transmit antenna during the k-th symbol interval from the beginning of a codeword. Then the m-th coding matrix $\mathbf{C}_M \in \mathcal{C}^{N_T \times n}$ could be expressed as

$$\mathbf{C} = \begin{bmatrix} c_{1,1}(m) & \cdots & c_{1,n}(m) \\ \cdots & \cdots \cdots \\ c_{N_T,1}(m) & \cdots & c_{N_T,n}(m) \end{bmatrix} \tag{8.54}$$

Here $1 \leq m \leq M$, M is the number of the code matrices.

If an additional constraint is imposed in the form of the orthogonality condition

$$\mathbf{C}_m \mathbf{C}_m^H = n\mathbf{I}_{N_T}, \quad m = 1, \cdots, M \tag{8.55}$$

then $\mathcal{G} = \{\mathbf{C}_m\}$ is called a *unitary code*. Here $\mathbf{I}_{N_T}$ is the unitary matrix of size $N_T \times N_T$.

A unitary group code $\mathcal{G}$ for the given constellation $\mathcal{C}$, $n \geq N_T$ is a group $\mathcal{G}$ of $n \times n$ unitary matrices $\mathbf{G}_m \in \mathcal{C}^{n \times n}$ such that $\mathbf{G}\mathbf{G}^H = \mathbf{G}^H\mathbf{G} = n\mathbf{I}$ for all $\mathbf{G} \in \mathcal{G}$. In addition a $N_T \times n$ matrix $\mathbf{D}$ such that $\mathbf{D}\mathbf{G} \in \mathcal{C}^{N_T \times n}$ must be chosen to initialize the code. One of the possible choices is to choose $\mathbf{G}_m$ as the permutation matrices (Hughes 2000, 2003).

Application of the the unitary group codes is a generalization of the differentially coded modulation for a SISO channel. At the initial step the transmitter sends the matrix $\mathbf{X}_0 = \mathbf{D}$ to initialize the code. On the k-th step the message encoded into the code matrix $\mathbf{G}_m$ is transmitted as

$$\mathbf{X}_k = \mathbf{X}_{k-1}\mathbf{G}_k, \quad k = 1 : K \tag{8.56}$$

Here K is a number of code words transmitted before the initialization matrix $\mathbf{D}$ is sent again.

If the $N_T \times n$ code matrix to be transmitted is $\mathbf{C}$ and the channel state is described by a $N_R \times N_T$ matrix $\mathbf{H}$, then the $N_R \times n$ received signal $\mathbf{Y}$ is given by

$$\mathbf{Y} = \mathbf{H}\mathbf{C} + \mathbf{N} \tag{8.57}$$

where $\mathbf{N}$ is $N_R \times n$ matrix of i.i.d. AWGN.

If $M = 2^p$ and $n = N_T$ then any group unitary code could be associated with the equivalent *cyclic group code* (or diagonal group code) constructed as follows (Hughes 2003). Let $0 < k_1 \leq k_2 \leq \cdots \leq k_{N_t} < M$ be a set of odd integer numbers.

The following group code $\mathcal{G} = \left\{ \mathbf{I_n}; \mathbf{G}_0, \mathbf{G}_0^2, \cdots, \mathbf{G}_0^{M-1} \right\}$ and

$$
\mathbf{G}_0 = \begin{bmatrix}
\exp\left(j2\pi \frac{k_1}{M}\right) & 0 & .. & 0 \\
0 & \exp\left(j2\pi \frac{k_2}{M}\right) & .. & 0 \\
\cdots & \cdots & .. & \cdots \\
0 & 0 & .. & \exp\left(j2\pi \frac{k_{N_T}}{M}\right)
\end{bmatrix}
\tag{8.58}
$$

It can be easily seen that the code operates in such a way that the only single symbol is transmitted by a single antenna k_m and the m-th time slot after the initialization symbol $\mathbf{D}$. The detection of the cyclic group code is also relatively simple (Hochwald et al. 2004; Hughes 2000). Indeed, let

$$
\mathbf{S} = \mathrm{diag}\mathbf{Y}_k^H \mathbf{Y}_{k-1} = \mathrm{diag}\{S_{n_r,n_r}\}
\tag{8.59}
$$

is the diagonal matrix obtained by retaining only diagonal elements of $\mathbf{Y}_k^H \mathbf{Y}_{k-1}$.

8.3.5   Performance of DTSM

Since the autocovariance algorithm is inherently non-coherent (Simon and Alouini 2000) its performance is worse than that of coherent ML algorithm by approximately 3 dB (Hughes 2000, 2003).

Let us consider the time interval of the k_i-th symbol counting from the training symbol $\mathbf{D}$, that is $(k_{i-1} - 1)T_s \le t \le k_i T_s$. Then in each of the receiver diversity branches one obtains

$$
y_{k_i}(t) = \mu s_{k_i}(t) + n_{k_i}(t)
\tag{8.60}
$$

where μ is the random gain coefficient for the flat fading channel. In the previous block, at the same position the received signal is just

$$
y_{k_i}(t) = \mu s_{k_i}(t) + n_{k_i}(t)
\tag{8.61}
$$

The let us represent $S_i(t)$ in the way

$$
S_i(t) = \frac{1}{\sqrt{N_T}} \sum_{q=k_1}^{k_2} a_{iq} \cos\left(q\frac{2\pi}{T_s}t\right) + b_{iq} \sin\left(q\frac{2\pi}{T_s}t\right)
\tag{8.62}
$$

where $k_2 - k_1 + 1 = F_s T_s$ and F_s is the bandwidth allocated for the channel. Similarly

$$
S_{i-1}(t) = \frac{1}{\sqrt{N_T}} \sum_{q=k_1}^{k_2} a_{i-1,q} \cos\left(q\frac{2\pi}{T_s}t\right) + b_{i-1,q} \sin\left(q\frac{2\pi}{T_s}t\right)
\tag{8.63}
$$

Also, in the very same manner the expansion could be obtained for the noise term

$$\eta_i(t) = \frac{1}{\sqrt{N_T}} \sum_{q=k_1}^{k_2} \alpha_{i,q} \cos\left(q\frac{2\pi}{T_s}t\right) + \beta_{i,q} \sin\left(q\frac{2\pi}{T_s}t\right) \tag{8.64}$$

where all the coefficients have zero-mean and the variance

$$\sigma_n^2 = \frac{N_0}{2T_s}$$

The representation of $\eta_i(t)$ is evident from that of (8.64).

Using the notations introduced above, the correlation between two sequential received symbols could be written as

$$y_i y_{i-1} = \langle y_i, y_{i-1} \rangle = \sum_{p=1}^{4} I_p \tag{8.65}$$

Here $\langle y_i, y_{i-1} \rangle$ is the scalar product of $y_i(t)$ and $y_{i-1}(t)$.

Recalling that we use differentially encoded M-PSK symbols one can derive expressions for the terms I_p, $p = 1, \cdots, 4$ as follows[5]

$$I_1 = \int_0^{T_s} S_i(t) S_{i-1}(t) dt = \mu^2 \frac{T_s}{N_T} \sum_{m=k_1}^{k_2} a_{im}^2 + b_{im}^2 = \mu^2 P_s T_s \tag{8.66}$$

$$I_2 = \int_0^{T_s} \eta_i(t) S_{i-1}(t) dt = \frac{1}{\sqrt{N_T}} \sum_{m=k_1}^{k_2} (a_{im} - b_{im})(\alpha_{im} - \beta_{im}) s \tag{8.67}$$

$$I_3 = \int_0^{T_s} S_i(t) \eta_{i-1}(t) dt = \frac{1}{sqrt N_T} \sum_{q=k_1}^{k_2} a_{iq}\alpha_{i-1,q} + b_{iq}\beta_{im} \tag{8.68}$$

$$I_4 = \int_0^{T_s} \eta_i(t)\eta_{i-1}(t) dt = \sum_{q=k_1}^{k_2} \alpha_{iq}\alpha_{i-1,q} + \beta_{iq}\beta_{im} \tag{8.69}$$

Then, using (8.65)–(8.68) one obtains the following condition for the correct decoding of a symbol pair:

$$\sum_{q=k+1}^{k_2} \left[(a_{iq} + \gamma_{iq})^2 + (b_{iq} + \gamma_{i-1,q})^2 \right] > \sum_{q=k+1}^{k_2} \gamma_{iq}^2 + \gamma_{i-1,q}^2 \tag{8.70}$$

where

$$\gamma_{iq} = \frac{\alpha_{iq} - \beta_{iq} + \alpha_{i-1,q}}{2}$$

$$\gamma_{iq} = \frac{\alpha_{iq} - \beta_{iq} + \alpha_{i-1,q}}{2}$$

---

[5] Remember that points of the constellation $\mathcal{C}_M$ are those of the unitary energy, that is, $P_S = 1$.

It is worth mentioning that this is a typical algorithm for a quadratic non-coherent receiver for block-decoding (Proakis 2001).

Let us use the notation $\mathcal{L}_1$ for the left-hand side of Equation (8.70) and the notation $\mathcal{L}_2$ for the right-hand side of it. Since $\mathcal{L}_2$ is a sum of squares of zero-mean independent Gaussian distributions of equal variance, distribution of $\mathcal{L}_2$ is the central χ^2 (Kendall and Stuart 1977) with $2N = 2F_s T_s N_T$ degrees of freedom. By a similar argument, distribution of $\mathcal{L}_1$ is non-central χ^2 with $2N$ degrees of freedom.

Recall that inequality (8.56) must be tested for all $M = 2^p$ symbols from a choosen constellation. Therefore, the probability of a symbol error could be defined as

$$P_0 = \int_0^\infty p(\lambda_1)d\lambda_1 \times \left[\int_{\lambda_1}^\infty p(\lambda_2)d\lambda_2 \right]^{M-1} \tag{8.71}$$

where

$$p(\lambda_2) = \frac{1}{2^N \Gamma(N) D} \left(\frac{\lambda_2}{D}\right)^{N-1} \exp\left(-\frac{\lambda_2}{2D}\right) \tag{8.72}$$

$$p(\lambda_1) = \frac{1}{2D} \left(\frac{\lambda_1}{\Delta}\right)^{N-1} \exp\left[-\frac{\lambda_1 + \Delta}{2D} I_{N-1}\left(\frac{\sqrt{\lambda_1 \Delta}}{D}\right)\right] \tag{8.73}$$

and $D = N_0/T_s$, $\Delta = P_s/N_T$. The variances of the random variables α_{iq} and β_{iq}

$$\sigma_{\alpha_{iq}}^2 = \sigma_{\beta_{iq}}^2 = \sigma_s^2 = \frac{N_0}{2T_s} \tag{8.74}$$

$$\sigma_{\gamma_{iq}}^2 = \frac{N_0}{4T_s} \tag{8.75}$$

Then it is easy to see, that the SNR at the output is

$$\gamma = 2\gamma_0 = 2\frac{2P_s T_s}{N_T N_0} \tag{8.76}$$

Introducing (8.72) into (8.70), it is possible, at least numerically, to evaluate the value of P_0. For analytical evaluation, the following options are available

• $M = 2$, N_R, $N = F_s T_s N_T$ arbitrary (Okunev 1997)

$$P_0 = \frac{1}{2} \exp(-\gamma_o) + \frac{\exp(-\gamma_0)}{2^N} \sum_{l=1}^{N-1} \left(\frac{\gamma_0}{l!}\right)^l \sum_{n=1}^{N-1} \frac{C_{n+N-1}^{n-l}}{2^n} \tag{8.77}$$

• $N = 1$, arbitrary M, N_R (SIMO case) (Proakis 2001)

$$P_0 = \sum_{l=1}^{M-1} \frac{(-1)^l C_{M-1}^l}{l+1} \exp\left(-\frac{2l\gamma_0}{l+1}\right) \tag{8.78}$$

In the case of high SNR $\gamma \gg 1$ and arbitrary $M = 2^p$ and N one can obtain the following asymptotic expressions (Okunev 1997)

$$\tilde{P}_0 \approx (M - 1)P_0 \tag{8.79}$$

where P_0 is calculated using Equation (8.77)

$$P_0 \approx 2F\left[\frac{\sqrt{\gamma_0}\sin(\pi M/2)}{\sqrt{1 + N/2\gamma_0}}\right], \quad N \leq 10 \tag{8.80}$$

which can be further simplified for $\gamma \gg N$ to produce

$$P_0 \approx 2F\left[\sqrt{\gamma_0}\sin(\pi M/2)\right] \tag{8.81}$$

It is also possible to find useful lower and upper bounds (Okunev 1997)

$$F\left[\frac{\sqrt{\gamma_0}\sin(\pi M/2)}{\sqrt{1 + N/2\gamma_0}}\right]$$
$$< P_0 < 2F\left[\frac{\sqrt{\gamma_0}\sin(\pi M/2)}{\sqrt{1 + N/2\gamma_0}}\right]\left[1 - \frac{1}{2}F\left[\frac{\sqrt{\gamma_0}\sin(\pi M/2)}{\sqrt{1 + N/2\gamma_0}}\right]\right] \tag{8.82}$$

At the moment we can utilize the fact that in the MIMO systems with differential codes described above, the receiver performs some sort of combining, either coherent or non-coherent. Sometimes, in the case of non-coherent combining, it is assumed that the phases difference in all branches is removed, albeit the total phase remains unknown. Such an assumption is known as the co-phased assumption (Hughes 2003).

Under the co-phased assumption, and the generalized Gaussian i.i.d. fading all branches, one can derive the following expression for symbol error probability

$$P_0 = \sum_{l=1}^{M-1} \frac{(-1)^{l+1}C_{M-1}^l}{l-1} \prod_{j=1}^{N_R} \frac{1}{\left(1 + \frac{2l}{1+l}\gamma_x\right)\left(1 + \frac{2l}{1+l}\gamma_y\right)}$$
$$\times \exp\left[-\frac{l}{1+l}q_j^2\gamma_j\left(\frac{\cos^2\phi_{p_j}}{1 + \frac{2l}{1+l}\gamma_x} + \frac{\sin^2\phi_{p_j}}{1 + \frac{2l}{1+l}\gamma_y}\right)\right] \tag{8.83}$$

where $\gamma_{x_j} = \sigma_x^2\gamma_0$, $\gamma_{y_j} = \sigma_y^2\gamma_0$, $\gamma_j^2 = \gamma_{x_j}^2 + \gamma_{y_j}^2$.

Furthermore, for large SNR $\gamma_0 \gg \max(1, N)$ one can obtain the following asymptotic result

$$P_0 \approx \frac{C_0}{\left(2\gamma \sin^2 \frac{\pi}{2M}\right)^{N_R} \prod_{j=1}^{N_R} \delta_j^2} \tag{8.84}$$

where

$$C_0 = C_{2N_R-1}^{N_R} \prod_{j=1}^{N_R}$$

$$\times \exp\left\{ -\frac{q_j^2 \left(1 + \beta_j^2\right)}{2\beta_j^2} \left(\cos^2 \phi_{p_j} + \beta_j^2 \sin^2 \phi_{p_j}\right) \right\} \frac{\left(1 + \beta_{2j}\right)\left(1 + q_j^2\right)}{2\beta_j}$$

(8.85)

Equations (8.83) and (8.84) can be easily obtained by the averaging of (8.77), (8.80) over the joint distribution of individual branch parameters. According to the model assumption such a distribution is just a product of $2N_R$ Gaussian distributions.

Unfortunately, non-coherent diversity combining can be treated analytically in the closed form only for the case of $M = 2$ and $N = 1$ and high SNR $\gamma_0 \gg 1$. We do not present these results due to space limitation.

The results of the calculations using Equations (8.77), (8.78), and (8.83) are shown in Figures 8.5–8.6. At this point we just would like to point out the fact that a decrease in P_0 with an increase of the number of antennas could be attributed to these factors: reduction of noise due to averaging over N_R independent signal measurements (accumulation of SNR or power gain) and reduction of the amount of fading due to receive diversity (hardening effect). In order to quantify these

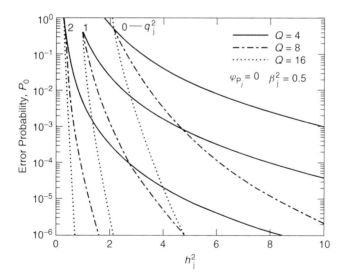

Figure 8.5   Simulation results.

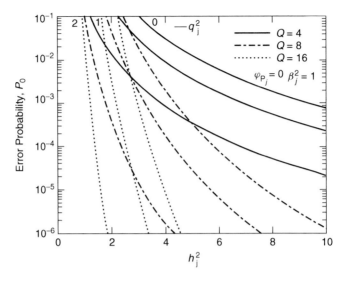

Figure 8.6   Simulation results.

effects one can consider the following quantity

$$\eta_{N_R}(P_0) = \frac{P_{N_R}(P_0)}{P_1(P_0)} \qquad (8.86)$$

where $P_{N_R}(P_0)$ is the power of the transmitted side needed to achieve the probability P_0 on the receive side after combining in N_R antennas. It is well known that $\lim_{N_R \to \infty} = \infty$ and mainly reflects accumulation of the SNR over large number of antennas. However, if one considers a ratio η_{N_R+1}/η_{N_R} then the effect of the fading reduction is isolated, since for large $N_R \gg 1$ accumulation of SNR is relatively small and the possible power gain comes from the reduction of amount of fading.

As an example one can consider expression (8.84) to show that

$$\eta_{N_R} = \left(\frac{\prod_{j=2}^{N_R}}{C_{2N_R-1}^{N_R}} \right)^{1/N_R} \frac{1}{N_R P_0 (1 - 1/N_R)} \qquad (8.87)$$

and, for

$$N_R > N_{R_{lim}} \geq \lfloor \frac{\ln P_0}{2 \ln \delta_{N_R}^2} \rfloor \qquad (8.88)$$

$P_0 \approx 10^{-4}$, $\delta_L^2 = 0.5$ one can simplify this even further

$$\eta_{N_R} = \left(\frac{\prod_{j=2}^{N_R}}{C_{2N_R-1}^{N_R}} \right)^{1/N_R} \frac{1}{N_R P_0 (1 - 1/N_R)} \qquad (8.89)$$

and

$$N_{R_{lim}} \geq \left\lceil \frac{\ln P_0}{\ln \delta_{N_R}^2} \right\rceil \tag{8.90}$$

that is it is approximately twice more than that in Equation (8.88).

8.3.6 Numerical results and discussions

The following conclusions can be drawn from the results of simulations

1. For M-PSK modulation $F_s T_s = 1$ and, therefore $N = N_T$.

2. While $M = 2^p$ increases, P_0 increases as well, which is typical for any decoding algorithm.

3. While $M = 2$ and N_T increases, the probability of error P_0 also increases. This happens because the error probability depends on the noise power and not just the noise power spectral density N_0. Therefore, the noise power grows as N_T increases (see (8.56)) while P_s is fixed.

4. For $N_T F_s T_s = 1$ performance loss in SNR for the autocovariance receiver is approximately 3 dB compared to the coherent decoding (Okunev 1997).

8.3.7 Some comments

1. Squared non-coherent combining of detection of DTSM represented by Equation (8.70) is the same as the non-coherent combining first suggested in (Fink 1970). Such coincidence is quite educational and opens the door to investigation of performance of detection algorithm using Fink's approach (Fink 1970):

2. For M-PSK constellation $F_s T_s = 1$ (narrowband signals), therefore the equivalent signal base in (8.69) is N_T. Therefore, there is a direct analogy between performance of DTSM and performance of transmission diversity schemes as generalization of fundamental properties of autocovariance receivers, described in Section 8.3: probability of error P_0 depends not only on SNR but also on the base of the signal. Thus, for a fixed P_s, the error probability P_0 increases with N_T.

3. Advantages of DTSM could be used in implementing a trade-off between diversity in the region of low and moderate SNR, where channel estimates are not very reliable. In these regions a given P_0 could be implemented by accumulating SNR in individual diversity branches, with multiplexing performed over individual channels if corresponding SNR allows.

4. Use of the generalized Gaussian model allows us to cover many significant realistic physical channels.

Bibliography

Abdel-Samad A, Davidson T, and Gershman A 2006 Robust transmit eigen beamforming based on imperfect channel state information. *IEEE Trans. Signal Processing* **54**(5), 1596–1609.

Abdi A and Kaveh M 2002 A space-time correlation model for multielement antenna systems in mobile fading channels. *IEEE J. Select. Areas Commun.* **20**(3), 550–560.

Abdi A, Tepedelenlioğlu C, Kaveh M, and Giannakis G 2001 On the estimation of the *k* parameter for the Rice fading distribution. *IEEE Commun. Lett.* **5**(3), 92–92.

Abou-Faycal I, Medard M, and Madhow U 2005 Binary adaptive coded pilot symbol assisted modulation over Rayleigh fading channels without feedback. *IEEE Trans. Commun.* **53**(6), 1036–1046.

Abramowitz M and Stegun I (ed) 1965 *Handbook of Mathematical Functions*. Dover, New York.

Agrawal A, Ginis G and Cioffi J 2002b Channel diagonalization through orthogonal space-time coding *Proc. IEEE ICC 2002*, pp. 1–5.

Akki A 1994 A statistical model for mobile-to-mobile landcommunication channel. *IEEE Trans. Veh. Technol.* **43**(4), 826–831.

Akki A and Haber F 1986 A statistical model for mobile-to-mobile land communication channel. *IEEE Trans. Veh. Technol.* **35**(1), 2–7.

Alcocer A, Parra R, and Kontorovich V 2005a An orthogonalization approach for communication channel modelling *Proc. VTC-2005, Fall*.

Alcocer A, Parra R, and Kontorovich V 2005b Some general properties of the covariation matrix for MIMO communication channels. *IST Mobile Wireless Communication summit*.

Alcocer A, Parra R, and Kontorovich V 2006 A MIMO channel simulator applying universal basis. *Proc. VTC-2006, Fall*.

Algans A, Pedersen K, and Mogensen P 2002 Experimental analysis of the joint statistical properties of azimuth spread, delay spread, and shadow fading. *IEEE J. Select. Areas Commun.* **20**(3), 523–531.

Almers P, Bonek E, Burr A, Czink N, Debbah M, Degli-Esposti V, Hofstetter H, Kyösti P, Laurenson D, Matz G, Molisch A, Oestges C, and Özcelik H 2007 Survey of channel and radio propagation models for wireless mimo systems. *EURASIP Journal on Wireless Communications and Networking*. **2007**(1), 56–70.

Wireless Multi-Antenna Channels: Modeling and Simulation, First Edition.
Serguei L. Primak and Valeri Kontorovich.
© 2012 John Wiley & Sons, Ltd. Published 2012 by John Wiley & Sons, Ltd.

Almustafa K, Primak S, Baddour K and Willink T 2009 On achievable data rates and optimal power allocation in fading channels with imperfect channel state information. *Wireless Personal Communications* **50**(1), 69–81.

Alpert 1967 *Propagation in Troposphere*. Chapman & Hall/CRC, New York.

Andersen J and Blaustaein N 2003 *Multipath Phenomena in Cellular Networks*. Artech House, 2002.

Arscott F 1964 *Periodic Differential Equations*. Pergamon Press, New York.

Asplund H, Glazunov A, Molisch A, Pedersen K and Steinbauer M 2006 The COST 259 directional channel model – part ii: Macrocell. *IEEE Trans. Wireless Commun.* **5**(12), 3434–3450.

Baddour K and Beaulieu N 2005 Robust Doppler spread estimation in nonisotropic fading channels. *IEEE Trans. Wireless Commun.* **4**(6), 2677–2682.

Baddour K and Willink T 2007 Improved estimation of the Ricean k factor from I/Q samples *IEEE Vehicular Technology Conference, Fall*, pp. 1228–1232.

Bai X and Shami A 2008 Two dimensional cross-layer optimization for packet transmission over fading channel. *Wireless Communications, IEEE Transactions on* **7**(10), 3813–3822.

Bapat R and Beg M 1989 Order statistics for nonidentically distributed variables and permanents. *Sankhyā: The Indian Journal of Statistics, Series A* **51**(1), 79–93.

Barakat R 1986 Weak-scatterer generalization of the k-density function with application to laser scattering in atmospheric turbulence. *J. Opt. Soc. Am. A* **3**(4), 401–409.

Barakat R 1987 Weak-scatterer generalization of the k-density function ii.probability density of total phase. *J. Opt. Soc. Am. A* **4**(7), 1213–1219.

Beaulieu N 2006 A useful integral for wireless communication theory and its application to rectangular signaling constellation error rates. *IEEE Trans. Commun.* **54**(5), 802–805.

Beckmann P and Spizzichino A 1963 *Scattering of Electromagnetic Waves from Rough Surfaces*. Pergamon Press, New York, USA.

Bello P 1963 Characterization of randomly time-variant linear channels. **11**(4), 360–393.

Bender C and Orszag S 1999 *Advanced Mathematical Methods for Scientists and Engineers*. Springer, New York.

Benedetto S and Biglieri E 1999 *Principles of Digital Transmission with Wireless Applications*. Kluwer Academic/Plenum, New York.

Bertoni H 2000 *Radio Propagation For Modern Wireless Systems*. Prentice Hall PTR, Upper Saddle River, NJ.

Biguesh M and Gershman A 2006 Training-based MIMO channel estimation: a study of estimator tradeoffs and optimal training signals. *IEEE Trans. Signal Processing* **54**, 884–893.

Blanz JJ and Jung P 1998 A flexibly configurable spatial model for mobile radio channels. *IEEE Trans. Commun.* **46**(3), 367–371.

Blaunstein N, Toeltsch M, Laurila J, Bonek E, Katz D, Vainikainen P, Tsouri N, Kalliola K, and Laitinen H 2006 Signal power distribution in the azimuth, elevation and time delay domains in urban environments for various elevations of base station antenna. *IEEE Trans. Antennas Propagat.* **54**(10), 2902–2916.

Bliss DW, Forsythe KW, Hero III AO and Yegulalp AF 2002 Robust MSE equalizer design for MIMO communication systems in the presence of model uncertainties. *IEEE Trans. Signal Processing* **50**(9), 1840–1852.

Bultitude R, Brussaard G, Herben M, and Willink T 2000 Radio channel modelling for terrestrial vehicular mobile applications *Proc. Millenium Conf. Antennas and Prop. Davos, Switzerland*, pp. 1–5.

Burr A 2003 Capacity bounds and estimates for the finite scatterers MIMO wireless channel. *IEEE J. Select. Areas Commun.* **21**(5), 812–818.

Cai X and Giannakis G 2005 Adaptive PSAM accounting for channel estimation and prediction errors. *IEEE Trans. Wireless Commun.* **4**(1), 246–256.

Cantrell P and Ojha A 1987 Comparison of generalized q-function algorithms. *IEEE Trans. Inform. Theory* **33**(6), 591–596.

Cavers J 1991 An analysis of pilot symbol assisted modulation for Rayleigh fading channels. *IEEE Trans. Veh. Technol.* **40**(4), 866–693.

Chuah C, Tse D, Kahn J, and Valenzuela R 2002 Capacity scaling in MIMO wireless systems under correlated fading. *IEEE Trans. Inform. Theory* **48**(3), 637–650.

Coldrey M and Bohlin P 2007 Training-based MIMO systems – Part I: Performance comparison. *IEEE Trans. Signal Processing* **55**, 5464–5476.

Coon JP and Sandell M 2007 Constrained optimization of MIMO training sequences. *EURASIP Journal on Advances in Signal Processing* **2007**(1), 1–13.

Correia, Ed. L 2007 *Mobile Broadband Multimedia Networks*. Wiley, UK.

Costa N and Haykin S 2006 A novel wideband MIMO channel model and McMaster's wideband MIMO SDR *Proc. ACSSC*, pp. 956–960.

Cover T and Thomas J 2002 *Information Theory* second edn. Wiley and Sons, New York.

Debbah M and Muller R 2005 MIMO channel modeling and the principle of maximum entropy. *IEEE Trans. Inform. Theory* **51**(5), 1667–1690.

Degli-Esposti V, Guiducci D, de'Marsi A, Azzi P, and Fuschini F 2004 An advanced field prediction model including diffuse scattering. *IEEE Trans. Antennas Propagat.* **52**(7), 1717–1728.

Dohler M, McLaughlin S, Laurenson D, Beach M, Tan CM, and Aghvami H 2007 Implementable wireless access for B3G networks - Part I: MIMO channel measurement, analysis and modeling. *IEEE Commun. Mag.* **45**(3), 85–92.

Dong M, Tong L and Sadler B 2004 Optimal insertion of pilot symbols for transmissions over time-varying flat fading channels. *IEEE Trans. Signal Processing* **52**, 1403–1418.

Falahati S, Svensson A, Sternad M and Ekman T 2005 Adaptive modulation systems for predicted wireless channels. *IEEE Trans. Commun.* **52**(2), 307–316.

Fechtel S 1993 A novel approach to modeling and efficient simulation of frequency-selective fading radio channels. *IEEE J. Select. Areas Commun.* **11**(3), 422–431.

Fernández-Durán J 2004 Circular distributions based on nonnegative trigonometric sums. *Biometrics* **60**(6), 499–503.

Fink L 1970 *Theory of Transmission of Discrete Messages (In Russian)*. Sov Radio, Moscow.

Fleury B 2000 First- and second-order characterization of direction dispersion and space selectivity in the radio channel. *IEEE Trans. Inform. Theory* **46**(6), 2027–2044.

Fleury B, Tschudin M, Heddergott R, Dahlhaus D, and Pedersen K 1999 Channel parameter estimation in mobile radio environments using the SAGE algorithm. *IEEE J. Select. Areas Commun.* **17**(3), 434–450.

Foschini G and Gans M 1998 On limits of wireless communications in a fading environment. *Wireless Personal Communications* **6**(3), 311–335.

Foschini G, Gitlin R, and Weinstein S 1974 Optimization of two-dimensional signal constellations in the presence of gaussian noise. *IEEE Trans. Commun.* **22**(1), 28–38.

Franklin J 2000 *Matrix Theory* second edn. Dover, Mineola, NY.

Gantmacher F 1959 *The Theory of Matrices*. Chelsea Pub. Co, New York.

Gesbert D, Bölcskei H, Gore DA and Paulraj AJ 2002 Outdoor MIMO wireless channels: Models and performance prediction. *IEEE Trans. Commun.* **50**(12), 1926–1934.

Gharavi-Alkhansari M and Gershman A 2005 On diversity and coding gains and optimal matrix constellations for space-time block codes. *IEEE Trans. Signal Processing* **53**(10), 3703–3717.

Giannakis GB, Liu Z, Ma X, and Zhou S 2007 *Space-Time Coding for Broadband Wireless Communications*. Wiley, New York.

Giorgetti A, Smith P, Shafi M, and Chiani M 2003 MIMO capacity, level crossing rates and fades: the impact of spatial/temporal channel correlation. *Journal of Communication and Networks* **5**(2), 104–115.

Goldsmith A and Varaiya P 1997 Capacity of fading channels with channel side information. *IEEE Trans. Inform. Theory* **43**(6), 1986–1992.

Golomb SW and Gong G 2005 *Signal Design for Good Correlation for Wireless Communication, Cryptography, and Radar: For Wireless Communication, Cryptography, and Radar*. Cambridge University Press, Cambridge, UK.

Golub G and van Loan C 1996 *Matrix Computations* third edn. John Hopkins University Press, Baltimor, MD.

Grenander U and Szegö G 1984 *Toeplitz Forms and their Applications*. Chelsea Pub. Co., New York.

Haghighi SJ, Primak S, Kontorovich V, and Sejdic E 2010 *Mobile and Wireless Communications Physical Layer Development and Implementation*. InTech Publishing. Chapter: Wireless Communications and multitaper analysis: applications to channel modelling and estimation, pp. 25–46.

Hassibi B and Hochwald B 2003 How much training is needed in multiple-antenna wireless links?. *IEEE Trans. Inform. Theory* **49**(4), 951–963.

Haykin S 1989 *Modern Filters*. MacMillan Publishing Company, New York.

Haykin S 2005 Cognitive radio: brain-empowered wireless communications. *IEEE J. Select. Areas Commun.* **23**(2), 201–220.

Helstrom C 1960 *Statistical Theory of Signal Detection*. Pergamon Press, New York.

Herdin M, Czink N, Özcelik H and Bonek E 2005 What makes a good MIMO channel model. *IEEE 61st Semiannual Vehicular Technology Conference (VTC 2005 - Spring)*, pp. 1–6, Stockholm, Sweden.

Hochwald BM, Marzetta TL, and Tarokh V 2004 Multiple-antenna channel hardening and its implications for rate feedback and scheduling. *IEEE Trans. Inform. Theory* **50**(9), 1893–1909.

Hughes B 2000 Differential space-time modulation. *IEEE Trans. Inform. Theory* **46**(7), 2567–2578.

Hughes B 2003 Optimal spa-time constellations from groups. *IEEE Trans. Inform. Theory* **49**(2), 401–410.

Hutchinson T and Lai CD 1990 *Continuous Bivariate Distributions, Emphasizing Applications*. Rumsby Scientific Publishing, Adelaide, Australia.

Ivrlač M and Nossek J 2003 Quantifying diversity and correlation in Rayleigh fading MIMO communication systems *ISSPIT 2003*, pp. 158–161.

Ivrlač M and Nossek J 2005 MIMO perfomance measures - a signal processing point of view. *ICECom 2005*, pp. 1–4.

Jakeman E and Pusey P 1976 A model for non-Rayleigh sea echo. *IEEE Trans. Antennas Propagat.* **24**(6), 806–814.

Jakeman E and Tough R 1987 Generalized K distribution: a statistical model for weak scattering. **4**(9), 1764–1772.

Jakes W 1974 *Microwave Mobile Communications*. IEEE Press, New York.

Jeruchim MC, Balaban P, and Shanmugan KS 2000 *Simulation of Communication Systems: Modeling, Methodology and Techniques*. New York: Springer.

Jingyu H, Bin S, and Xiaohu Y 2005 The phase probability distribution of general Clarke model and its application in Doppler shift estimator. *IEEE Antennas Wireless Propagat. Lett.* **4**(4), 373–377.

Kaiserd T, Bourdoux A, Boche H, Fonollosa J, Andersen J and Utschick W 2006 *Smart Antennas: State of the Art*. Hindawi, NY.

Kay S 1993 *Statistical Signal Processing: Estimation Theory*. Prentice Hall PTR, Upper Saddle River, NJ.

Kendall M and Stuart A 1977 *The Advanced Theory of Statistics* fourth edn. Macmillan, New York.

Kermoal J, Mogensen P, Jensen S, and Andersen J 2000 Experimental investigation of multipath richness for multi-element transmit and receive antenna arrays. *Proceedings of the IEEE Vehicular Technology Conference, Spring*, pp. 2004–2008.

Kermoal JP, Schumacher L, Pedersen KI, Mogensen PE, and Frederiksen F 2002 Capacity bounds and estimates for the finite scatterers MIMO wireless channel. *IEEE J. Select. Areas Commun.* **20**(6), 1211–1226.

Klovski D 1982a *Digital Data Transmission over Radiochannels*. Sviaz, Moscow, USSR, (in Russian).

Klovski D 1982b *Processing of Time-Space Information in Communication Channels*. Sviaz, Moscow, USSR, (in Russian).

Kontorovich V 2006a The MIMO channel orthogonalizations applying universal eigenbasis. *Seventh Int. Conf. Mathematics in Signal Processing*, pp. 9–13.

Kontorovich V 2007 2-D RAKE receiver for the MIMO channel: some generalizations *Proc. 16th IST Mobile and Wireless Communications Summit*.

Kontorovich V Norway, 2006b Orthogonalization approach to channel modeling. *International Summer Workshop: Propagation and Channel Modelling*.

Kontorovich V and Primak S 2008 2D RAKE receiver for MIMO channels: Optimum algorithm with minimum complexity. *Stochastic Models*. **24**(Suppl. 1), 194–217.

Kontorovich V, Primak S, Alcocer-Ochoa1 A, and Parra-Michel R 2008 MIMO channel orthogonalisations applyinguniversal eigenbasis. *Proc. IET Signal processing* **2**(2), 87–96.

Korn A and Korn T 1967 *Manual of Mathematics*. McGraw-Hill, New York.

Kovacevic J and Chebira A 2007 Life beyond bases: The advent of frames (part i). *IEEE Signal Processing Mag.* **24**(4), 86–104.

Kuchar A, Rossi J, and Bonek E 2000 Directional macro-cell channel characterization from urban measurements. *IEEE Trans. Antennas Propagat.* **48**(2), 137–146.

Kuo PH and Smith P 2005 Temporal behavior of MIMO channel quality metrics *Proc. ICWNCMC*.

La Frieda J and Lindsey W 1973 Transient analysis of phase-locked tracking system in the presense of noise. *IEEE Trans. Inform. Theory* **19**(2), 155–165.

Lamahewa TA, Kennedy RA, Abhayapala TD, and Betlehem T 2006 MIMO channel correlation in general scattering environments *Proc. 7th Australian Communications Theory Workshop*.

Lancaster H 1958 Structure of bivariate distributions. *The Ann. of Math. Stat.* **29**(3), 719–736.

Larsson EG and Stoica P 2003 *Space-time Block Coding for Wireless Communications*. Cambridge University Press, Cambridge, UK.

Lee WC 1997 *Mobile Communications Engineering* second edn. McGraw-Hill, California.

Li KYQ and Ho M 2005 Measurement investigation of tap and cluster angular spreads at 5.2 GHz. *IEEE Trans. Antennas Propagat.* **53**(7), 2156–2160.

Liu H 2005 High-rate transmission scheme for pulse-based ultra-wideband systems over dense multipath indoor channels. *Proc. IEE* **150**(2), 235–240.

Loyka S 2005 Multiantenna capacities of waveguide and cavity channels. *IEEE Trans. Signal Processing* **54**(3), 863–872.

Loyka S and Levin G 2007 On finite-SNR diversity-multiplexing tradeoff. *GLOBECOM'07*, Washington, DC, pp. 1456–1461.

Lozano A and Tulino A 2002 Information capacity of a random signature multiple-input multiple-output channel. *IEEE Trans. Inform. Theory* **48**(12), 3117–3128.

Luo J, Zeidler J and Proakis JG 2002 Error probability performance for WCDMA systemas with multiple transmit and receive antennas in correlated Nakagami fading channels. *IEEE Trans. Veh. Technol.* **51**(6), 1502–1516.

Mallat S 1999 *A Wavelet Tour of Signal Processing*. Elsvier, Amsterdam.

Mardia K and Jupp P 2000 *Directional Statistics*. Wiley, New York.

Medard M 2000 The effect upon channel capacity in wireless communication of perfect and imperfect knowledge of the channel. *IEEE Trans. Inform. Theory* **46**(5), 933–946.

Mehta M 1991 *Random Matrices*. Academic Press, New York.

Messier G and Hartwell J 2009 An empirical model for non-stationary Ricean fading. *IEEE Trans. Veh. Technol.* **58**(1), 14–20.

Middleton 1960 *Introduction to Statistical Communications Theory* first edn. McGraw-Hill, New York.

Miller DAB 2000 Communicating with waves between volumes: evaluating orthogonal spatial channels and limits on coupling strengths. *Appl. Optics* **39**(11), 1681–1699.

Misra S, Swamy A, and Tong L 2006 Optimal training for time-selective wireless fading channels using cutoff rate. *EURASIP J. Appl. Sig. Proc.* pp. 1–15.

Molisch AF 2004 A generic model for MIMO wireless propagation channels in macro- and microcells. *IEEE Trans. Signal Processing* **52**(1), 61–71.

Molisch AF, Asplund H, Heddergott R, Steinbauer M, and Zwick T 2006 The COST 259 directional channel model – part i: Overview and methodology. *IEEE Trans. Wireless Commun.* **5**(12), 3421–3433.

Moustakas A, Simon S, and Sengupta A 2003a Mimo capacity through correlated channels in the presence of correlated interferers and noise: a (not so) large n analysis. *IEEE Trans. Inform. Theory* **49**(10), 2545–2561.

Moustakas AL and Simon SH 2007 On the outage capacity of correlated multiple-path mimo channels. *IEEE Trans. Inform. Theory* **53**(11), 3887–3903.

Moustakas AL, Simon SH, and Sengupta AM 2003b MIMO capacity through correlated channels in the presence of correlated interferers and noise: A (not so) large *n* analysis. *IEEE Trans. Inform. Theory* **49**(10), 2545–2561.

Muller R 2002 A random matrix model of communication via antenna arrays. *IEEE Trans. Inform. Theory* **48**(9), 2495–2506.

Müller R 2004 Random matrices, free probability, and the replica method *Proc. EUROSIPCO*, Vienna.

Narasimhan R 2006 Finite-SNR diversity-multiplexing tradeoff for correlated Rayleigh and Rician MIMO channels. *IEEE Trans. Inform. Theory* **52**(9), 3965–3979.

Narasimhan R and Cox DC 1999 Speed estimation in wireless systems using wavelets. *IEEE Trans. Commun.* **47**(9), 1357–1364.

Norklit O and Andersen JB 1998 Diffuse channel model and experimental results for array antennas in mobile environments. *IEEE Trans. Antennas Propagat.* **46**(6), 834–840.

Oestges C and Clerckx B 2007 *MIMO wireless communications: from real-world propagation to space-time code design*. Academic Press, New York.

Oestges C, Erceg V, and Paulraj A 2003 A physical scattering model for MIMO macrocellular broadband wireless channels. *IEEE J. Select. Areas Commun.* **21**(5), 721–729.

Øien GE, Holm H, and Hole KJ 2004 Impact of channel prediction on adaptive coded modulation performance in Rayleigh fading. *IEEE Trans. Veh. Technol.* **53**(3), 758–769.

Okunev Y 1997 *Phase and Phase-difference Modulation in Digital Communications* first edn. Artech House, New York.

Ord J 1972 *Families of Frequency Distributions*. Griffin, London.

Özcelik H, Czink N, and Bonek E 2005 What makes a good MIMO channel model *IEEE 61st Semiannual Vehicular Technology Conference (VTC 2005 – Spring)*, pp. 1–6, Stockholm, Sweden.

Papoulis A 1991 *Probability, Random Variables, and Stochastic Processes* third edn. McGraw-Hill, Boston, MA.

Parra R, Kontorovich V and Orozco-Lugo A 2002 Modelling wide band channels using orthogonalizations. *IEICE Trans.* **E-852**(3), 544–551.

Patzold M 2002 *Mobile Fading Channels*. John Wiley & Sons, New York.

Paulraj A, Nabar R and Gore D 2003 *Introduction to Space-Time Wireless Communications*. Cambridge University Press, Cambridge, UK.

Pedersen K, Andersen J, Kermoal J and Mogensen P 2000 A stochastic multiple-input-multiple-output radio channel model for evaluation of space-time coding algorithms *Proceedings of the IEEE Vehicular Technology Conference*, pp. 893–897.

Percival D and Walden A 1993 *Spectral Analysis for Physical Applications: Multitaper and Conventional Univariate Techniques*. Cambridge University Press, New York.

Petrus P, Reed JH and Rappaport TS 2002 Geometrical-based statistical macrocell channel model for mobile environments. *IEEE Trans. Commun.* **50**(3), 495–502.

Pitsianis N and van Loan C 1993 Approximation with Kronecker products. *Linear Algebra for Large Scale and Real Time Applications*.

Poor V 1994 *An Introduction to Signal Detection and Estimation* 2nd edn. Springer, Berlin.

Primak S 2008a On role of scattering environment and the antenna parameters on quality of MIMO channel prediction. part i: infinite past. Technical report, ECE, UWO.

Primak S 2008b UIU model revisited. Crc report, UWO.

Primak S and Kontorovich V 2009 On sum of specular components model of mimo channels. *Proceedings of WTS 2009*, Prague.

Primak S and Sejdic E 2008 Application of multitaper analysis to wireless communications problems *Proc. of ISABEL'08*, Aalborg, Denmark.

Primak S, Kontorovich V and Lyandres V 2004 *Stochastic Methods and their Applications to Communications: SDE Approach*. John Wiley & Sons, Chichester.

Proakis J 1997 *Digital Communications* third edn. McGraw-Hill, New York.

Proakis J 2001 *Digital Communications* fourth edn. McGraw-Hill, New York.

Rappaport T 2002 *Wireless Communications: Principles and Practice*. Prentice Hall, Upper Saddle River.

Rényi A 1970 *Probability Theory*. North Holland, Amsterdam.

Salo J, El-Sallabi HM, and Vainikainen P 2006 Statistical analysis of the multiple scattering radio channel. *IEEE Trans. Antennas Propagat.* **54**(11), 3114–3124.

Salz J and Winters J 1994 Effect of fading correlation on adaptive arrays in digital mobile radio. *IEEE Trans. Veh. Technol.* **43**(4), 1049–1057.

Sarmanov O 1958a Maximum correlation coefficient (non-symmetric case). *Dokl. Akad. Nauk SSSR* **121**(3), 52–55.

Sarmanov O 1958b Maximum correlation coefficient (symmetric case). *Dokl. Akad. Nauk SSSR* **120**(3), 715–718.

Sayeed A 2002 Deconstructing multiantenna channels. *IEEE Trans. Signal Processing* **50**(10), 2563–2579.

Schreier P and Scharf L 2003a Second-order analysis of improper complex random vectors and processes. *IEEE Trans. Signal Processing* **51**(3), 714–725.

Schreier P and Scharf L 2003b Second-order analysis of improper complex random vectors and processes. *IEEE Trans. Signal Processing* **51**(3), 714–725.

SCM Editors 2002 3GPP2 TSG-C WG3 spatial channel model AHG, 1× EV-DV spatial channel model evaluation methodology *Tech. report*.

SCM Editors 2006 UMTS: Spatial channel model for MIMO simulations *Tech. report 25.996*.

Sejdic E, Luccini M, Primak S, Baddour K, and Willink T 2008 Channel estimation using DPSS frames *Proc. of ICASSP08*.

Sen I and Matolak D 2008 Vehicle–vehicle channel models for the 5-GHz band. *IEEE Trans. Intell. Transport. Syst.* **9**(26), 235–245.

Shilov GE and Gurevich B 1978 *Integral, Measure, and Derivative: A Unified Approach*. Dover, New York.

Simon M 2002 *Probability Distributions Involving Gaussian Random Variables* first edn. Kluwer Academic Publisher, Boston.

Simon M and Alouini MS 2000 *Digital Communication over Fading Channels: A Unified Approach to Performance Analysis*. John Wiley & Sons, New York.

Slepian D 1965 Permutation modulation. *Proc. IEEE* **53**(3), 228–236.

Slepian D 1978 Prolate spheroidal wave functions, Fourier analysis, and uncertainty. V-The discrete case. *Bell System Technical Journal* **57**, 1371–1430.

Slepian D, Landau H, and Pollak H 1961 Prolate spheroidal wave functions. *Bell Systems Tech. Jour.* **40**(1), 43–64.

Smith P, Kuo PH, and Garth L 2005 Level crossing rates for mimo channel eigenvalues: implications for adaptive systems *Proc. of ICC*, pp. 2442–2446.

Stewart G and Sun JG 1990 *Matrix Perturbation Theory*. Academic Press, San Diego, CA.

Su W and Xia X 1981 Tables of sphere packings and spherical codes. *IEEE Trans. Inform. Theory* **27**(3), 327–338.

Su W and Xia X 2004 Signal constellations for quasi-orthogonal space-time block codes with full diversity. *IEEE Trans. Inform. Theory* **50**(10), 2331–2347.

Suzuki H 1977 A statistical model for urban radio propagation. *IEEE Trans. Commun.* **25**(7), 673–680.

S. Vaihunthan, S. Haykin and M. Sellathurai 2005 MIMO Channel Capacity Modeling using Markov Models *Proc. of IEEE VTC*, Fall.

Svantesson T and Swindlehurst A 2003 A performance bound for prediction of a multipath MIMO channel *Proc. of 37th Asilomar Conf. Signals, Systems and Computers*, Pacific Grove, CA. pp. 233–237.

Svantesson T and Swindlehurst A 2006 A performance bound for prediction of MIMO channels. *IEEE Trans. Signal Processing* **54**(2), 520–529.

Szegö G 1939 *Orthogonal Polynomials*. AMS, Providence, Rhode Island.

Teal PD, Abhayapala TD, and Kennedy RA 2002 Spatial correlation for general distributions of scatterers. *IEEE Signal Processing Lett.* **9**(10), 305–308.

Teletar E 1999 Capacity of multi-antenna gaussian channels. *European Transactions on Telecommunications* **10**(6), 585–595.

Tepedelenlioğlu C, Abdi A, and Giannakis G 2003 The Ricean k factor: estimation and performance analysis. *IEEE Trans. Wireless Commun.* **2**(3), 799–810.

Tepedelenliŏglu C and Giannakis GB 2001 On velocity estimation and correlation properties of narrow-band mobile communication channels. *IEEE Trans. Veh. Technol.* **50**(4), 1039–1052.

Thomson D 1982 Spectral estimation and harmonic analysis. *Proc. IEEE* **70**(9), 1055–1096.

Toeltsch M, Laurila J, Kalliola K, Molisch A, Vainikainen P, and Bonek E 2002 Statistical characterization of urban spatial radio channels. *IEEE J. Select. Areas Commun.* **20**(3), 539–549.

Tong L, Sadler B, and Dong M 2004 Pilot-assisted wireless transmissions: general model, design criteria, and signal processing. *IEEE Signal Processing Mag.* **21**, 12–25.

Tse D and Viswanath P 2005 *Fundamentals of Wireless Communication*. Cambridge University Press, New York.

Tulino A and Verdú S 2004 Random matrix theory and wireless communications. *Foundations and Trends in Communications and Information Theory* **1**, 1–182.

Tulino A, Lozano A, and Verdu S 2005a Impact of antenna correlation on the capacity of multiantenna channels. *IEEE Trans. Inform. Theory* **51**(7), 2491–2509.

van Trees H 2001 *Detection, Estimation, and Modulation Theory: Part I* first edn. John Wiley & Sons, New York.

van Trees H 2002 *Detection, Estimation and Modulation Theory: Optimal Array Processing* first edn. Wiley-Interscience, New York.

Vatalaro F and Forcella A 1997 Doppler spectrum in Mobile-to-Mobile communications in the presence of three-dimensional multipath scattering. *IEEE Trans. Veh. Technol.* **46**(1), 213–219.

Verdu S 1998 *Multiuser Detection*. Cambridge University Press, UK.

Vucetic B and Yuan J 2002 *Space-Time Coding*. Wiley, New York.

Vuokko L, Vainikainen P, and ichi Takada J 2005 Clusters extracted from measured propagation channels in macrocellular environments. *IEEE Trans. Antennas Propagat*. **53**(12), 4089–4098.

Wallace J and Jensen M 2002 Modeling the indoor mimo wireless channel. *IEEE Trans. Antennas Propagat*. **50**(5), 591–599.

Wallace J and Maharaj B 2007 The accuracy of tensor and directional methods for MIMO channel modeling *Antennas and Propagation, 2007. EuCAP 2007. The Second European Conference on*, pp. 1–6.

Wallace J, Jensen M, Swindlehurst L, and Jeffs B 2003 Experimental characterization of the MIMO wireless channel: Data acquisition and analysis. *IEEE Trans. Wireless Commun*. **2**(2), 335–343.

Weichselberger W 2004 On the decomposition of the MIMO channel correlation tensor *ITG Workshop on Smart Antennas*.

Weichselberger W, Herdin M, Ozcelik H, and Bonek E 2006 A stochastic MIMO channel model with joint correlation of both link ends. *IEEE Trans. Wireless Commun*. **5**(1), 90–100.

Weichselberger W, Ozelik H, Herdin M and Bonek E 2005 A novel stochastic MIMO channel model and its physical interpretation *IEEE Trans. Wireless Communications*, **5**(1), 90–100.

Willink T, Primak S, Baddour K, Almustafa K, Luccini M, and Ahmadi A 2006 Modelling and prediction of MIMO channels CRC UWO contract No. Research Contract.

Xiao H, Burr A, and White G 2005 Simulation of time-selective environment by 3GPP spatial channel model and analysis on the performance evaluation by the CMD metric *Proc. International Wireless Summit (IWS 05)*, pp. 446–450, Aalborg, Denmark.

Xu H, Chizhik D, Huang H, and Valenzuela R 2004 A generalized space-time multiple-input multiple-output (MIMO) channel model. *IEEE Trans. Wireless Commun*. **3**(3), 966–975.

Yip K and Ng T 1997 Karhunen-Loeve expansion of the WSSUS channel output and its application to efficient simulation. *IEEE J. Select. Areas Commun*. **15**(4), 640–646.

Yoo T and Goldsmith A 2006 Capacity and power allocation for fading MIMO channels with channel estimation error. *IEEE Trans. Inform. Theory* **52**(5), 2203–2214.

Yu K and Ottersten B 2002 Models for MIMO propagation channels: a review. *Wireless Communications and Mobile Computing* **2**(7), 653–666.

Zatman M 1998 How narrow is narrowband?. *IEE Proc. Radar, Sonar and Navigation* **145**(2), 85–91.

Zemen T and Mecklenbrauker C 2005 Time-variant channel estimation using discrete prolate spheroidal sequences. *IEEE Trans. Signal Processing* **53**(9), 3597–3607.

Zhang L, Burr A, and Pearce D 2006 Capacity of MIMO wireless channel with finite scattering in cellular systems. *IEE Proc. Comm*. **153**(4), 489–495.

Zhang M, Smith PJ, and Shafi M 2005 A new space-time MIMO channel model. *AusCTW*.

Zhang X and Ottersten B 2003 Performance analysis of V-BLAST structure with channel estimation errors. *IEEE Workshop on Signal Processing Advances in Wireless Communications*, pp. 487–491.

Zhao G and Loyka S 2004a Impact of multipath clustering on the performance of MIMO systems. *Proc. Wireless Communications and Networking Conference*, pp. 1–5.

Zhao G and Loyka S 2004b Performance study of MIMO systems in a clustered multipath channel. *Canadian Conference on Electrical Engineering*, pp. 453–456.

Zhao X, Kivinen J, Vainikainen P, and Skog K 2002 Propagation characteristics for wideband outdoor mobilecommunications at 5.3 ghz. *IEEE J. Select. Areas Commun.* **20**(3), 507–514.

Zheng L and Tse D 2002 Communication on the Grassman manifold: A geometric approach to the noncoherent multiantenna channel. *IEEE Trans. Inform. Theory* **48**(2), 359–383.

Zhong C, Wong KK, and Jin S 2009 Capacity bounds for MIMO nakagami- m fading channels. *IEEE Trans. Signal Processing* **57**(9), 3613–3623.

Zwick T, Fischer C, and Wiesbeck W 2002 A stochastic multipath channel model including path directions for indoor environments. *IEEE J. Select. Areas Commun.* **20**(6), 1178–1192.

Index

Wireless Multi-Antenna Channels: Modeling and Simulation, First Edition.
Serguei L. Primak and Valeri Kontorovich.
© 2012 John Wiley & Sons, Ltd. Published 2012 by John Wiley & Sons, Ltd.

CPSIA information can be obtained at www.ICGtesting.com
Printed in the USA
BVOW051738171011

273610BV00003B/18/P